For the Moon and Mars

N-1

A Reference Guide to the Soviet Superbooster

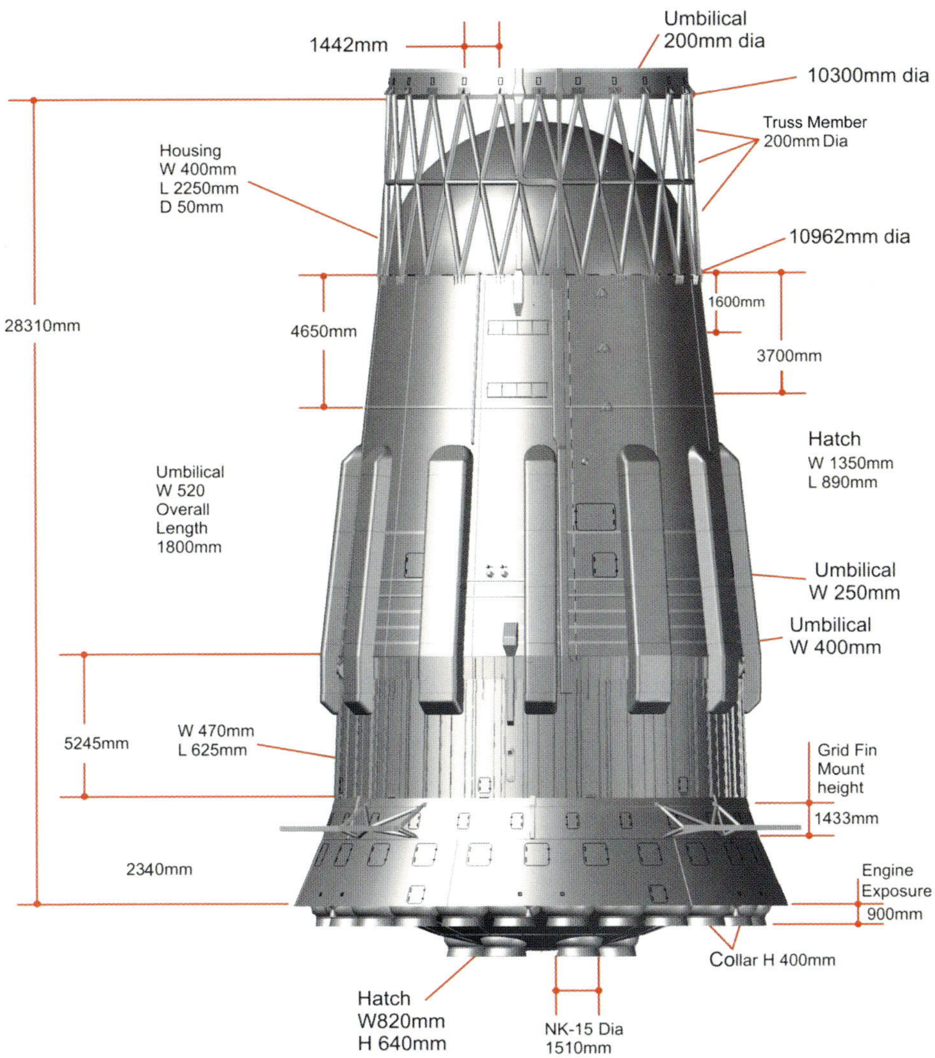

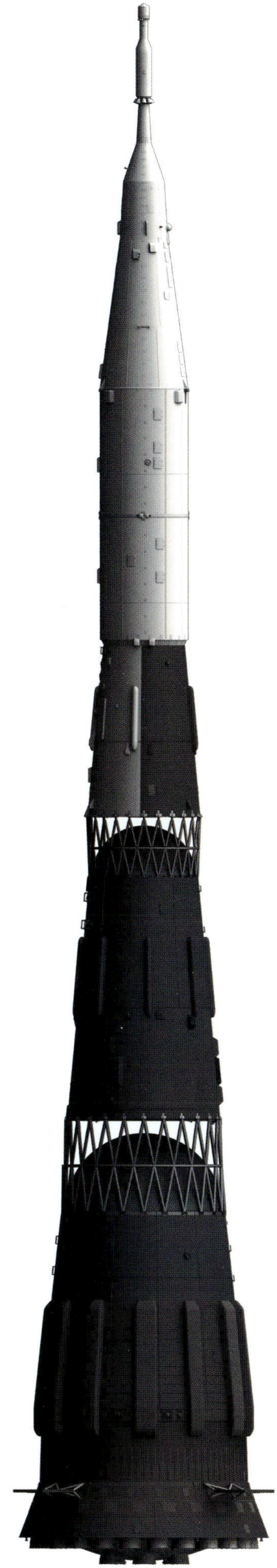

by Matthew Johnson and Nick Stevens with

Alexander Shliadinsky, Igor Bezyaev and Vladimir Antipov

Edited by Jack Hagerty

ARA Press
www.arapress.com

Copyrights

N-1: For the Moon and Mars, A Guide to the Soviet Superbooster (1st Edition)

ISBN 978-0-9899914-0-7

Printed and bound in the United States by:
ARA Press
785 Jefferson Avenue
Livermore, CA 94550-3137
www.arapress.com

The body text is 12 pt Times New Roman
Page headers and footers are 12 pt Ravel Medium
Chapter titles are 18 pt Arial
Photo and Figure captions are 10 pt Arial
Cover and Title Page are multiple weights of Ravel Medium

Table of Contents

Dedicated to my parents, Marv and Donna Lee, who supported my hobby and interests in both stages of my life. Thank you for the freedom and encouragement to pursue it

—Matt Johnson

2004: Born Again

Ten months after diagnosis, I watched my father wither away, shrink and disappear from colon cancer at the age of 69. I sat back and sought a way to capture the good old, possibly innocent, days with my father. A tribute and panacea. With an immediate response, I zeroed in on the early days of the space program and our work together on one Estes model after another. Like so many middle aged men with no fathers, I quickly became a "Born Again Rocketeer."

1968: Apollo and TV Table Rockets

The living room, after dinner, after Star Trek, during Apollo. With a TV table at the couch as our modeling surface, my dad would carefully open the plastic bag of the latest Astron Estes kit and sort the parts. As it was late and near bedtime for my 5 year old frame, I asked, "Can we start like we always do? With the engine mount? That always takes time to dry" (damn the white glue). He says yes and we begin yet another rocket together. It is a 1/200 Saturn V. My dad says he will finish up after I go to bed.

The next morning I wake up and search the house for the newly born Saturn. Hanging in the garage is a freshly spray painted Saturn V. It is painted red. There is a big thumb print just under the second stage transition. Even at this young age, I am struck with the notion that the red Saturn V hanging before me was a product of parental vigor and the desire to make a connection with the little dorky kid who (thought) knew too much science for his own good. My older brother (by six years) was a popular athlete and probably reflected my father's genes a bit more than I, as dad loved sports and was a coach.

I proudly flew our red Saturn V that summer at the park in town where the local hobby shop had organized public Sunday launches. As my dad was deep into the school administration, we also had "Rocket Day" at the school where we would launch a batch of Estes rockets for the class, watch a few crash, watch a few fly great, and go home and fix the few that were stepped on by too eager little feet. So for at least one day a year, the dork would be pretty popular at school.

My dad taught me how to build rockets from that age on. After puberty, for some reason, the rockets trail off. After high school, after college, rocketry was a fond memory. With no children of my own to spark the parental vigor, it took the desire to reconnect with my father to prompt into being the 45 year old "Born Again Rocketeer." I buy kits for kids on the field and give away engines whenever I can. (A kid can never have too many engines... oh how I always needed an engine.) I credit my father with my current Level 2 NAR High Power standing. How I wish he could see me launch a "K" engine now.

What I would not give now to see for a red 1/200th Saturn V, with a big thumb print just below the second stage transition, hanging in the garage — MJ

Foreword

For those of us who have spent a lifetime digging at the edges of the Soviet space program, the N-1 rocket has been a kind of holy grail. When I was in high school back in the mid-eighties, I would stay up late reading the works of various British and American sleuths who were claiming—contrary to many nay-sayers—that, yes, the Soviets did indeed have their own "Superbooster," one that was designed to compete with NASA's Saturn V. Not knowing its real name, Westerners at the time called it the "G-1" (as the next in a series of Soviet launch vehicles given Latin alphabet designators). There were bits and pieces of data floating around about this quasi-mythical rocket, including some possible launch dates. All we really knew was that the rocket never delivered anything to orbit successfully, let alone to the Moon. Then in August of 1989, a long essay by a journalist named Sergey Leskov in the Soviet newspaper *Izvestia* revealed that, yes, there had been such a rocket, and that Soviets had called it the "N-1."

As the months rolled by, more and more revelations appeared in the Soviet (and soon, Russian) press. A long interview with Vasily Mishin, the man who succeeded the legendary Sergey Korolev as "chief designer" of the major Soviet design bureau, appeared in *Pravda* in October 1989. He revealed how the history of the N-1 had been marred by all sorts of infighting and disagreements among Soviet designers. There was a palpable sense of sadness at the final outcome of the project: the program was suspended in 1974 and eventually all its hardware was destroyed, as if to blot out its very existence from history. More information about the rocket appeared in the 1990s in Soviet journals such as *Aviatsiya i kosmonavtika* ("Aviation and Cosmonautics") and *Kryl'ya rodiny* ("Wings of the Motherland") authored by writers such as Igor Afanas'yev, Mikhail Rebrov, and others. Among those who sought to reconstruct the history of this giant rocket were Russians Vladimir Antipov and Alexander Shliadinsky, both of whom are contributors to this wonderful volume on the technical aspects of the N-1.

In the nearly quarter of a century since the secret of the N-1 was revealed, there has been a deluge of information in the media about the rocket. You can find dozens of websites, articles, and books about it. Because of the proliferation of information about the rocket, there is also a lot of conflicting information circulating, especially about various technical aspects of the N-1. What were the differences between each of the four N-1s launched? What were the performance specifications of each of its elements? What were the intricate details of its L-3 payload, designed to take a single Soviet cosmonaut to the surface of the Moon? Principal authors Matt Johnson and Nick Stevens have found the answers to all of these questions and assembled them in one handy volume, profusely illustrated with both period photographs and original renderings. I have no doubt that this book will be seen as an ideal reference work for modelers who are interested in making their own N-1 rockets. Kudos to Jack Hagerty for putting this whole project together. When one considers that until 1989, the N-1 rocket was as much of a mystery as the Loch Ness monster, what you hold in your hands is no less than astonishing.

— Asif A. Siddiqi

Author of

Challenge to Apollo: The Soviet Union and the Space Race, 1945-1974

The Red Rocket's Glare

Series Editor, English version of Boris Chertok's 4-volume history of the Soviet

space program: *Rockets and People*

Editor/Publisher's Preface

I am a child of the Space Age. Some of my earliest memories are of sitting in front of the 13" Philco at age three watching Disney's "Man in Space" series in black and white. I was one month shy of nine years old when Yuri Gagarin made his historic flight inaugurating humanity's presence in space. I vividly remember getting up at 4:30 AM while in middle school for most of the Gemini launches since the coverage started about an hour before their 9 AM (eastern) liftoff (the only time I really regretted living in California). Neil Armstrong took his famous "one small step" in the summer between my Junior and Senior years of high school, and Gene Cernan took the last step on the moon (so far) when I was about halfway through my engineering degree.

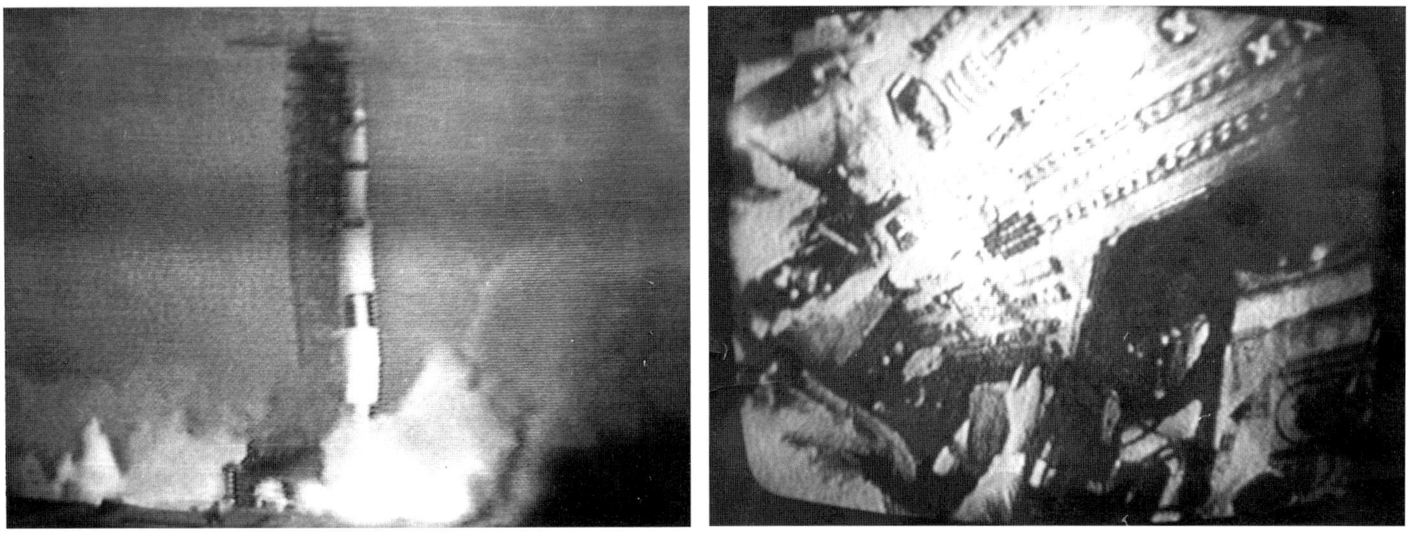

Two shots of the Apollo 8 mission photographed directly from the TV screen. A desperate attempt by the writer to capture this historic mission using the only tools he had as a 16-year-old living in 1968.

By the time I graduated, it was all over, so it seemed. Not only was the Apollo program long since completed, but Skylab was also in the past; and we wrapped up all the Apollo hardware flights with the Apollo-Soyuz Test Project the month after I graduated from U.C. Berkeley.

I did manage to follow my passion with jobs in the space program, but the only one hiring was the military. After a stint with GE Aerospace as ground support for an Air Force reconnaissance satellite, and another with Lockheed working on the Trident missile, I decided that aerospace was more satisfyingly pursued as a hobby rather than a profession. I was able to re-start my model rocketry hobby in the early '90s, which was just about the time we started hearing about the giant Soviet moon rocket that had been kept secret for so long. After a few years of limited information and rampant speculation, we all "knew" what had happened to the Soviet moon program, and it went something like this:

The Russians had a huge lead in the early space race thanks to their bigger boosters that were developed to carry their larger (read cruder) atomic weapons. After Kennedy gave his "We-will-go-to-the-moon" speech that started the Apollo program, the Soviets squandered their lead with more propaganda flights: the first dual mission (two vehicles in orbit at the same time), the first woman in space, the first multi-manned ship and the first spacewalk. By the time they got serious about competing on the moon race they had some catching up to do. The monolithic Soviet bureaucracy swung into action and directed its scientists and engineers to copy the American effort, and do it faster to beat us to the moon. Lacking the industrial infrastructure to build really big engines, like the Saturn's F-1, they opted for a large number of smaller engines that were difficult to control. They were also unable to develop the more efficient hydrogen-oxygen type engine so had to use the lower-energy kerosene-oxygen propellants in all the booster stages. This limited their payload to much smaller spacecraft and a crew of two cosmonauts, only one of which would go down to the surface.

Then their mysterious "Chief Designer" died of a heart attack and the Soviet politburo in its infinite wisdom, transferred the project to a different design bureau which had to start from scratch to understand the vehicle. Even though the Americans had their own setback with the *Apollo 1* fire, they were still in the lead so the Russians skipped ground testing of the massive, multi-engine first stage and went for broke trying to fly the entire stack live for the first time. When one of the many engines caught fire, the primitive control system mistakenly shut off all the other engines and the rocket crashed. A second all-up attempt exploded on the pad killing many of the engineers and technicians crucial to the program. At that point they knew they had lost the race, but tried a couple more face-saving launches after *Apollo 11*, and when those also failed the program was cancelled. The whole thing had an air of desperation about it.

Well [ahem], back to reality. Except for the late start and the directive to beat the Americans, most of that was, I'm happy to say, way off base. The actual story is far more involved and far more interesting. The Soviet space program was anything but monolithic and the political maneuvering that went on unseen by the west reads more like a Tom Clancy novel than the history of a technology program. The N-1 was hardly a slapped-together attempt to be first on the moon; its origins go back much further and its goals far more ambitious. I'm thrilled to be able to present the story here, probably for the first time to most readers.

Jack Hagerty
Livermore, CA
November 2013

Note on the translation of hardware terms

Many of the hardware diagrams in this book are reproduced directly from the Russian sources. Having a background in aerospace, I've translated the callouts freely based on the function of the device rather than literally from the Russian. More often than not this means using terms from the US space program which, in many cases I'm sure, is nowhere close to a literal translation. I've left the original Russian callouts on every diagram that had them so that native speakers can get the original meaning in context.

Note on technical drawings
(If you don't recognize the figures below you can skip this note)

I have spent my whole engineering career working in the US where we use Third Angle projections. The N-1 was, of course, designed in the Soviet Union (Europe and Asia) where they use First Angle projections. Some of these design drawings, and others of European origin, have been reproduced in this book. For editing purposes, if they were in First Angle I rearranged them into Third Angle so that I could visualize them more easily to be sure they were correct. My apologies to my European and Asian readers.

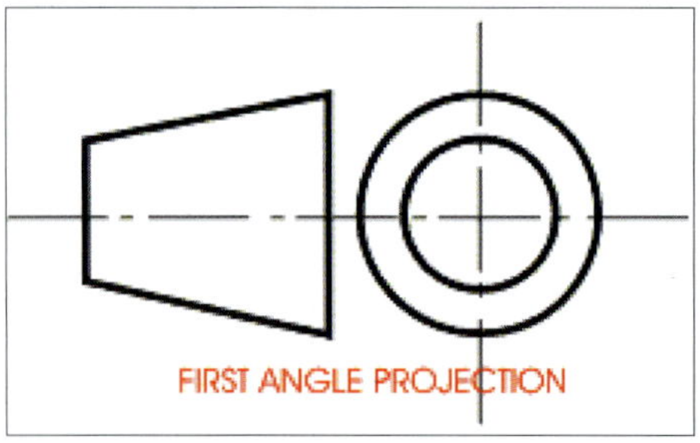

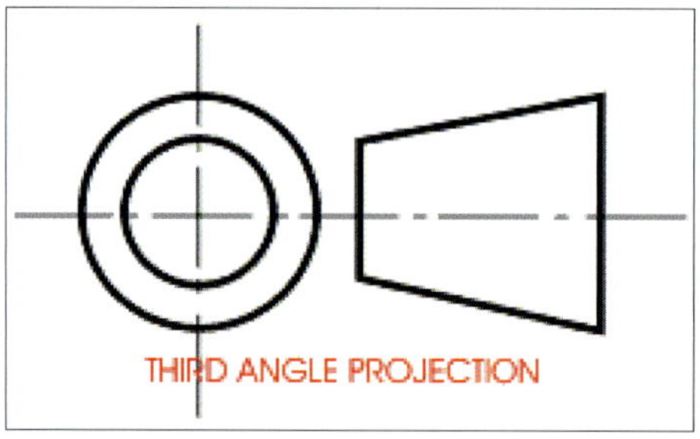

Introduction

For a modeler, the appeal of the N-1 is instantaneous. Who could not be drawn to the sheer audacity of the design? The N-1 has thirty engines in the first stage alone. The first three stages sprout forty-two engines, twenty-four fairings and twenty telemetry antennae! Then there are all those panels, jets, lattice fins and separation rockets scattered over the body, not to mention the two huge interstage trusses holding the thing together. The N-1 is a modeler's dream of detail and a rocketeer's dream of engineering.

After the fall of the Berlin Wall in 1989, information began to trickle out of the new Russian Federation about the long-denied moon landing project: a giant rocket had existed! By 1993, enough documentation had been gathered to allow engineer-historian Alex Shliadinsky to produce his famous set of mechanical drawings of the mysterious moon rocket; and by 2005, Alex had created his AutoCAD files of the N-1. This was an achievement of herculean proportions, considering the level of detail required for a rocket that had been dismantled and scattered all over the debris fields of Baikonur.

After becoming a Born Again Rocketeer in 2006, I began research in earnest on the N-1 with the desire to create a thirty-engine flying model that would be NAR contest-worthy. Over three years, that research yielded international online friendships with Alex Shliadinsky in St. Petersburg, Vladimir Antipov in Moscow, Igor Bezyaev in Obninsk, and Nick Stevens in London, as well as the late Mike Dorffler of Estes Rocket design fame.

This book uncovers odd bits of documentation, old photographs, videos, scientists' dreams and a heart wrenching story of near success and loss. While it started out primarily geared toward a modeler's reference to the rocket, as Tolkien said in the introduction to Lord of the Rings, "the tale grew in the telling." My aim was to share the discoveries I made while gathering the necessary information needed to build an N-1 model as accurately as possible; but the sheer volume of detail provided by the contributors mentioned above meant that this would be far more than a modelers' source book. Other books and websites exist to tell the N-1 story but none can claim the depth of experience represented by my collaborators. While my goal was to create a flying model, the material in this book will aptly suit any static modeler's needs all the way to museum quality.

In addition to Alex's drawings and Nick's wonderful CGI renderings, we also have two chapters written by Alex covering his personal story of discovery regarding the N-1 and his observations on the rocket's design and history. These priceless chapters, as well as other Russian documents on the project, were translated by Igor Bezyaev. Igor has included two chapters of his own, one on the planned moon mission using the N-1 and a second on the booster's original purpose for missions to Mars and Venus.

Matthew Johnson
Grand Rapids, MI USA
July 2013

Part I

History and Specifications

1: A Brief History of the N-1

1: Краткая история Н-1

A Path to the Moon—by Matthew Johnson

By tossing down the space-race gauntlet in May 1961, President Kennedy set two nations on the path to the moon. Even after the United States emerged triumphant on 20 July 1969, the Soviets maintained a stark silence on the matter. It would be decades before we would discover, publicly at least[1], just how close that race for the moon truly was. On 21 February 1969, less than five months before the flight of Apollo 11, the Soviet Union had the first N-1 moon rocket, designated N1-3L, on the pad and ready for launch at 09:18 GMT.

[1] The US CORONA spy satellite program provided proof of a very large moon-capable rocket on the pad at Baikonur in 1969. See page 56

N1-3L in the assembly hall of Site 112

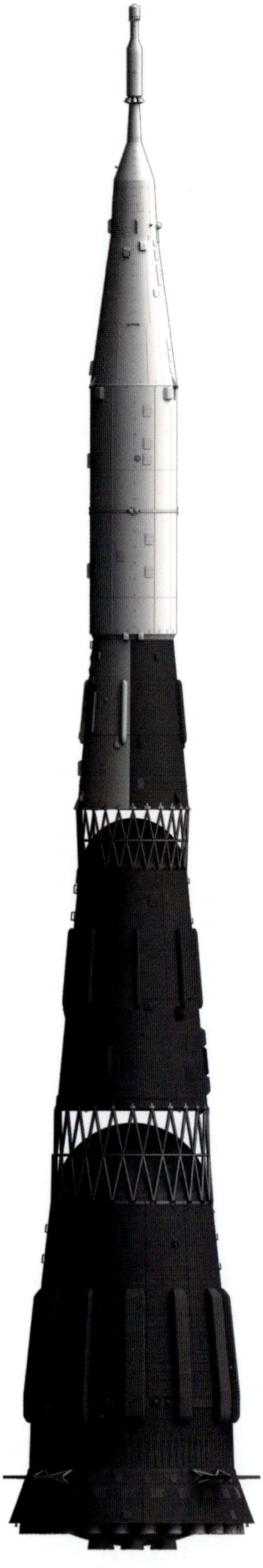

N1-3L

First Launch

The flight of that first N-1 would only last for 68.7 seconds before engine failure and fire in the first stage would cause an abort destruction of the craft.[2] One can only imagine how the course of history would have been different had the flight succeeded.

It is well known that the Soviet approach to solving the engineering problems of such a complex rocket as the N-1 was, either by design or necessity, to build a reasonable prototype, fly it and analyze the results, even if those results were disastrous.[3] Obviously, the operation of thirty engines in the first stage would be a major concern. The Soviet engineers devised a control system called KORD[4] (Literally: Control {KO} of Rocket {R} Engines {D}) to control the ignition and subsequent operation of the thirty NK-15 engines[5]

A total of twelve R&D flights were planned to sort out this immensely complex vehicle and it's amazing that they came as close to success as the did by the fourth flight.

---

[2] Vladimir Antipov and Alex Shliadinsky report that, knowing the flight was a failure, the engineers decided to fire the second stage engines early to gather more telemetry. This is where the information was gathered about the burn through of the first stage fuel dome, which resulted in the addition of the blast deflectors on the subsequent versions of the N-1.

[3] The explosion of the N1-5L (next page) remains the single biggest explosion in the history of rocketry and one of the largest non-nuclear explosions ever.

[4] While it had difficulty controlling the engines on other flights, the KORD system did shutdown the rocket correctly on this first flight and fire the LES (Launch Escape System), saving the mock-up of the lunar craft from destruction.

[5] The Soviets were unable to develop large engines like those on the Saturn V in time for the moon program (see page 47). However, they were already masters of smaller engines and managed to create the powerful NK-15 for Block-A, the NK-15V (vacuum version of the NK-15) for Block-B and the smaller NK-19 for Block-V. The high efficiency of these engines came from the design being a closed system; piping all of the liquid oxygen propellant through the turbines, creating an oxygen-rich exhaust that was fed directly into the main combustion chamber. This captures all of the energy from the impeller side of the turbine that normally is wasted out an exhaust port. It was also a task was balked at by western engineers as too difficult, if not impossible (this is covered in more detail in Chapter 4). The NK-15's were to be replaced by the improved NK-33 in Block-A, the NK-43 vacuum version in Block-B and the NK-39 in Block-V on later flights, but all four of the flown versions used the NK-15/19. The fifth flight, N1-8L, was the first one built with the NK -33/43/39, but the program was cancelled before it was flown.

NK-33

N1-3L Liftoff

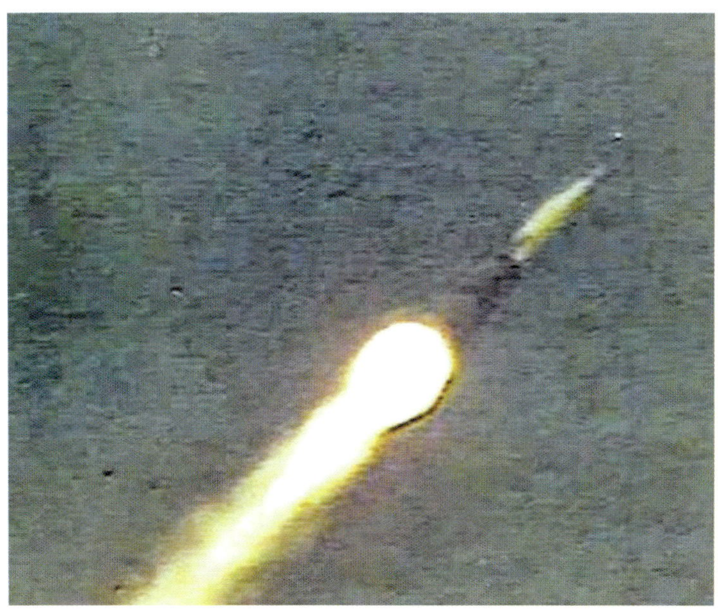

N1-3L In Flight

Second Launch

The second N-1 flight, designated N1–5L, was launched on 3 July 1969 at 20:18 GMT. At 0.25 seconds into the launch, either a bolt or a fragment of welding slag (accounts differ) was sucked into a fuel pump causing that engine to catch fire and shut down. The N-1 fell back onto the pad at a 45° angle and exploded, killing scores of engineers.[6] Occurring only two weeks before the launch of Apollo 11, the failure of this N-1 insured that the Soviet Union would not be the first to land a man on the moon.

N1-5L LES Fires as the booster explodes

[6] This launch failure caused the KORD vehicle control system to fire the Launch Escape System (LES) as designed and carry the mockup moonship away from the pad. These two LES firings are not the only time the escape system was used under launch conditions in the Soviet space program. On 26 September 1983 Vladimir Titov and Gennady Strekalov were waiting in *Soyuz T-10a* preparing to launch to the *Salyut 7* space station when a fueling accident caused the booster to catch fire. The LES was activated remotely by ground control and the crew endured 15 or more g's for five seconds as they were pulled away from the collapsing booster. They landed some 4 km downrange and were badly bruised after the high acceleration, but were otherwise unharmed. This remains the only time an LES was used to save a human crew.

In the US program, the first two Mercury-Redstone flights had LES firings. In MR-1 (unmanned) a badly programmed controller jettisoned the escape tower just a few seconds before booster ignition. MR-2, with the chimp "Ham" on board had the booster shut down six seconds early which caused the LES to (correctly) fire and pull the capsule away. Ham got a 16-g jolt and sailed 35 miles higher in his trajectory than planned, but was unharmed. Finally, the third Mercury-Atlas test (MA-3) had a booster guidance failure where it never pitched over and just continued to ascent vertically. Range safety had to destroy it, but the LES pulled the (unmanned) capsule free as designed—Ed.

Third Launch

After research, investigation and upgrades (including a screen on each propellant pump inlet), a third N-1, designated N1–6L, was launched on 26 June 1971 at 23:15 GMT. Fraught with roll stabilization problems from the moment of lift off, the rocket was aborted and destroyed by ground command at 50.1 seconds into the flight just as the L3 moonship section started to break off of the booster. This launch experienced a severe "pogo" effect, a large oscillating vibration along the axis of the rocket. This is something that also affected the second test launch of the Saturn V.

Launch of N1-6L

Fourth Launch

On 23 November 1972 at 6:11 GMT a fourth launch of the N-1, designated N1–7L, was attempted after numerous modifications were carried out on the design to increase payload capacity and correct design flaws. This last N-1 would fly for 106.93 seconds before engine failure in the first stage once again caused a shutdown and a ground-commanded destruction of the rocket, less than 10 seconds from the planned staging.

Launch of N1-7L

An Ignominious End

The next launch, N1-8L, was scheduled for late in 1974. There had been a large number of improvements from N1-7L including having the final configuration of NK-33/43/39 engines in Block-A, B and V, respectively. The vehicle had been completed, moved out to the launching pad and was undergoing fueling tests when the program was cancelled in May of that year.

In 1976 the Director General of Energia, V.P. Glushko, ordered all remaining N-1 hardware and equipment destroyed. In spite of this order, nearly 100 of the NK-33 engines survived in storage. And, proving the Soviet penchant for utilitarianism, many of the other pieces have existed right into the 21st century as store signs, gazebos, sheds, sandboxes, pig troughs, bus stops and a field of wreckage around Baikonur.

Lunar Architecture Gallery

When V.P. Glushko ordered all of the existing pieces of the N-1 program destroyed, he didn't mean ground up into little pieces. All of the boosters and components that had been built were dismantled and scrapped as fairly intact assemblies. It wasn't long before the ever-resourceful Soviet citizens assimilated the pieces into the surrounding cities and countryside. For hundreds of kilometers around Baikonur you can find reminders of the moon-program that almost was. Below and on the next two pages are examples of what Vladimir and his researchers have found.

Parts from the enormous Block-A were quite useful. Above, a fuel tank cover was used to create a rural hut. The image to the right is the view from inside a structure made from a base ring fairing looking out of one of the engine access hatches.

Right, another tank cover, along with some of the Block-A/B interstage truss, is used to make some large play equipment.

Note also the Block-V engine heat shield on the extreme left being used as a sun shade. This is the same structure as shown at the bottom of page 9.

The Lattice Fins were prized for their exotic look. Above left, four of them are being used as open walls on a picnic cover. Left, a half dozen form a bus stop shelter, and above, half of a fin is used for a roadside sign.

Many N-1 pieces wound up in playgrounds and parks. Below, segments of a Block-A engine heat shield are used as planters (although there are no flowers in this winter photo). To the right, the end segment of a Block-B fuel line cover becomes a toy boat. Do you think that the children playing here imagining this to be a real boat had any idea that it was actually part of a real moon rocket?

The huge expanses of sheet metal that went into each N-1 made the cast-offs especially valuable as sheds and covers. Above, a Block -A's unrolled skin is used as a garage roof (the roof is 70 m (230 ft) long!). The grey area in the center outlines where the fuel line cover was. The black rectangles cover the holes where the fuel lines exited the body to go down the side around the LOX tank. The attachment fittings for the lines are still there.

Top right, a Block-V skin is split open lengthwise for use as a storage shed. Note the hatch and visible rivet pattern.

Right are segments of the L3 fairing being used on a dairy farm as storage sheds (upper) and, along with some of the Block-B/V interstage truss, cattle feeding covers.

Below is the Block-V engine heat shield being used as a shade (with four holes in it).

A Surviving Legacy

The silver lining of this dark cloud is that some brave and enterprising engineers provided a gift to the future. In 1997, after the fall of the Soviet Union and the death of Glushko (coincidentally both in 1989), the "veritable forest" of ninety-four NK-33 engines stored in the warehouse was offered for sale to western aerospace companies. This surviving legacy of the Soviet moon race was sold to the American companies Rocketplane Kistler,[7] and Aerojet General. Unfortunately, Rocketplane Kistler filed for Chapter 7 bankruptcy on 15 June 2010, but Aerojet sells the engines as the AJ26. An upgraded version of the NK-33 called the NK-33A, is back in production by Roscosmos for use by Energia on a new family of Soyuz boosters, designated Soyuz 1.

---

[7]For a description of the Kistler *K-1* project that had planned to use the NK-33 engines, see *Spaceship Handbook* from ARA Press.

Once upon a time...

A. The "forest" of NK-33 engines hidden for twenty years
B. Nikolai Dmitriyevich Kuznetsov, designer of the N-1's engines
C. AeroJet in the US tests an NK-33
D. Model of an NK-33 in the Soyuz 1
E. N1-3L on the factory floor
F. Mighty thirty engine rear view of N1-3L
G. N1-3L being raised onto launch pad
H. N1-M1 (Weight Model) on pad
I. N1-5L on the "Grasshopper" transporter
J. N1-6L Liftoff
K. N1-7L on pad

Dear Reader: These blank pages, as well as the ones between 21-22 and 67-68, are not a printing error, they are a layout error that was not caught at the proof stage. There are no missing pages.

2: Specifications and Clarifications on the N-1

2: Технические данные Н-1 и пояснения

Main text by Matthew Johnson

N-1 Designations and Color Schemes

Flight Round Numbers

"N" is an abbreviation for *Nositel* or *Raketa-nositel* ("Rocket-Booster")

Flight rounds numbered 3L thru 7L plus the M1:

- N1-3L: Launched 21 February 1969
- N1-4L: Partway through assembly redesignated as N1-M1 (see below) for use as pad fit-check
- N1-5L: Launched 3 July 1969
- N1-6L: Launched 26 June 1971
- N1-7L: Launched 23 November 1972
- N1-M1: Also called GVM ("Overall Weight Model"). It worked out all the changes introduced into the construction of the flying versions.

Spacecraft Designations

- L3 Lunar Complex: Everything above Block-V (see below). The lunar exploration craft were designed for "Lunar Plan L3." There were plans for lunar flights up to "L5" involving a lunar base.
- LOK (Lunar Orbital Craft): Designation for the Soyuz 7K-L3 spacecraft.
- LK (Lunar [Landing] Craft): Designation for the lunar lander.

Block Designations[1]

- Block-A: Designation for the first stage.
- Block-B: Designation for the second stage.
- Block-V: Designation for the third stage.
- Block-G and D: Designations for L3 Complex stages: Block-G (Trans-Lunar Trajectory stage) and Block-D (Lunar Insertion and LK de-orbit stage).
- Block-E: Designation for LK lander engines.
- Block-I: Designation for the LOK engine used for Trans-Earth Trajectory.

Color Schemes

As a victim of factory fluorescent lighting, European cross-cultural film developing, the whim of uninformed photographic developers and aged photographs, the color scheme of the N-1 has long been a much debated issue.[2] The battleship grey and white color of the N-1 series has been confirmed by Alex Shliadinsky and Vladimir Antipov[3] through exhaustive research, examination of remaining fragments of the N-1 at Baikonur and recollections of eye witnesses.

Many publications feature a green colored N-1 rocket. This is a gross error due to the fact that the majority of photographs used for reference are of poor quality. Professionals who printed the photos, more often than not, achieved the color in accordance with their views. I have seen some pictures of the Soyuz rocket, which is the same grey color, presented green, grey and even blue. By the time the printed images of the N-1 rocket were available, the declassified images of R-1, R-2 and R-11 were standard. These missiles had a short range and were secretly transported to the launch sites. They were painted green to mask their appearance and this influenced the printing of other photos. Long-range missiles like the N-1 did not require such a disguise. Besides, one should bear in mind that the N-1 rocket was launched from Baikonur where there is very little greenery, the landscape consists mainly of grey sand. It would be very strange to place a green rocket against the background of grey sand. In terms of concealment this is nonsense.

- Vladimir Antipov, 2011

The rocket N1-3L was painted only in grey and white colors; green was not used. In photos of Soviet rockets, a green color is often visible. Actually it appears in photos because of the color rendition of the film during processing. [For exam-

[1] Note that these Block designations are the direct translations from the Greek-based Russian alphabet into our Latin-based one. Once you get past "B" the letters don't really line up consistently, so don't be concerned that the Blocks do not seem to be in alphabetical order, they actually are—Ed.

[2] This color is often rendered in light green. The error is so prevalent that it can be found in the National Association of Rocketry's *N-1 Scale Data Pack*, and on the Smithsonian Institution's model of the N-1—Matt Johnson

[3] Confirmed by Mike Dorffler of Estes, 2008—Matt Johnson

N1-3L

ple,] I was present [at the launch] of a rocket at Baikonur in 1991. The rocket was grey. I have been very surprised, when seeing albums devoted to this [launch], that there are photos of a green rocket.

- Alex Shliadinsky, 2008

Their consensus on the different N-1 color schemes is as follows:

- N1-3L—Block-A and Block-B (stages 1 and 2) painted grey. Block-V (stage 3) painted half grey, half matte white. L3 payload fairing painted matte white.
- N1-5L and 6L—Blocks A, B and V (stages 1, 2 and 3) painted matte white with the band on the service tower side painted grey on all three. Interstage trusses and fuel tank domes were painted grey. L3 payload fairing painted matte white.
- N1-7L—Completely matte white.

Beginning with the launch of N1-8L, the rockets were to be painted completely glossy white. The transition from grey to white color was associated with the fact that the rockets became very hot in the Baikonur summers and the inside compartments sometimes reached temperatures up to 70°C (158°F).

- Vladimir Antipov, 2011

N1-5L

This widely available photo is often flipped. This is the correct orientation.

N1-6L

N1-7L

Red Objects

Many red objects can be seen on the hull of the N-1 rockets in various photographs ranging from the factory floor to the launch site. These objects were dust covers and pressure sustaining devices (the "red cans" often seen on the third stage and L3 section). All red objects were removed from the N-1 before flight - Alex Shliadinsky, 2008

Block-B Engine Variations for N1-3L

The NK-15V ("V" for high altitude in Russian shorthand) second stage engines were not ready in time for the flight of rocket N1-3L in February 1969. The standard NK-15 first stage engines were used. As the first flight of the N-1 had a payload that weighed much less than on the subsequent flights, the standard NK-15 engines were sufficient -Alex Shliadinsky, 2008

NK-15 on 3L Block-B

NK-15 Engine

NK-15V Engine

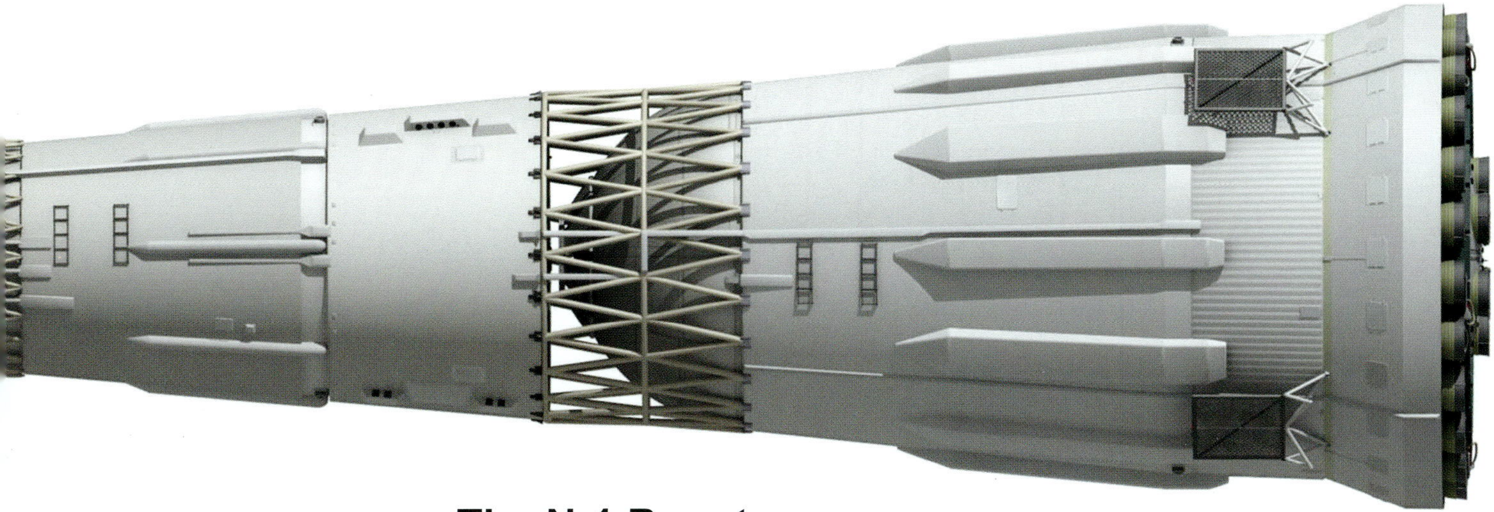

The N-1 Booster

This is what an operational N-1 would have looked like without payload

Block-B Block-V Block-G Block-D Block-E (LK) Block-I (LOK)

The N-1 Unstacked

the N-1 and L3 Complex stages at the same relative scale

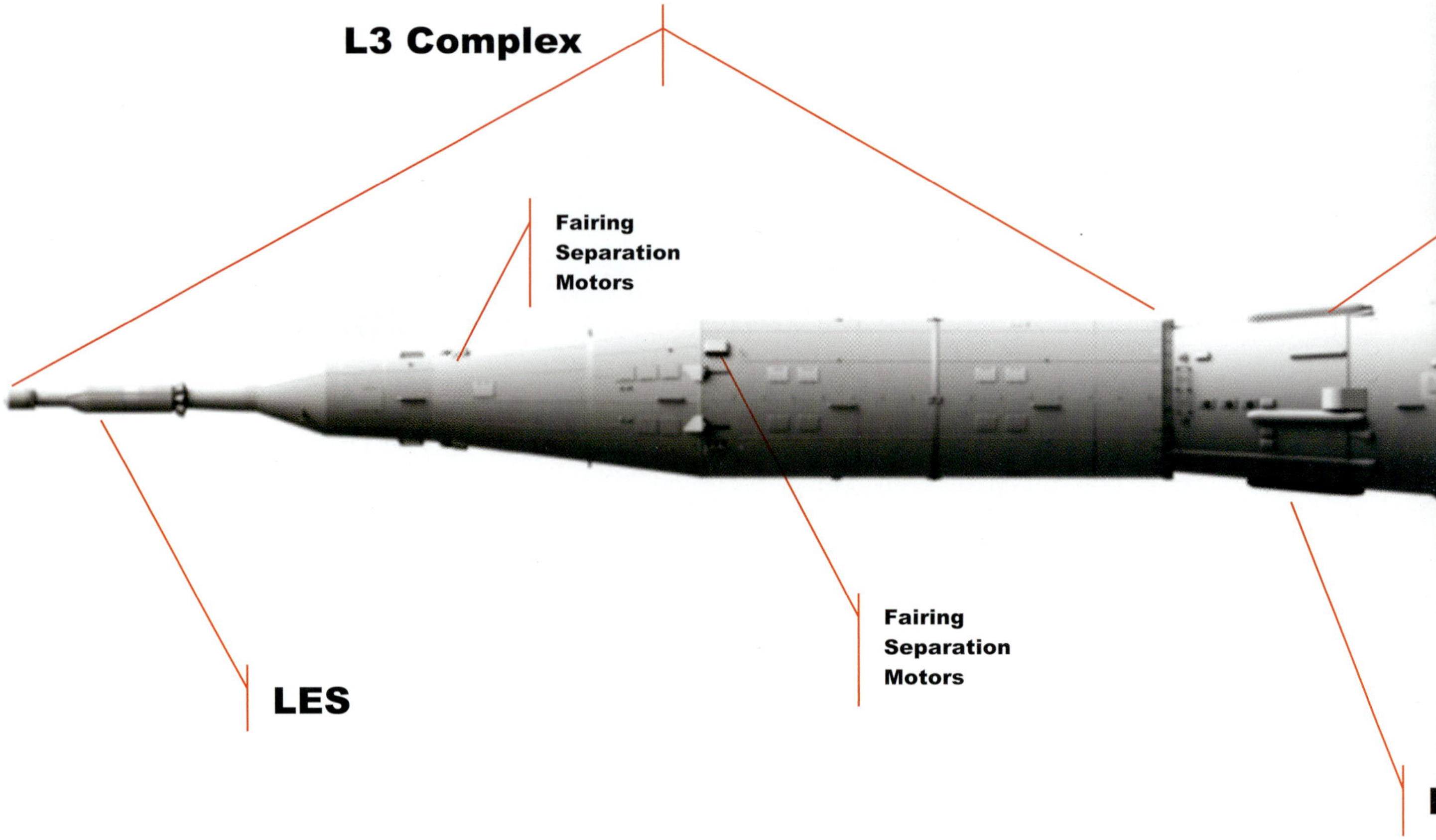
L3 Complex
Fairing Separation Motors
LES
Fairing Separation Motors

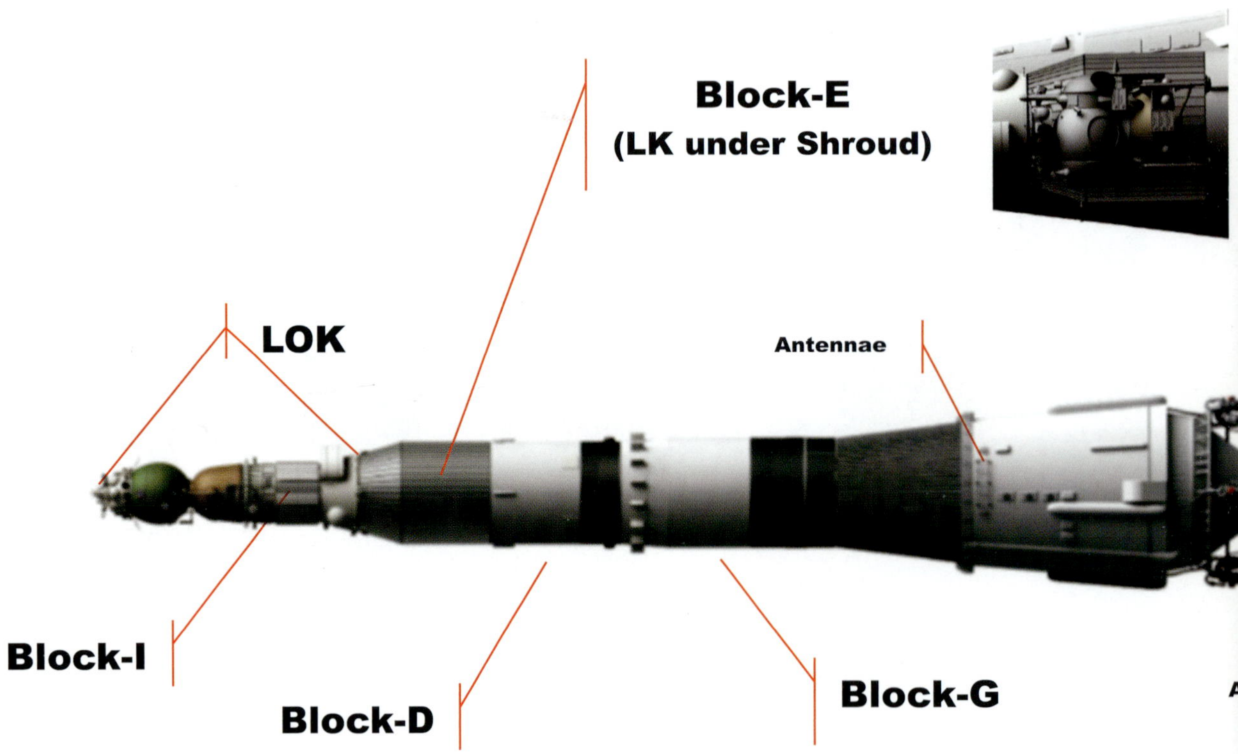
Block-E
(LK under Shroud)
LOK
Antennae
Block-I
Block-D
Block-G

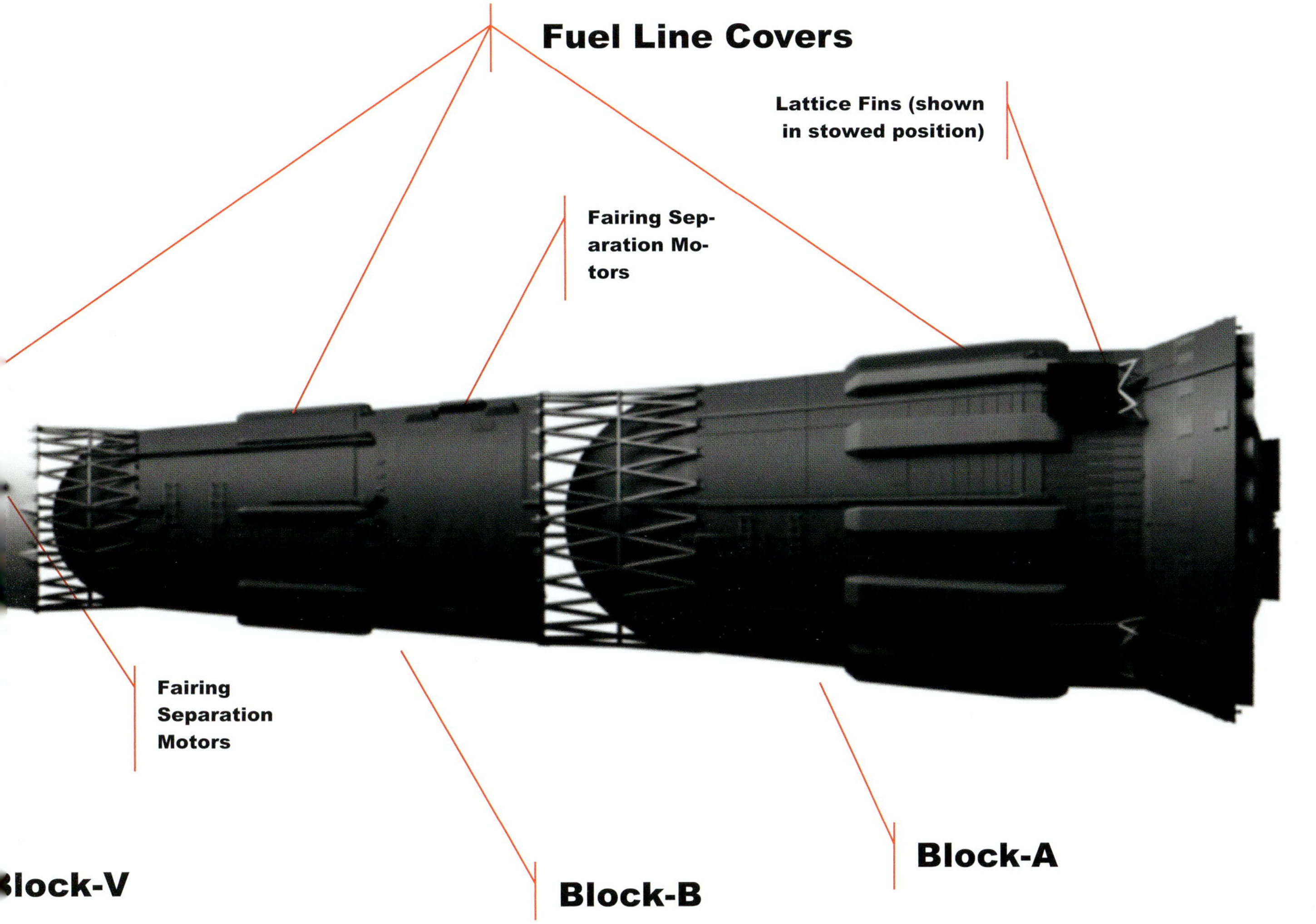
Fuel Line Covers
Lattice Fins (shown in stowed position)
Fairing Separation Motors
Fairing Separation Motors
Block-V
Block-B
Block-A

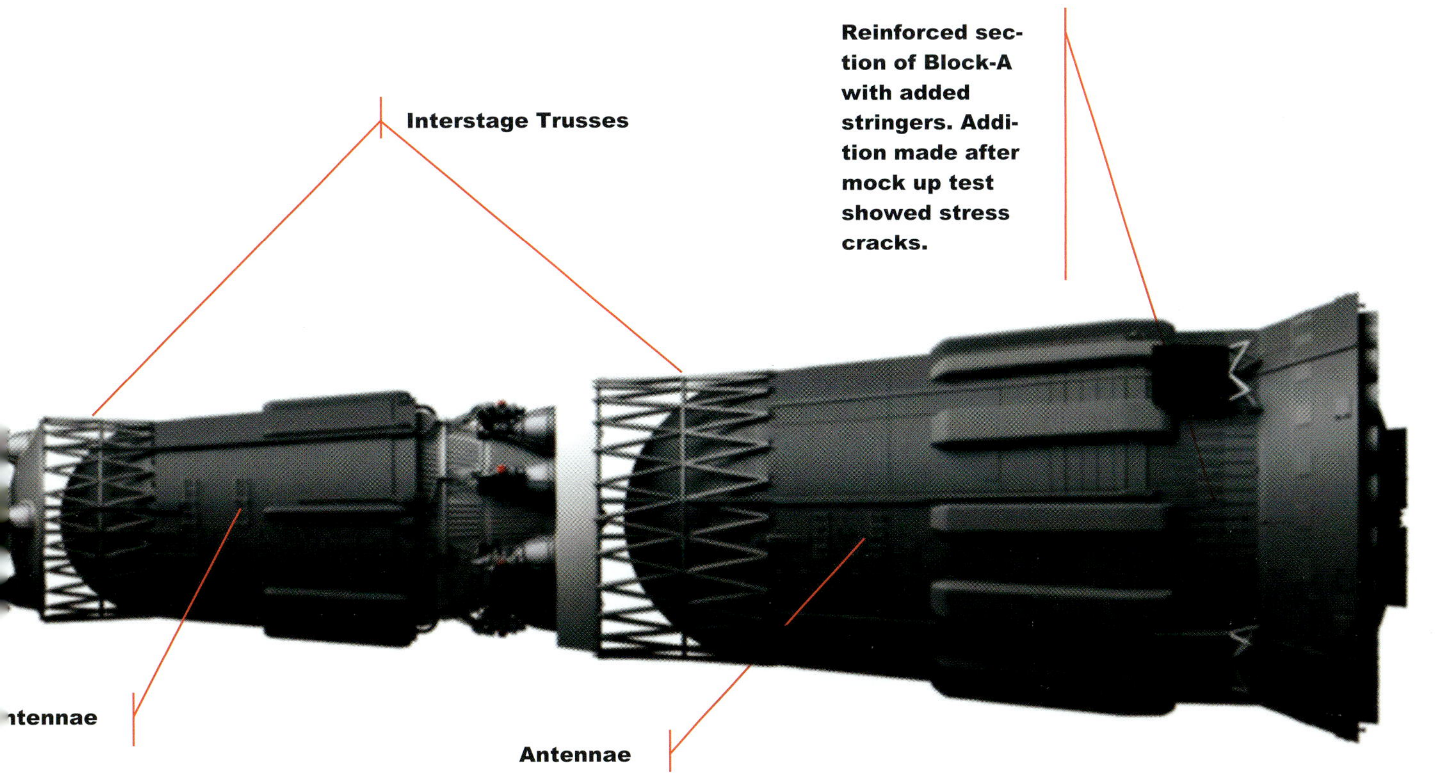
Interstage Trusses
Reinforced section of Block-A with added stringers. Addition made after mock up test showed stress cracks.
Antennae

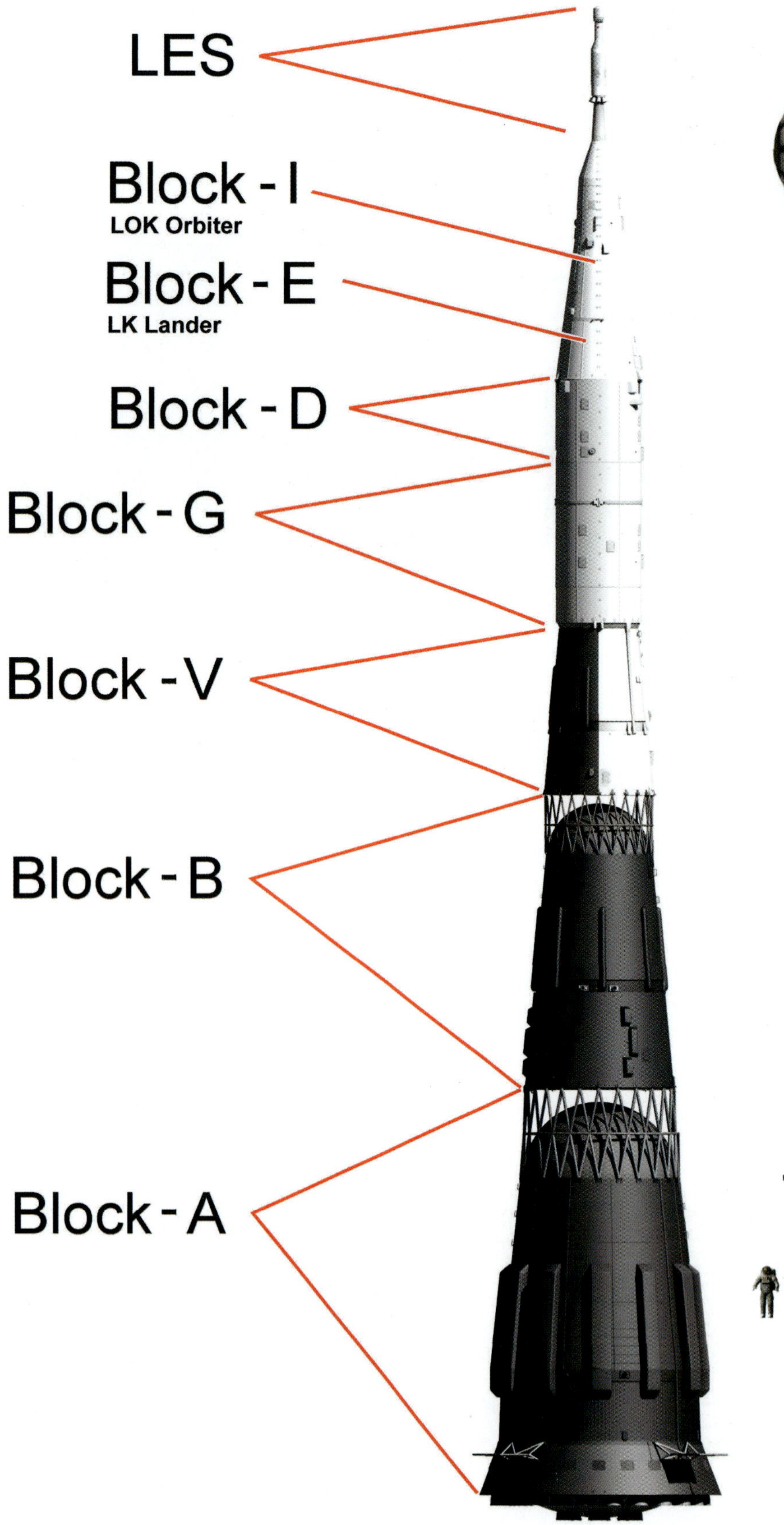

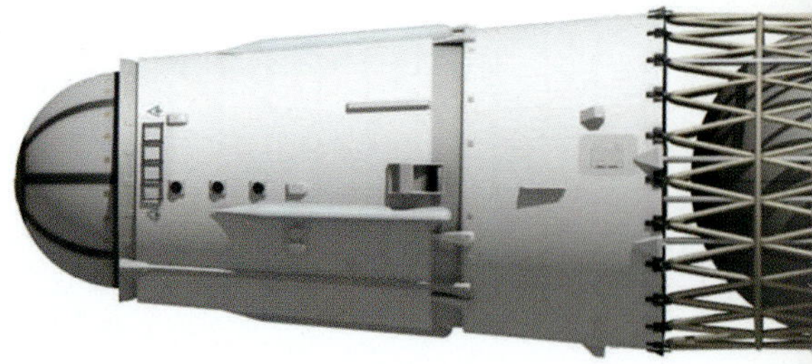

Block-A

All of

Infamous Block-V "Pipe Organs"

The N-1 rocket sported three rather large curved umbilical lines along the vertical axis of the third stage, often referred to as the "pipe organs" for obvious reasons. These are so prevalent in extant photographs that it has become a common belief that these were part of the N-1 itself, even famously appearing on the scale Lunar Models' resin kit. These were actually jettisoned at launch.

The designated elements are systems for preparing the rocket for launch. Similar systems were used on Energia and Zenit rockets. These devices are arranged on the vertical axis of the rocket. During the moment of ignition they were shot away from the rocket. They were present in all flights of the N-1.

—Alex Shliadinsky, 2008

Umbilical connections, factory floor.

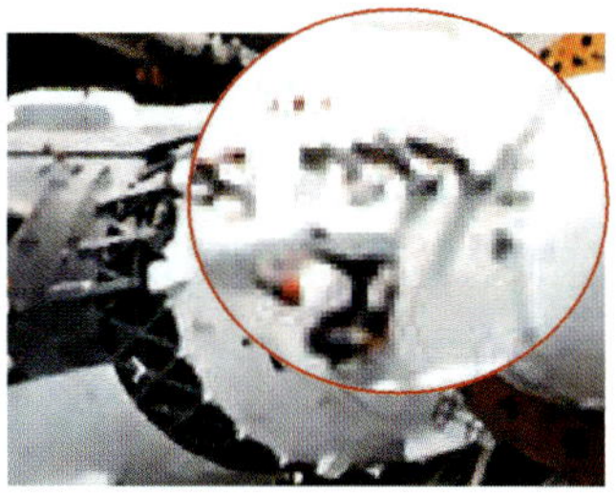

Umbilical connections with "Pipe Organs", during transport.

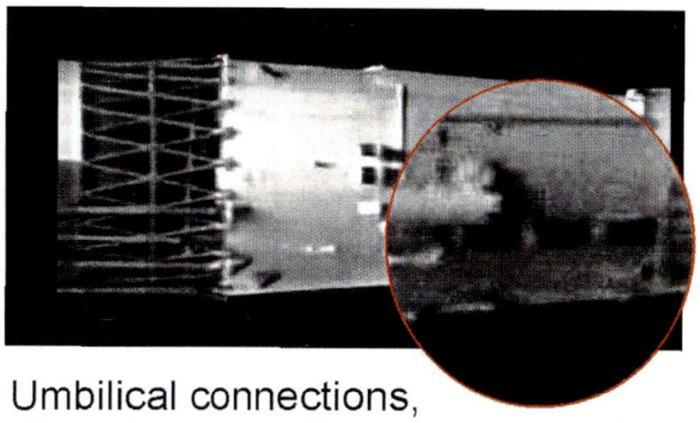

Umbilical connections, "Pipe Organs" jettisoned at launch.

Block-V Cover

The nature of a major object on the surface of the third stage became the focus of disputed interpretation over the years, as neither the mechanical drawings nor the AutoCAD files could explain its surface completely. The question was whether or not the object was a smooth cover or a large latching device with a roll-bar. Luckily, an amazing photograph, albeit in black and white, taken during the launch of the 5L, clearly shows a solid surface

—Alex Shliadinsky, 2011

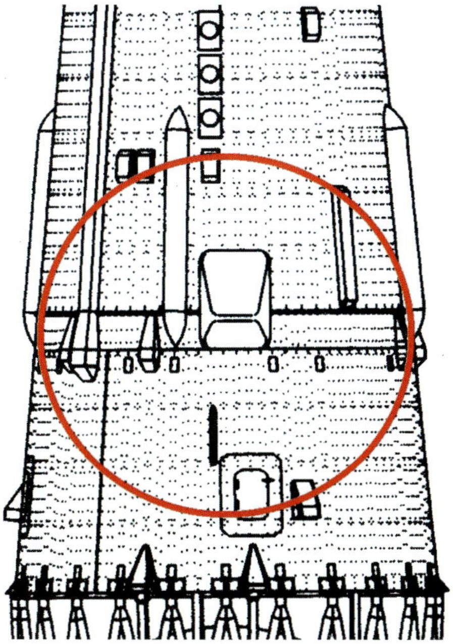

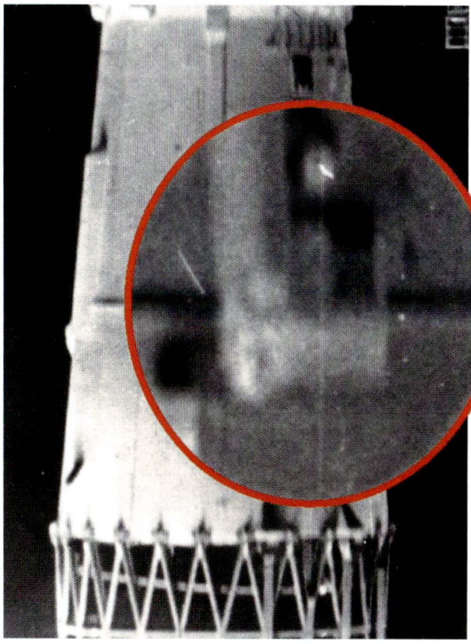

N-1 Variants

Block-A Reinforcing Stringers

The surface of the lower section of the first stage has caused confusion for researchers because its structure changed over time. The first flight version, N1-3L, was smooth except for some bracketing along the upper edge where it connected to the conical section. However, Alex Shliadinsky discovered that as the N-1 was fit-tested on the pad, cracks appeared and additional ribbed supports (stringers) were deemed necessary. These were added before the launch of the N1-3L, as seen in the early drawings. These stringers can also be seen on N1-5L and N1-6L. By N1-7L, the structure was modified to strengthen it internally so the stringers were not necessary..

Before making a flight version of the N-1 rocket, a full size mockup was created (GVM - Overall Weight Model, N1-M1). This was necessary for testing launch pad contraction and other systems. The main difference between the flight rocket and the mockup was that on the top of the nose there was no Launch Escape System (LES). During the test of lifting the rocket to the vertical position, cracking was detected on the transition section and the main load-bearing frame. The decision was made to make those elements stronger by installing stringers on the transition section. This change consisted of metal straps with thickness of 10 mm and a width of 50 mm. Some of them passed along all of the transition section and some of them stretched from the main load-bearing frame down to about 720 mm. On the main load-bearing frame there were installed metal L-bars 70x70x750 mm. The presence of stringers is easily seen in photographs to help distinguish the mockup from the first three launched rockets.

—Alex Shliadinsky, 2011

Stringers on N1-5L

Block-A Fuel Line Covers

As the design of the N-1 was refined over time, there were significant changes made to the first stage. By the time N1-7L appeared, the upper end of the fuel line covers became pointed and the rear end tapered towards the hull. The base skirt of the first stage also lost its truncated conical shape and became a straight cylindrical rim instead.

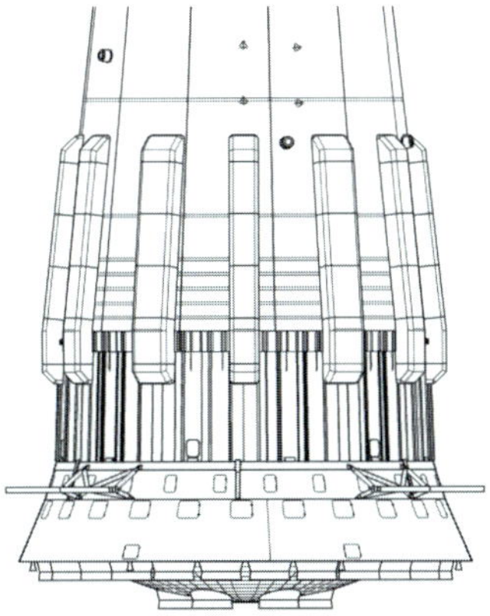

N1-3L

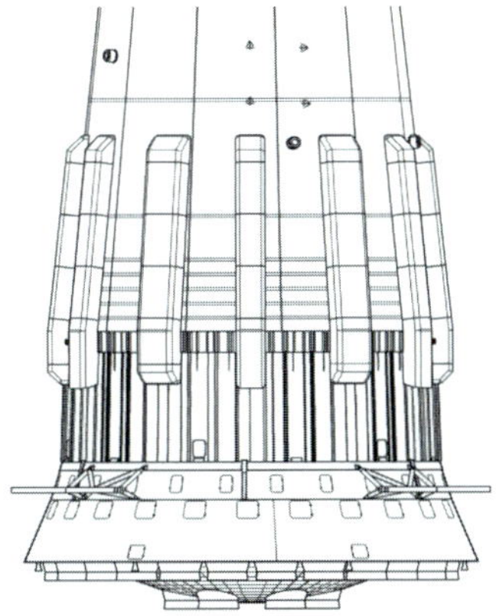

N1-5L

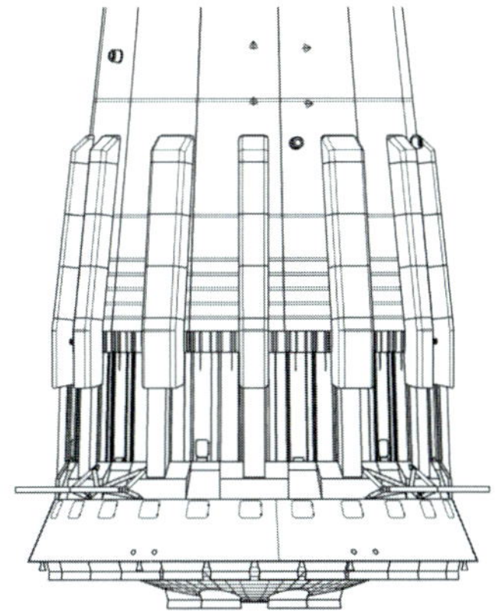

N1-6L

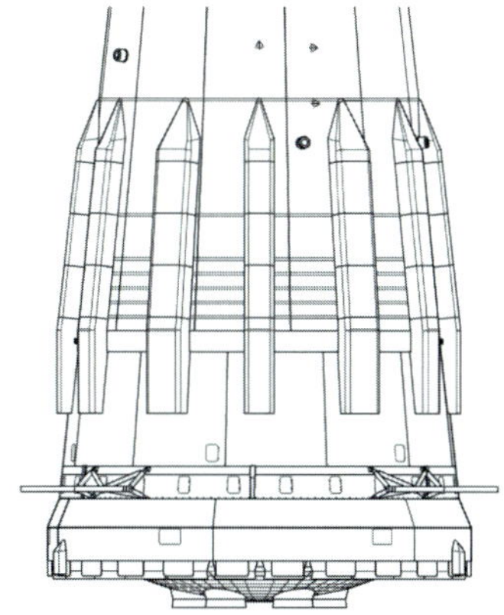

N1-7L

N1-M1—No Stringers

N1-6L—Stringers Visible Above Lower Skirt

N1-7L—No Stringers and white lattice fins

Block-A Heat Shield

The white heat shield protecting the bottom of the fuel tank appeared on all N-1's after the first 3L flight when it was determined that the exhaust from the NK-33 engines could pose a problem on future flights. The numerous dots are the black plastic attachment bolts.

N1-3L, No Heat Shield

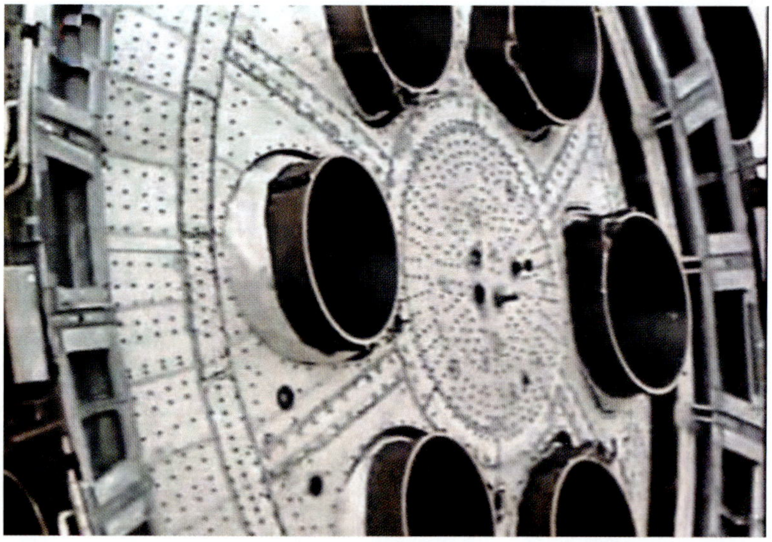

N1-5L, 6L, 7L With Heat Shield

N-1 M1 "Weight Model"

The N1-M1 weight model stands in place at launch complex 110 West. The N1-M1 is always detectable in photos by its lack of an LES tower. The N1-M1 proved invaluable in determining weak points on the N-1 structure and was responsible for the addition of the metal "stringers" on the exterior of the base of Block-A. By the time they got to the 7L, this structure has become smooth again with a higher degree of strength built in.

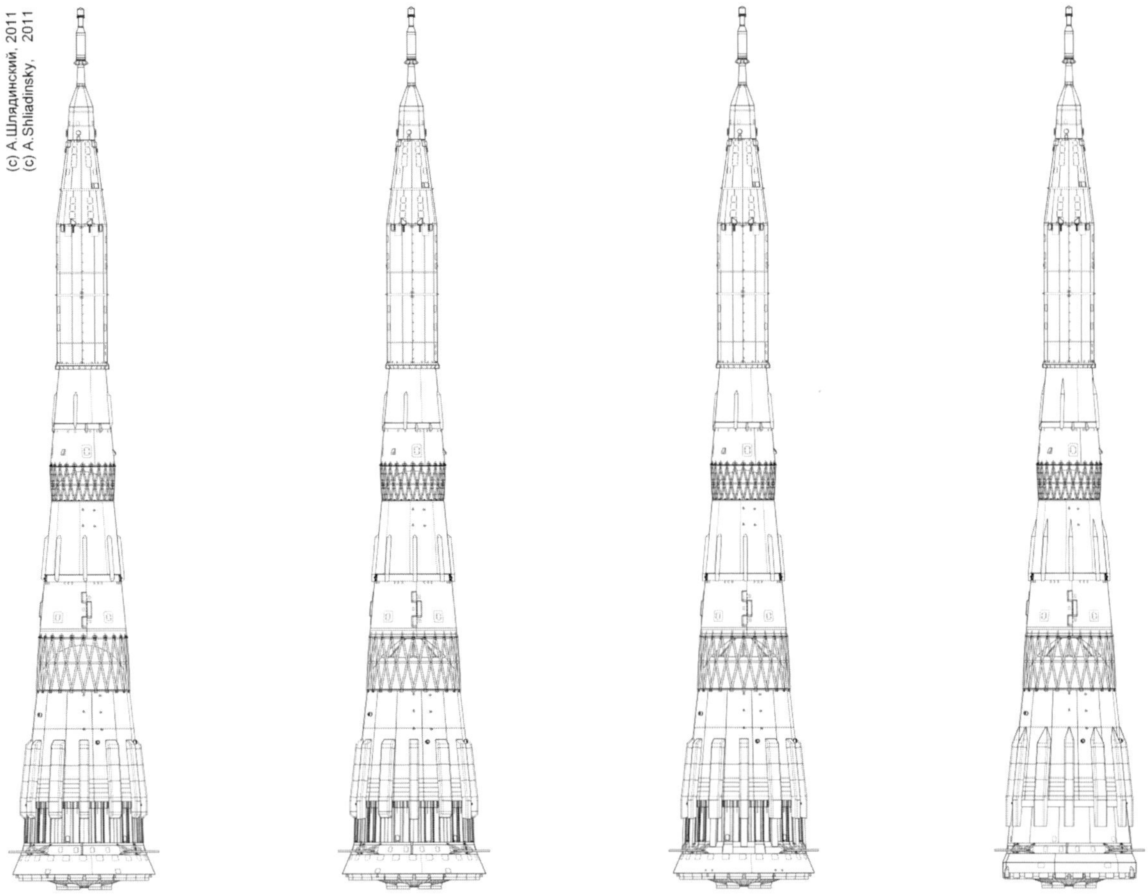

3Л 5Л 6Л 7Л

N-1 Variant Features

Feature	N1-3L	N1-5L	N1-6L	N1-7L
Block-A Color	Grey	90° Grey/ 270° White	120° Grey/ 240° White	White
Block-B Color	Grey	90° Grey/ 270° White	120° Grey/ 240° White	White
Block-V Color	180° Grey/ 180° White	White	White	White
Block-A Tank Dome Color	Grey	Grey	Grey	White
Block-B Tank Dome Color	Grey	Grey	Grey	White
Block-A Bottom Skirt	Conical	Conical	Conical	Cylindrical
Block-A Roll-Control Jets	Yes	Yes	Yes	No
Block-A Heat Shield	No	Yes	Yes	Yes
Lattice Fins	Black in Grey Frame	Grey	Grey	White
Fuel Line Cover Shape	Blunt Nose	Blunt Nose	Blunt Nose	Streamlined
Block-A Fire Extinguishers	No	No	Yes	No
Block-A Narrow Fuel Line Covers	Closed	Open	Open	Open
Block-A Wide Fuel Line Covers	Closed	Closed	Open	Open
Block-B Fairing Base	Exposed Latches	Streamlined	Streamlined	Streamlined

3: The N-1 Moon Mission

3: Н-1 для лунной программы

By Igor Bezyaev

This description of the L3 Soviet Manned Lunar Expedition is based on the book *Recollections of the Lunar Ship* ("Воспоминания о лунном корабле") by Filin Vyacheslav (Филин Вячеслав) published in 1992.[1] Vyacheslav was an engineer at NPO Energia in the late 1960's and early 1970's who worked on the LK development. I have also added some information regarding that Lunar Expedition Plan from other sources. This final version might be somewhat different than Vyacheslav's, but it shows to us the basic vision of the Soviet Manned Lunar Expedition.

In 1968 the Soviet L3 Manned Lunar Expedition Program consisted of six vehicles: two probes, three unmanned spaceships and one manned spaceship. Before beginning the main phase of the Lunar Expedition there was much preparatory work done. Many portions of this work were eventually performed as noted below.

The automatic probes E-8LC were sent to take high resolution photos of potential manned landing sites. These vehicles were actually flown as missions *Luna-19* (launched September 1971) and *Luna-22* (May 1974).

Next in the plan were two remote controlled lunar rover vehicles, also carried by the E-8 series spacecraft. These were flown as missions *Luna-17* (November 1970) and *Luna-21* (January 1973). They were launched on UR-500K rockets (now called the *Proton* booster) and the rovers they deployed were called "Lunokhod." The tasks for the Lunokhods were to explore two landing sites close to each other: 1) A site for the spare LK Lander, (RLK) without a cosmonaut and 2) A site for the manned LK lander.

Unfortunately, the two reconnaissance probes and the two rovers were as far as the plan got in actual flown hardware, but had it continued, here's how the rest of it would have gone:

Operation and control of Lunokhods was from Earth by radio signals and they had a major role in implementation of the Soviet Manned Lunar Expedition. In addition to exploring and checking the landing sites, the Lunokhods also acted as radio beacons for LK landings. The propellant reserve in LK landers was very small and side maneuvers were limited. So for a safe landing, the LK has to land in automatic mode, accurately following Lunokhod's radio beacon to a precise safe landing spot. It is interesting to consider this approach in view of the difficulties Neil Armstrong had when landing the Apollo 11 LM, *Eagle*.

One month later an N-1 would launch an L3 Complex, which contained a standard manned LOK and reserve RLK. The RLK was to land on the first landing site. After the RLK had landed, the cosmonauts in the LOK were to take high resolution photos of the lunar surface and then return to the Earth. On the surface, the Lunokhod would approach the RLK and send photos of each side of the RLK to the Earth. These pictures and the RLK telemetry data would provide conclusive proof of the good condition of the RLK. Only after this was confirmed could the main stage of program begin – the manned lunar landing.

A second L3 Complex would launch at the next available window, one month after the Reserve LK was in place. The duration of the Lunar Expedition was planned to be about 11-12 days. A detailed description of the mission sequence follows.

[Editor's Note: The numbers in Igor's sequence are only approximately followed in Serge Gracieux's drawings in the foldout illustration.

[1] Here is the link to the book: http://astronaut.ru/bookcase/books/filin/filin.htm. The link was still active as of press time, but the site is all in Russian – Ed.

Both describe the same mission, but don't line up exactly since they were done separately—JH]

1. Launch the L3 Complex into Earth orbit. Orbit altitude is 200 km. Crew would be flying without spacesuits, just in tracksuits.
2. The L3 Complex would spend about 25 hrs. doing 17 orbits around the Earth. During that time crew checks the status of all systems and makes preparation to start the flight to the Moon. On the 17th orbit, the rocket engine of Block-G, for Trans-Lunar Injection (TLI) burn. Block-G is jettisoned after this step.
3. The L3 assembly fires the Block-D engine to reach the correct speed and trajectory to get to the Moon.
4. The flight to the Moon takes about 101 hours. During that flight there are two mid-course corrections, the first 8-10 hours after TLI and the second 10-21 hours before the braking operation to enter lunar orbit.
5. The Block-D engine fires to slow the L3 and put it into a circular orbit around the Moon. The orbit would be 150 km above the moon, with an orbit taking 122 minutes.
6. Time in lunar orbit was planned to be 77 hours, (39 orbits). During that time there were two trajectory correction operations planned using Block-D. The goal of these operations was to achieve an elliptical trajectory with an apolune (high point) 100 km above the Moon, and a perilune (low point) 20 km above the surface. The first correction would be after 4 orbits, and the second after 14 orbits. Also required were two measurements of spacecraft altitude, the first on the 11th orbit, (before second correction burn) and the second on the 20th orbit, after reaching targeted orbit. It was also necessary to measure the relative position of the Lunokhods, so the correct descent trajectory can be calculated.
7. Preparation for landing. One cosmonaut spacewalks from the LOK to the LK lander. The Utility Module of the LOK would be used as an airlock. The cosmonaut in LK, still attached to the Block-D, separates from the LOK. The L3 Complex had the LK lander placed inside a load-bearing shroud. For landing, the LK slides out from inside the fairing.[2] To do this safely, there are guides on the LK Lander, which help it move along the inside of the shroud without damaging the many systems on the outside of the LK lander. After this operation, the supports are jettisoned and the landing legs unfold.
8. The LK adopts the correct orientation and at the specified time fires the Block-D engines, to start the descent. At a height of about 3-4 km the empty Block-D is jettisoned and LK fires its descent engine to get some distance from the dropped Block-D. This maneuver is needed to prevent the landing radar from seeing the empty Block-D and confusing it for the lunar surface. If any problems are detected during descent, the LK can immediately drop the landing stage, fire the ascent engine and return to orbit. Note that for a normal landing this would be an automatic system, but in an emergency the cosmonaut can take control.

At the moment of touchdown, four small "anchor rocket" motors fire upwards to hold the lander firmly on the surface and prevent it from bouncing.

Time on the lunar surface was planned for a minimum of 6 hours to a maximum 24 hours if there was some problem. After landing the cosmonaut has to check the LK's status to be sure there is no damage. While he is doing that, one or two Lunokhods approach the LK lander and help to survey LK exterior. If the LK is in good condition the cosmonaut gets permission to leave the lander and walk on the surface of the Moon. If the LK was damaged such that it cannot take off to reach lunar orbit, the cosmonaut can use the Lunokhod to carry him to the RLK, and use that one instead. The Lunokhod has oxygen stored on board for the cosmonaut and the "Krechet" (Falcon) space suit has a hose that can connect to the oxygen supply on the Lunokhod. There's a special platform on the front of the Lunokhod for the cosmonaut to stand on while being carried to the reserve lander (with the rover being controlled from Earth). In front of cosmonaut there's a control panel so he can drive the Lunokhod himself if necessary. Estimated speed was about 1.2 km/hour.

The plans for this part of the mission only reached draft stage and, like so much N-1 material, are no longer available. But they confidently concluded that a Lunokhod could serve well for carrying cosmonauts across the moon's surface.

[2] Note that in Serge Gracieux's drawing #6 he shows this as a typical split faring which is not correct according to Igor's research – Ed.

9. It is also worth noting that the engineers worried about the cosmonaut in that stage of the expedition. He would be completely alone on the lunar surface. There could be problems, including psychological problems. So engineers considered a special system consisting of ropes and pulleys to drag the cosmonaut into the cabin if the cosmonaut lost consciousness. It is difficult to imagine how such a system might work, but it was considered. There were also concerns that a cosmonaut might fall over backwards and be unable to stand again. This led to a design for a half-hoop that would attach to the rear of the cosmonaut's space suit immediately after stepping onto the surface of the moon. If the cosmonaut fell backwards he could roll and then use his arms to lift himself up again.

   The plan was to spend 2 hours at a time outside the lander with the possibility to refill air supplies between excursions. The planned actions were:

   - Erect the flag of the USSR
   - Deploy scientific equipment
   - Collect samples of lunar surface.
   - Take photos
   - Make a report to a television camera.

   When the cosmonaut finished all of his tasks he would go back into the LK and fill the cabin with air. Only after that he can open spacesuit face plate and eat. Note that during all operations, from entering the LK in orbit until that moment, the cabin of LK would be unpressurized and the cosmonaut would be in his spacesuit. This was done to save weight in the life support system that provides breathable air. After eating, the cosmonaut would check the lunar rock samples, check himself and take a rest.

   During lunar surface activity, the LOK in orbit takes measurements of its precise trajectory and the relative position of the LK and LOK. This needs to be completed by the second orbit. The data are required to calculate the flight path required for the LK ascent stage to rendezvous with the LOK. After this is done, the LOK takes photos of lunar surface.

   The lower part of the lander remains on the Moon, similar to the Apollo LM, to serve as a launch platform. Before launch, connections between the upper and lower parts, including cabling and hydraulic connections, are severed. Unlike Apollo, however, the ascent stage has two engines (Block-E) for improved safety since only one is required to successfully achieve lunar orbit and rendezvous.

   Detection of which engine to use was automatic and the nozzles of the ascent stage engines are angled so that the thrust is delivered in the correct direction regardless of which one was used. The ascent stage would then ascend smoothly into lunar orbit to rendezvous with the LOK.

10. Rendezvous and docking were achieved with a passive lander and the LOK doing the maneuvering using the Block-I engine (although the LK can rotate to get the correct orientation). Docking was achieved with the probe fingers on the LOK latching into a perforated plate on the top of the LK lander. After docking the cosmonaut would spacewalk back to the LOK taking the samples with him.

11. The cosmonauts then have some time to rest before preparing to fire the Block-I engine for the flight back to Earth. For this they transfer lunar samples to the Descent Module and take their places in the chairs there. The connected LK, Docking Module and Utility Module are jettisoned on 38th orbit. The LOK adopts the required orientation and makes the Trans-Earth (TE) burn on the 39th orbit.

12. The return flight takes about 82 hours. During that time two mid course corrections are scheduled 24 and 44 hours after the TE burn.

13. Two hours before entering Earth's atmosphere, the LOK rotates to the correct angle for re-entry and the Descent Module separates from the Instrumentation Module. The capsule descends much faster than one from Earth orbit and for this reason it will dip into the atmosphere twice. The first encounter helps reduce speed and the capsule skips outward before descending at lower speed into the atmosphere for a second time. This maneuver requires great precision in the trajectory.

    Finally the parachutes open and the mission is over. Landing locations on the ground in Soviet territory or in the Indian Ocean were considered.

N-1: For the Moon and Mars

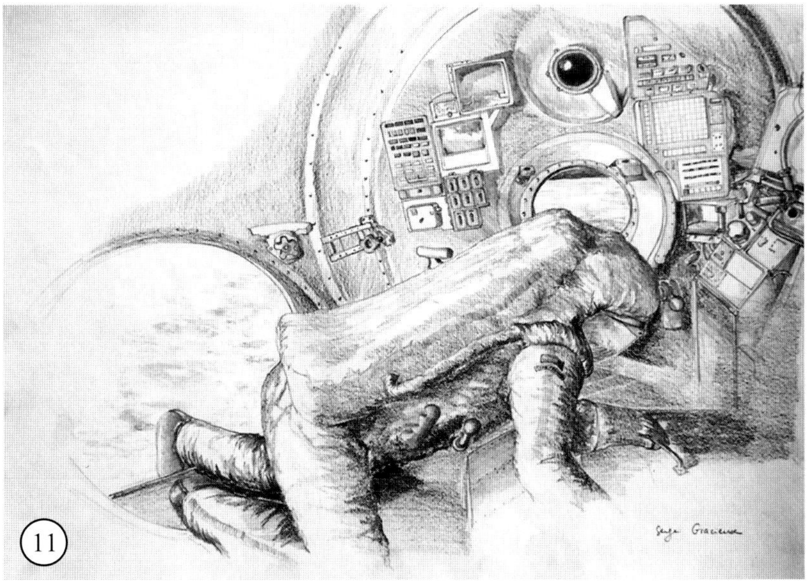

11

12

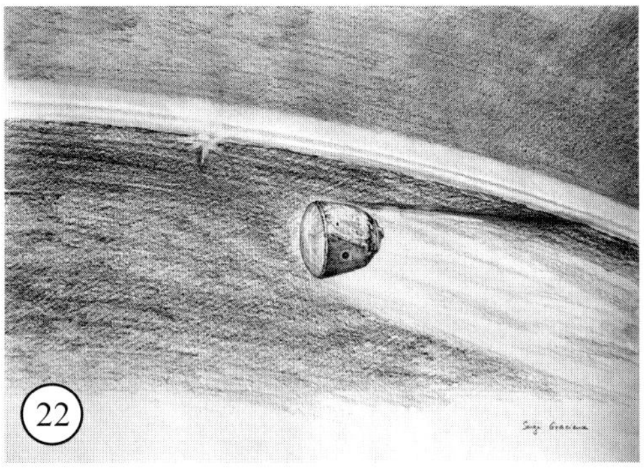

22

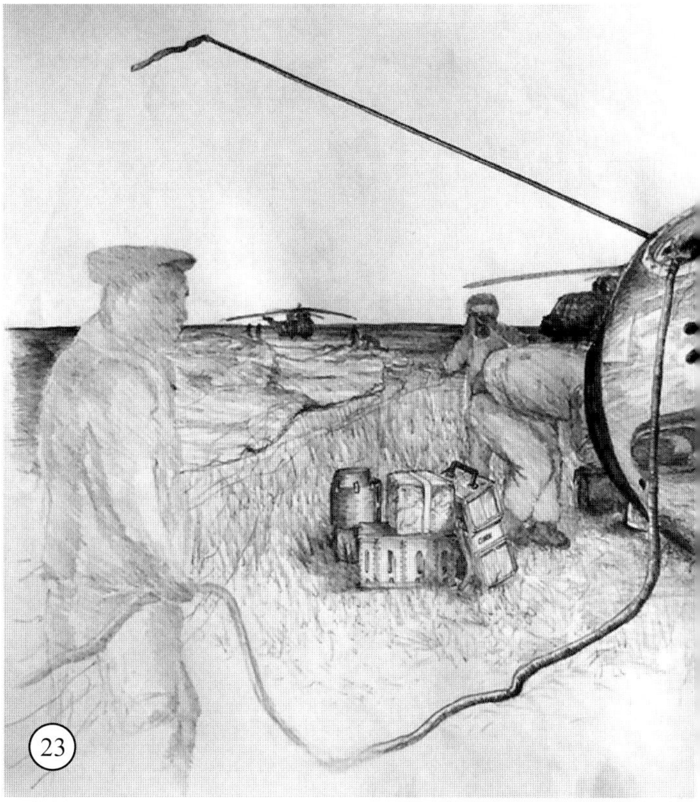

23

21

20

19

en I saw Serge's artwork of the N-1 mission, I knew that that was the appropriate medium to show n, drawn by Serge Gracieux, Object and Museum Program Studies Manager at the ***City of Space***

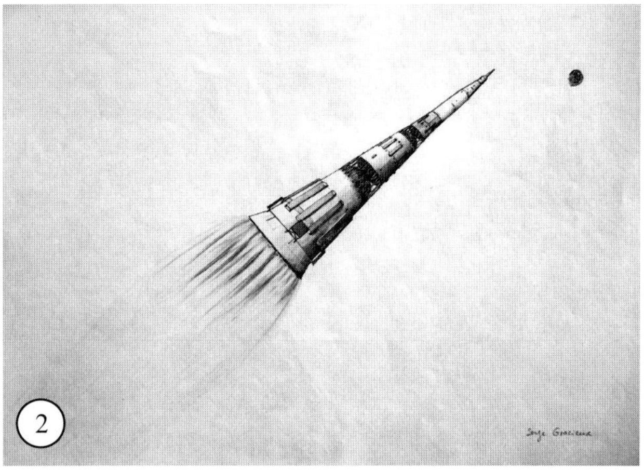

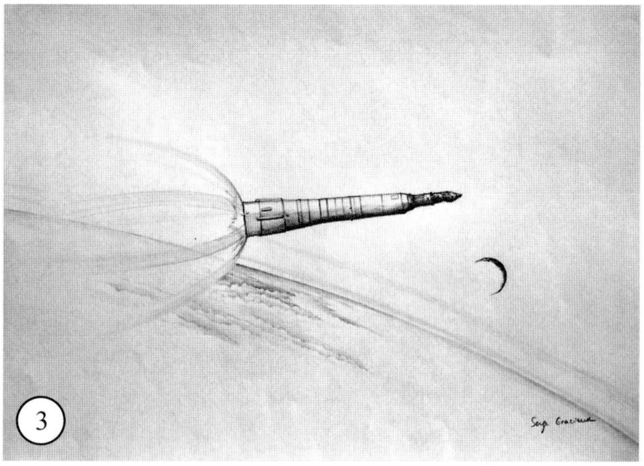

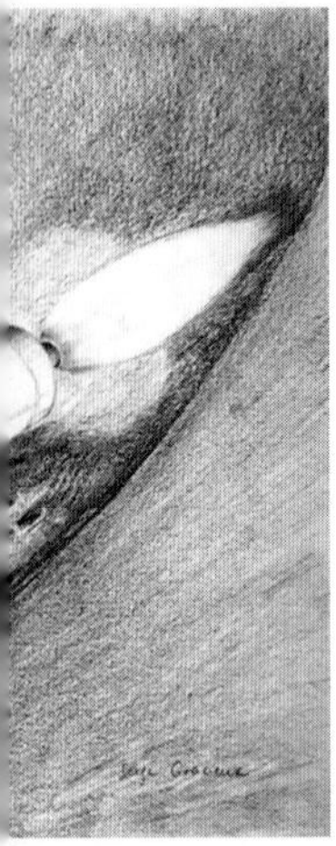

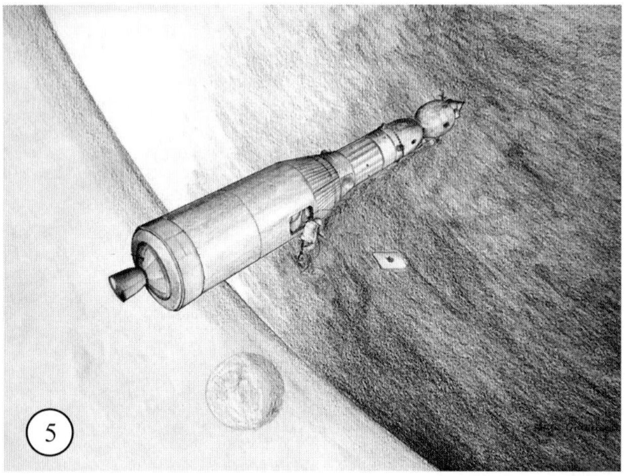

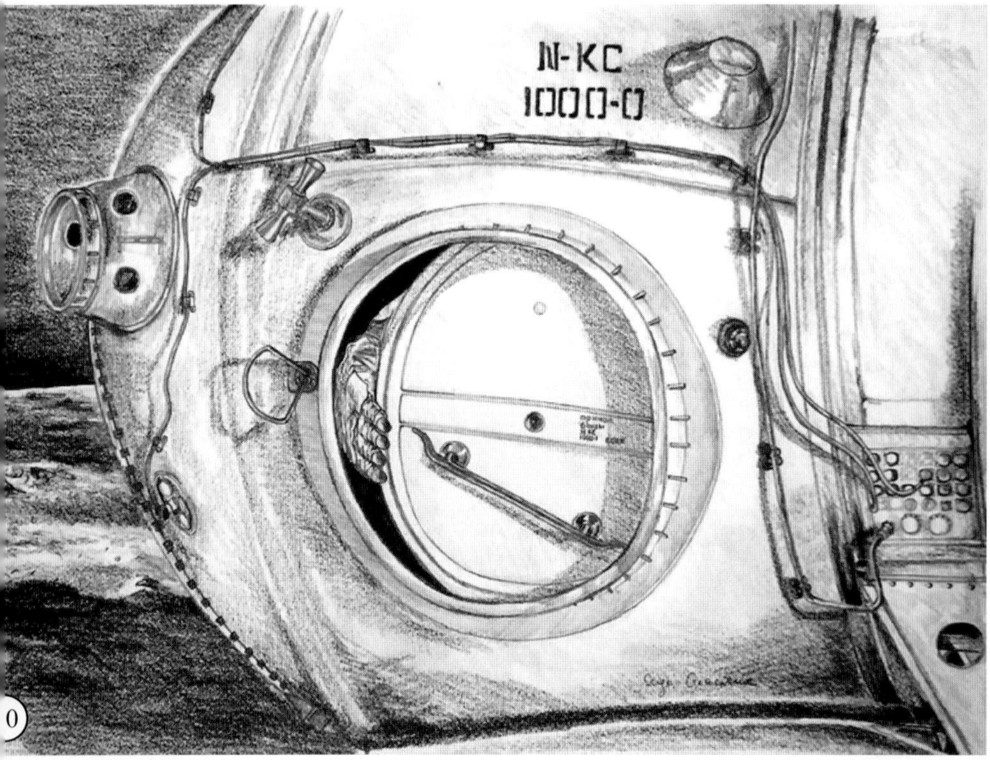

1 - Departure
2 - Launch
3 - Lunar Trajectory
4 - Lunar Orbit
5 - Transfer to LK
6 - LK Separation*
7 - LK Descent
8- Block-D Jettison
9 - Landing
10 - Hatch Open

*Research by Igor Bezyaev has confirmed that the LK slid out of a 1-piece cylindrical fairing, not 2-pieces as depicted here. Once the LK had moved a safe distance away, the fairing was jettisoned from the LOK

N-1: For the Moon and Mars

Nothing in this modern CGI world comes across with quite the passionate impact of the human touch. W
the possible future that never was. Presented here are twenty-four frames capturing the N-1 moon missi
museum in Toulouse, France—Matt Johnson

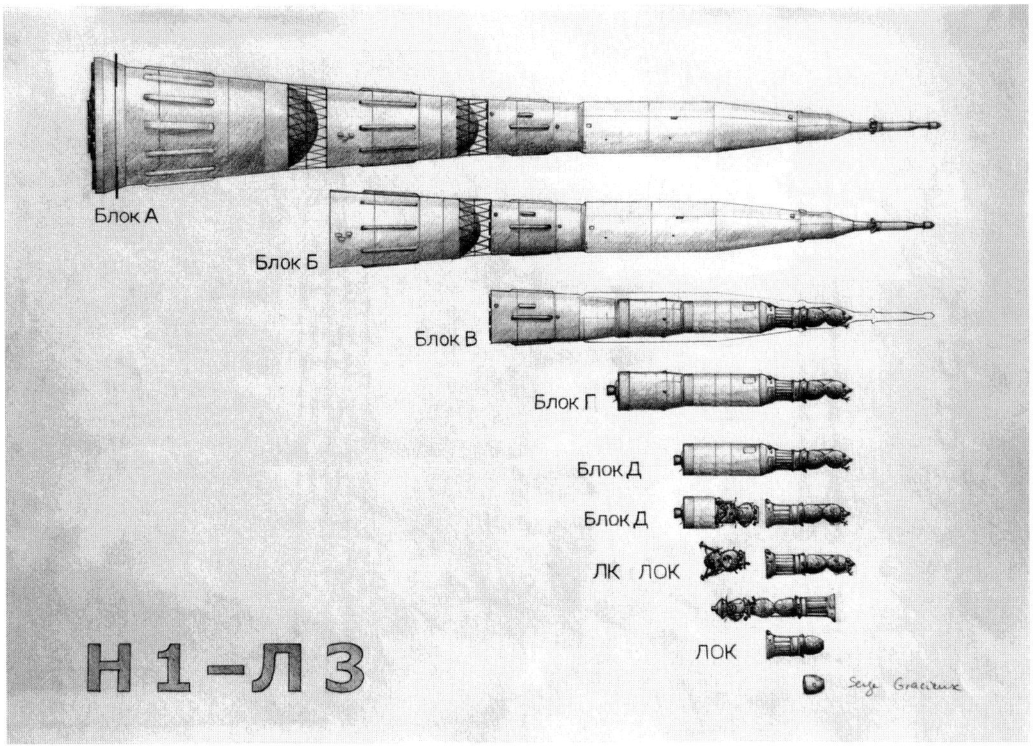

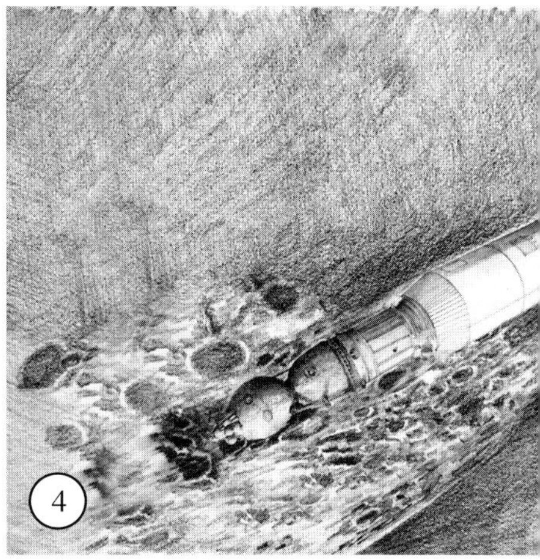

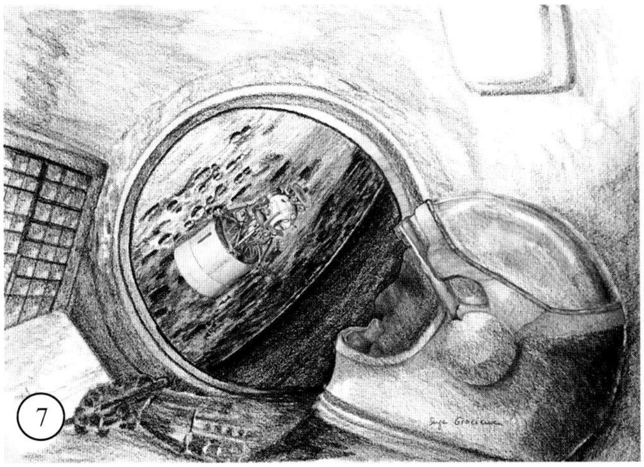

3: The N-1Moon Mission Profile

11 - Cosmonaut exit
12 - Cosmonaut descends ladder
13 - Cosmonaut On Moon
14 - Cosmonaut travels with Lunokhod
15 - Lift Off to Orbit
16 - Dock with LOK
17 - Transfer to LOK
18 - Transfer close-up
19 - Cosmonaut enters LOK
20 - LK and LOK utility module jettisoned
21 - LOK on Earth trajectory
22 - LOK breaks for re-entry
23 - Safe landing

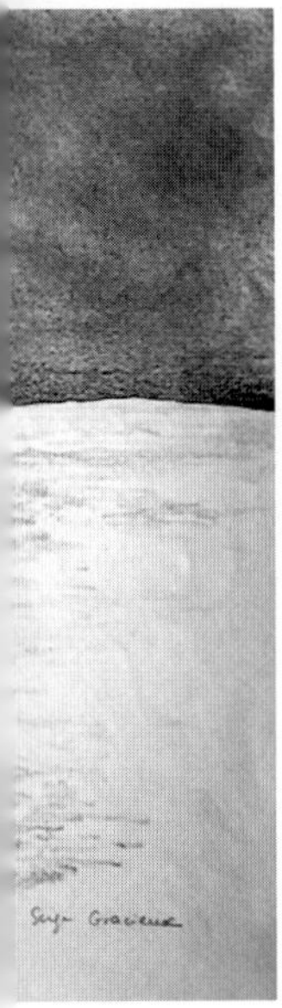

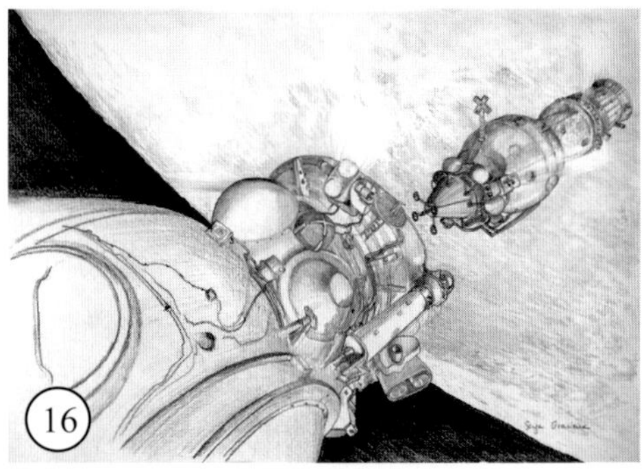

The LK Lander

Lean and Spartan, the LK lander represents the entire Soviet lunar program by its very nature. Lacking an Apollo-style docking tunnel between the craft, a lone cosmonaut (from the crew of two) would spacewalk to the LK, still under the shroud and attached to the top of Block-D. In order to enter the LK, he must open a port in the shroud to gain access to the LK hatch. After undocking, the LK would start its initial descent toward the lunar surface using the Block-D engine. The Block-D would then be jettisoned to crash into the moon as the LK performed the final descent with its own engine (Block-E). To ensure a firm grip on any possibly angled terrain, solid fuel "anchor rockets" (attached to each landing leg with their nozzles facing up) would fire on the moment of contact with the lunar surface, pinning the LK to the ground. After his sojourn on the moon, the cosmonaut would re-enter the LK, lifting off using the same engine that he landed with (although a backup engine was included),[3] leaving only the landing legs behind as a launch platform. Once in orbit, the LK would dock with the orbiting LOK and the cosmonaut would once again spacewalk to the safety of his companion's spacecraft. This mission would be quite an arduous task while remaining in the same spacesuit for the duration! - Matt Johnson

[3] Note that both the Descent and Ascent engines (with their propellant tanks) are in Block-E. As can be seen in the Bottom View on the next page and Figure "C" on page 32, there are actually three engines: the larger Descent engine nozzle is in the center while the two Ascent engines are the smaller angled nozzles on either side (one engine was primary and the other an identical backup in case of failure). They are angled so that no matter which engine was used, the thrust will be aimed through the LK's center of gravity. The four still-smaller nozzles are exhaust ports for the pump turbines.—Ed

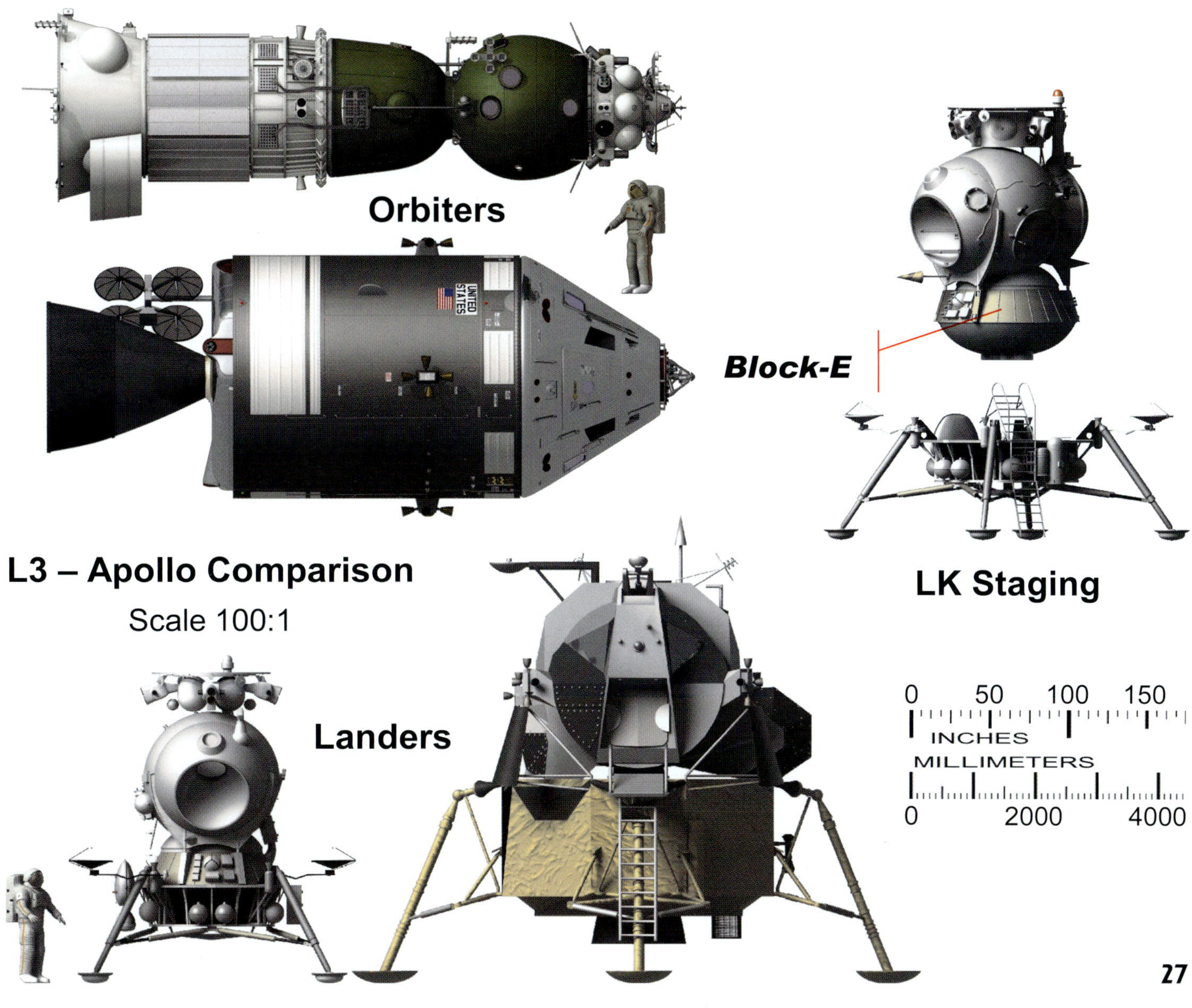

Лунный корабль 11Ф94

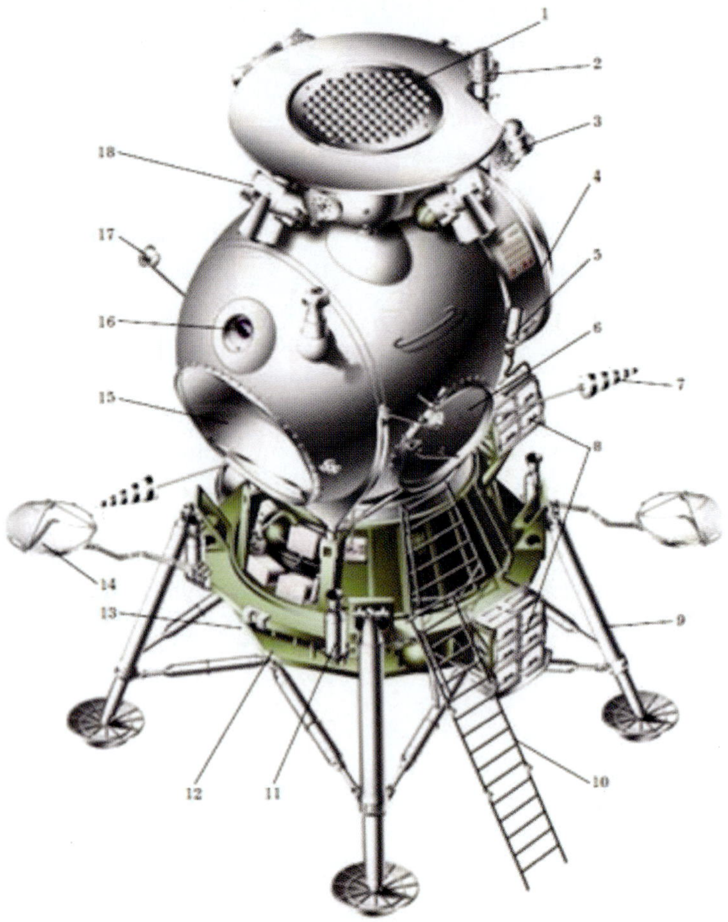

1 – стыковочный узел; 2 – датчик прицеливания; 3 – юстировочные датчики; 4 – приборный отсек; 5 – телекамера; 6 – выходной люк; 7 – всенаправленная антенна; 8 – источники питания; 9 – опорная стойка с амортизатором; 10 – трап; 11 – РДТТ прижатия; 12 – лунный посадочный агрегат; 13 – двигательная установка блока Е; 14 – остронаправленная антенна (2 шт.); 15 – вогнутость для иллюминаторов; 16 – иллюминатор для наблюдения за стыковкой; 17 – антенны системы сближения; 18 – блок двигателей ориентации

Lunar Craft (LK) 11F94

1 – Docking Plate
2 – Targeting Sensor
3 – Alignment Sensor
4 – Instrument Compartment
5 – TV Camera
6 – Hatch
7 – Omni Antenna (2)
8 – Power Supply
9 – Shock Absorbing Landing Strut (4)
10 – Ladder
11 – Solid–fuel Anchor Rocket (4)
12 – Landing Gear Assembly
13 – Block-E Engine (underneath)
14 – High Gain Antenna (2)
15 – Recess for Window
16 – Docking Window
17 – Rendezvous System Antenna
18 – Reaction Control System (4)

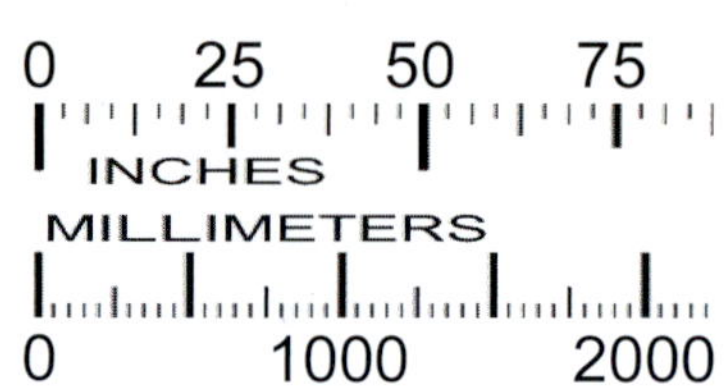

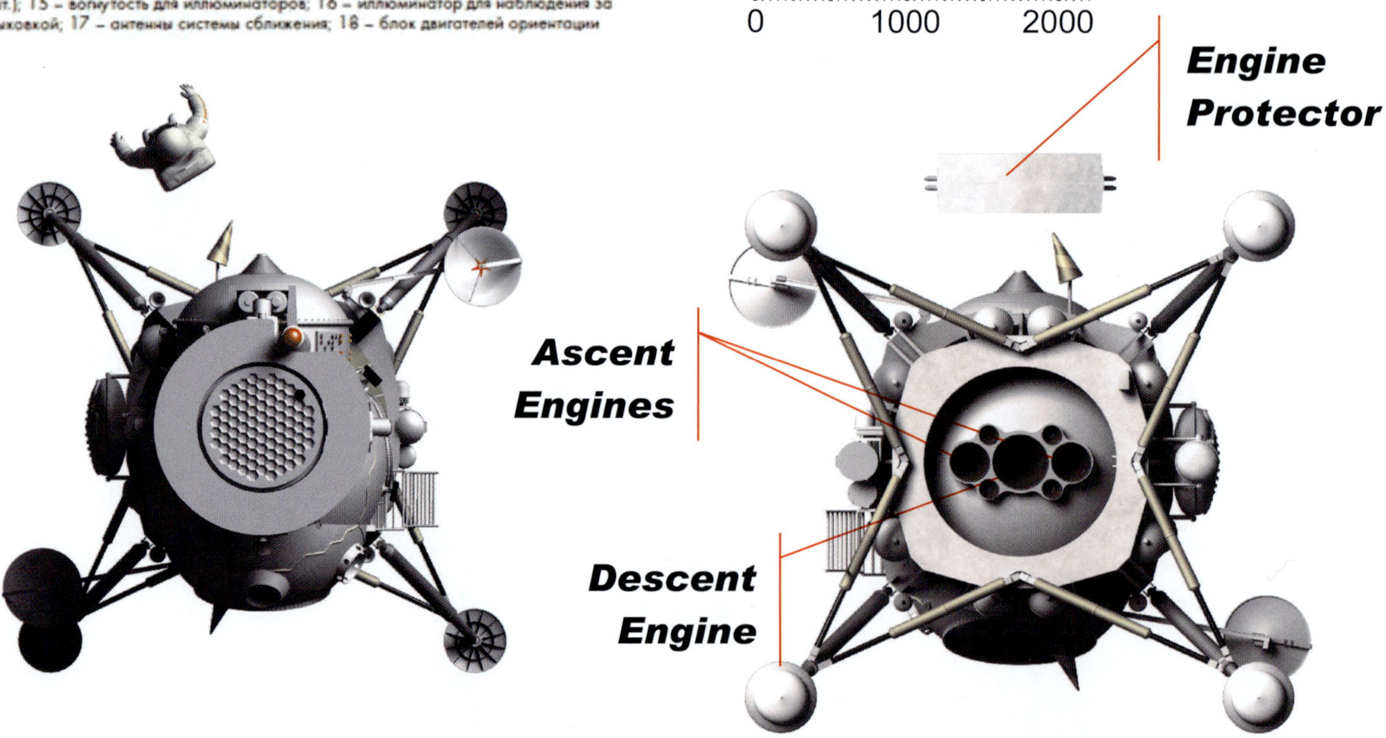

LK Top View **Scale 50:1** **Bottom View**

LK — 45° Rotations Scale 50:1

0°

45°

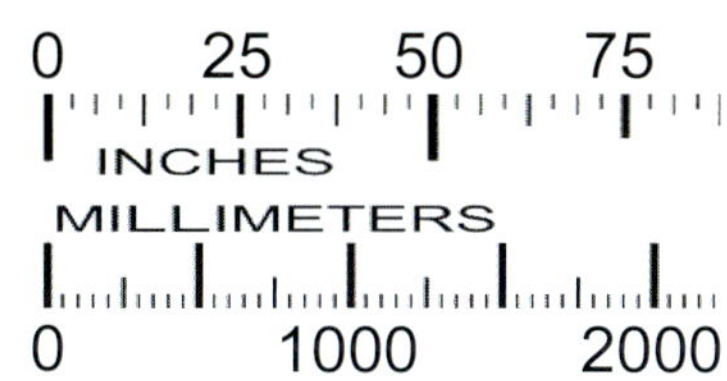

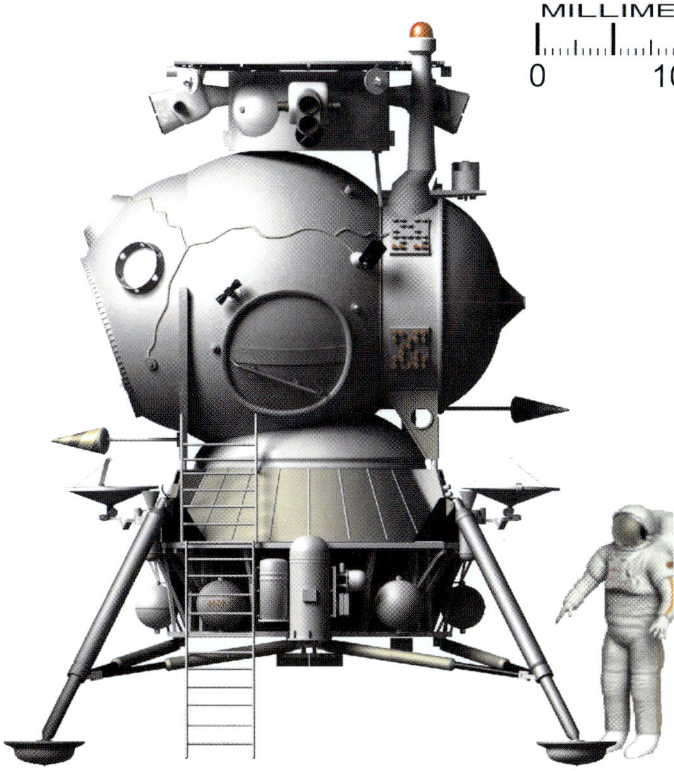

90°

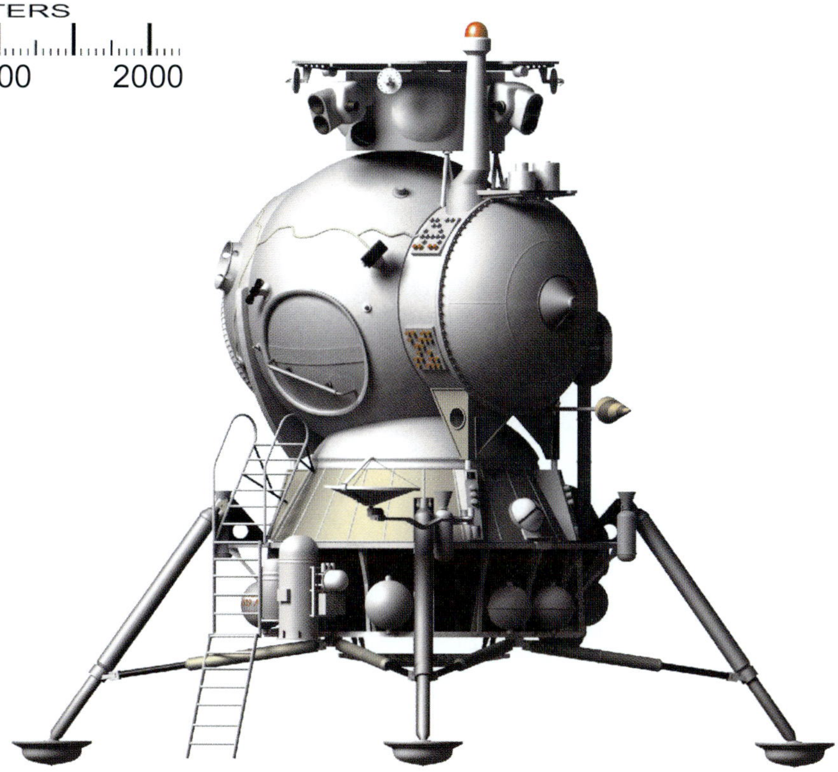

135°

LK — 45° Rotations Scale 50:1

180°

225°

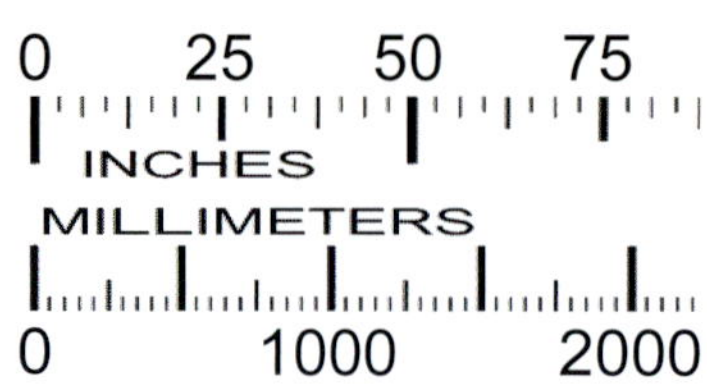

270°

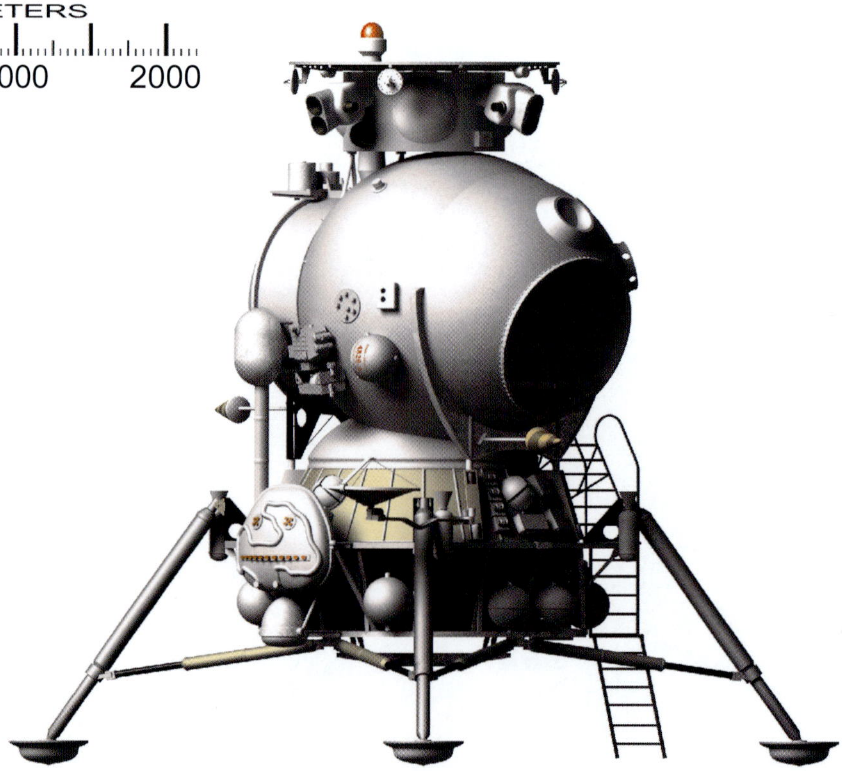

315°

LK On Display at NPO Energia Museum

A- Docking Plate

B- The *Kretchet (Falcon)* Lunar Space Suit

C- The Block-E Engines on bottom of the LK

D- The "Anchor Rocket" test of an LK boilerplate, pushing the unit down upon landing on a very uneven surface.

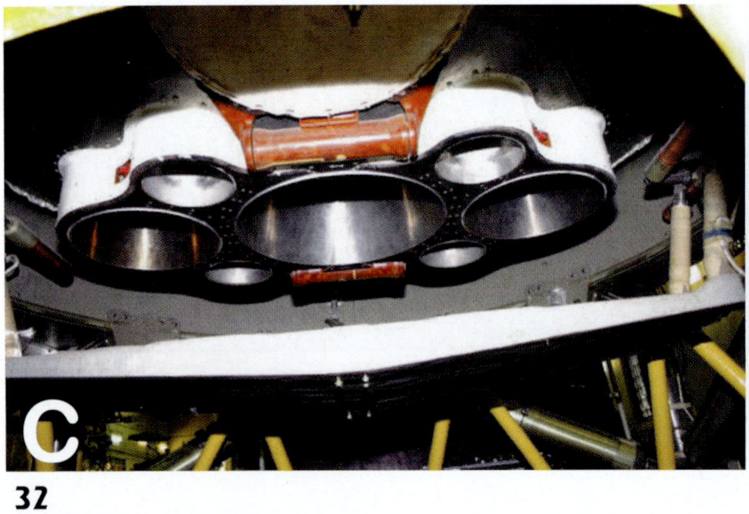

LOK (Lunar Orbiting Craft) — Soyuz 7K-L3

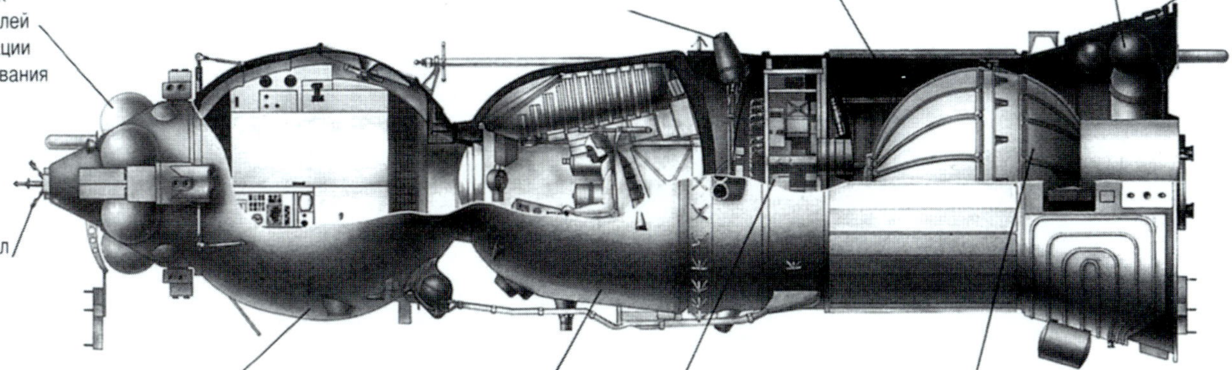

Left: The complex Docking Module with the radar and optical aids for rendezvous and the LK Capture Node (cone at the top). Some of the gas storage spheres are missing from this unit.

Right: the two occupied modules of the LOK. The Utility Module (black) used as living space outbound and for the lone Cosmonaut on-orbit around the moon, and the Reentry Module (brown) used as living space on the return trip in addition to reentry into the atmosphere

Лунный орбитальный корабль 11Ф93
(Lunar Orbiting Craft (LOK) 11F93)

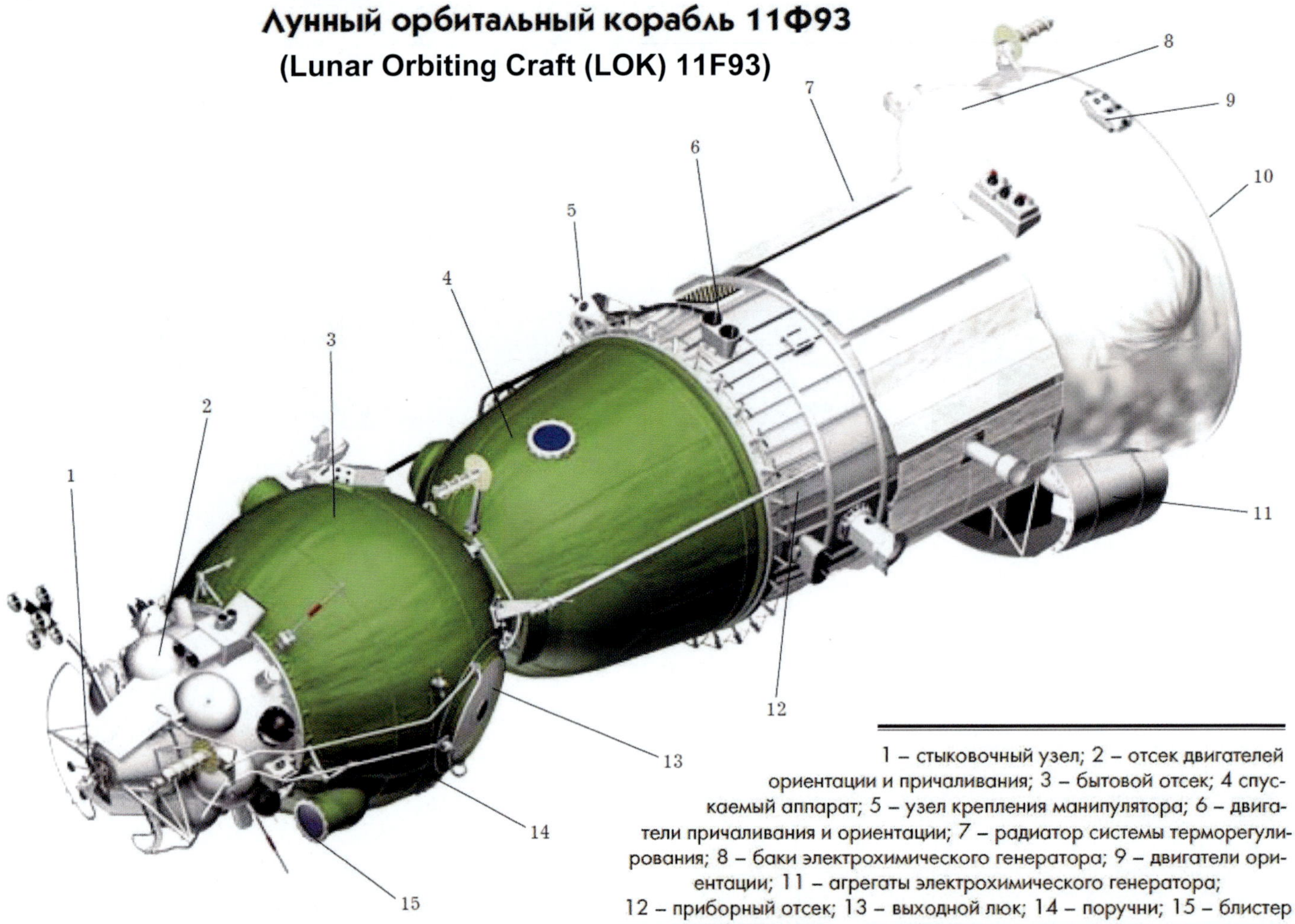

1 – стыковочный узел; 2 – отсек двигателей ориентации и причаливания; 3 – бытовой отсек; 4 спускаемый аппарат; 5 – узел крепления манипулятора; 6 – двигатели причаливания и ориентации; 7 – радиатор системы терморегулирования; 8 – баки электрохимического генератора; 9 – двигатели ориентации; 11 – агрегаты электрохимического генератора; 12 – приборный отсек; 13 – выходной люк; 14 – поручни; 15 – блистер

1 – Capture Node, 2 – Docking and Orientation Module, 3 – Utility Module, 4 – Reentry Module, 5 – Grapple Fixture, 6 – Docking and Maneuvering Thrusters, 7 – Thermal Control System, 8 – Fuel Cell Tanks, 9 – Maneuvering Thrusters, 10 – Block-I engine, 11 – Fuel Cells, 12 – Equipment Module, 13 – EVA Hatch, 14 – Handrails, 15 – Docking Window

The impressive mass of pipes and tubing that is the Block-I engine. The two larger drum-like objects are the main engine nozzles (which provide the thrust to get back to Earth) with the smaller orbital maneuvering nozzle (which is used to adjust the lunar orbit up or down) located between them. This arrangement allows the maneuvering engine to be directly on the spacecraft's centerline while still balancing the thrust from the main engines. Also visible is one of the RCS (attitude control) thrusters. It's counterpart is lost in the lower shadow.

After the gigantic engines of the booster stages, these are a little easier to imagine. The main nozzles on this engine are only 400mm (15") in diameter and are spaced 800mm (30") center-to-center. Still, they put out a combined thrust of 30 kN (almost 7,000 lbs).

LOK Rear View

Scale 40:1

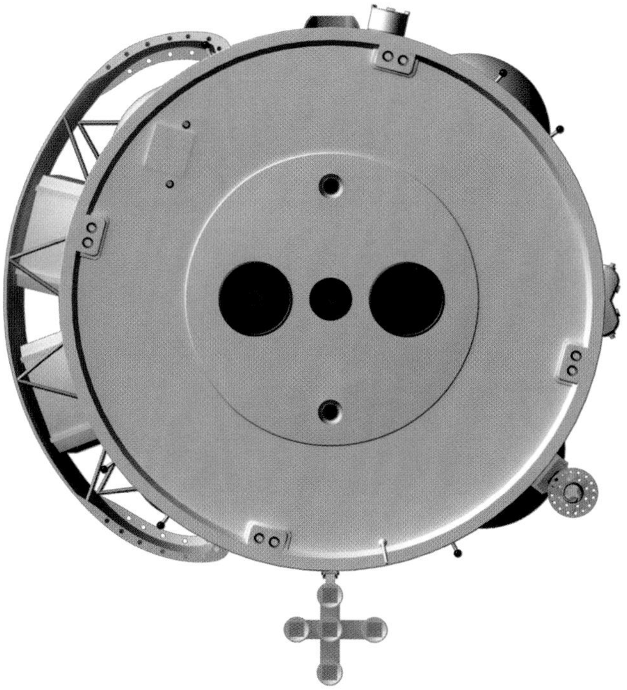

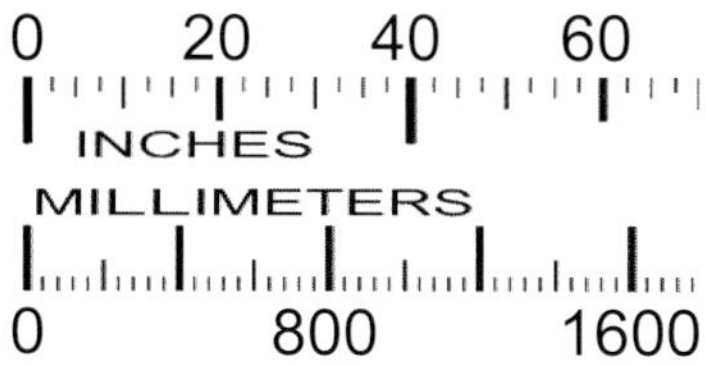

LOK — Soyuz 7K-L3

LOK Front View

Scale 40:1

LOK — 90° Rotations Scale 60:1

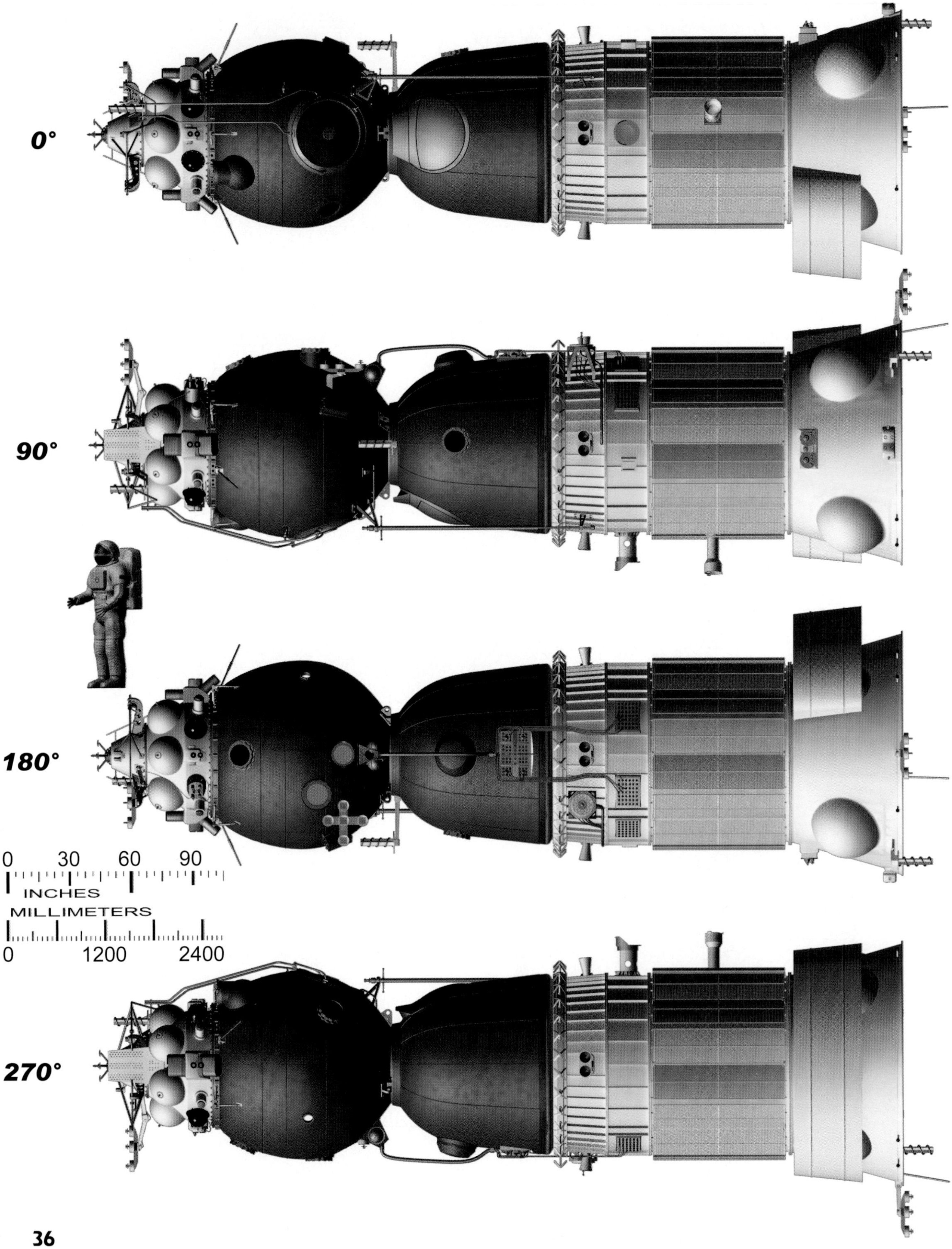

4: The N-1 for Mars and Venus

4: Н-1 для Марса и Венеры

The Greatest Space Project of the 20th Century

By Igor Bezyaev

Why was the N-1 rocket created? I think many would say it was a Moon rocket and it was created to take a Soviet cosmonaut to the Moon. But is this true? Some say that Sergey Korolev lost the Moon Race because the technical design of the N-1 rocket was flawed. Others say the Moon Race was lost because of the quarrel between Sergey Korolev and Valentin Glushko over the choice of propellants for the rocket. Still others blame failure on Korolev's decision not to build an expensive test stand, so the fully assembled first stage could be tested. Then there's the suggestion that this unique rocket was designed by Korolev in 1960 without a clear purpose and that it was only in 1964 the Soviet Union found a use for it, to beat the Americans to the Moon.

Historical Background

So what actually occurred during this period and why did Korolev, who achieved both the first artificial satellite and the first manned spaceflight, lose the race to the Moon? Is it fair to call it a race when Korolev only had half the time of the USA to get to the moon? Remember that the USA began work in 1961 and succeeded in 1969, eight years later.

Sergey Korolev

Korolev was only formally instructed to start the Moon Landing project in September 1964, with a landing required by 1967-1968 in order to beat the USA. It meant that work on lunar expeditions began three years later in the USSR, with a target date three years earlier!

The Early Days

In days of Konstantin Tsiolkovsky and Robert Goddard interplanetary flight was the very essence of spaceflight, with the clear aim of reaching the other planets of Solar system.

Konstantin Tsiolkovsky

In the autumn of 1931 a group of spaceflight enthusiasts created a study group in Moscow. This was the Group for the Study of Reactive Motion (GIRD).[1] In the group were Friedrich Zander, Sergey Korolev, Michael

Korolev (arrow) and members of GIRD pose with the ГИРД-Х (GIRD-10)

[1] Note that this essentially translates as "Jet Propulsion Laboratory"– Ed.

Tihonravov and others. The head of group was Zander, who dreamed of flights to Mars. Later Sergey Korolev was chosen as the group's leader and showed his skills as a gifted organizer, coordinating the many activities of the members.

Michael Tihonravov

On 17 August 1933 on range situated near Moscow in Nahabino, the flight of first Soviet liquid fueled rocket, designed by Tihonravov, took place: The GIRD-09. The rocket flew for 13 seconds and rose to a height of 400 meters.

Dmitri Ustinov

After the Great Patriotic War (World War II—Ed.), development of long range ballistic missiles began using captured German V-2 rockets as a starting point. Under the military control of Commissar Dmitri Ustinov, Korolev was appointed chief designer of long-range missiles at the OKB-1 design bureau at the secret NII-88 research center. Korolev demonstrated his organizational abilities in this new facility, keeping a dysfunctional and highly compartmentalized organization operating.

The first effort of the team was to produce a direct replica of the V-2. This was designated the R-1 and was first tested in October 1947. Korolev's group then began working on more advanced designs with improvements in range and throw weight. The R-2 doubled the range of the V-2 and was the first design to use a separate warhead. This was followed by the R-3, which had a range of 3,000 kilometers (1,900 mi).

In 1952 work began on the R-5, which had a more modest 1,200 kilometer (750 mi) range. It completed a first successful flight by 1953. The first true intercontinental ballistic missile (ICBM), not only in USSR but in the whole world, was the R-7 Semyorka. This was a two-stage rocket with a maximum payload of 5.4 tons, sufficient to carry the Soviet Union's bulky nuclear bomb an impressive distance of 7,000 kilometers (4,300 mi). After several test failures, the R-7 was successfully launched on August 1957, sending a dummy payload to the Kamchatka Peninsula. One and a half months later, on 4 October 1957, this rocket launched the first satellite, Sputnik ("traveler"), into orbit and began the space age.

R-1

R-5

Three years later, on 12 April 1961, a three stage R-7 rocket took the first man into space, Yuri Gagarin. To the outside world, Sergey Korolev was known only as "the chief designer," his name was a state secret (only after his death in 1966 was his identity revealed to the world). Still with him from the GIRD days was Tihonravov and together they continued to think of daring plans for interplanetary flight.

So mankind had made the first steps into space. However just going into space was not the goal – it was rather the fundamental goal of astronautics: to send men to new worlds! But the R-7 rocket was not powerful enough for piloted interplanetary flight. A new, much more powerful rocket was needed to take men to Venus and Mars.

So it is clear that the need for a heavy rocket began long before the official N-1 project started.

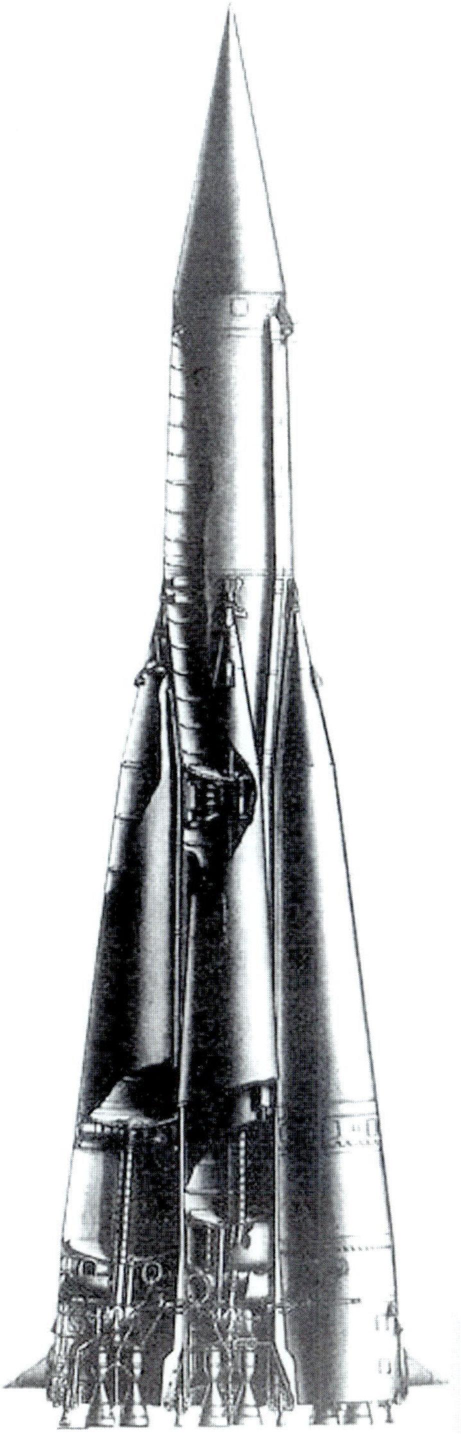

R-7 ICBM

R-7 Vostok Launcher

Heavy Rockets – the beginning

Studies of heavy rocket designs began in 1954 and the first proposals were presented at meeting of chief designers on 15 July 1957. It was agreed that the work was sound and should continue.

At that time nuclear power seemed to solve all energy problems. Nuclear submarines, icebreakers and even planes were designed. Rockets were no exception and first heavy rocket designs were also nuclear powered.

In December 1959, the OKB-1 engineering bureau prepared outline sketches of three rockets with nuclear engines. The weight of the payload launch to Earth orbit under these outline plans were:

- Variant 1: 2.6 - 4 tons
- Variant 2: 35 - 40 tons
- Variant 3: 150 tons, the "Superbooster"

However further studies showed it was possible to make heavy rockets using existing liquid propellant engines and work on using nuclear engines was not taken forward.

Bureau OKB-1 1959-1962

OKB-1 was separated from the military in April 1957, with Sergey Korolev in charge. Today it is known as Space-Rocket Corporation *Energia.*

As in the days of GIRD, OKB-1 had various departments specializing in different aspects of space flight. The most legendary was Department 9, supervised by Michael Tihonravov who had been with Korolev since the days of GIRD. Department 9 had two main areas of work:

First, designing the spaceships Vostok ("East"), Voskhod ("Dawn") and Soyuz ("Union"). Konstantin Feoktistov was the chief of this section. The plan was to launch these spaceships using modified R-7 rockets.

Совет главных конструкторов: В.П. Глушко, Н.А. Пилюгин, В.И. Кузнецов, С.П. Королёв (председатель), М.С. Рязанский, В.П. Бармин, Б.Е. Черток (не член совета)

Meeting of Chief Designers, July 1957. From left to right:
V.P Glushko, N.A. Pilugin, V.I. Kuznetsov, S.P. Korolev (chairman),
M.S. Razancky, V.P. Barmin, B.E. Chertok (not a board member)

Konstantin Feoktistov

Second, designing manned interplanetary ships and automatic interplanetary probes for flights to Moon, Mars and Venus. Two sections were involved in this work. In the first section Boris Adamovich was the chief. The work was to develop electrical ion rocket engines and a version of the Tyazhely Mezhplanetny Korabl, ("Heavy Interplanetary Ship") known as TMK, with electric ion engines for manned expeditions to Mars and Venus.

Gleb Maksimov

In the second section Gleb Maksimov was the chief. The work of that section was to develop a version of TMK using liquid-propellant rocket engines and to design automatic probes for flights to the Moon, Mars and Venus. Probes could be launched on modified R-7 rockets but for the TMK, the N-1 super booster was required.

There was no dedicated program for manned lunar expeditions at that time. Speculative versions of lunar missions were discussed, but the sole purpose of lunar missions was as a step to manned Mars expeditions. For example, flights to Moon could test elements of TMK and the landing module. Automatic interplanetary probes were required for the study of interplanetary space environment and testing TMK systems.

Preliminary studies of TMK variants became the basis of the N-1 requirements. For a flyby of Mars or Venus, (the logical first step in interplanetary space travel), the required weight in Earth orbit should be about 75 tons. This includes the TMK weighing 15 tons and the booster rocket, which would accelerate TMK to interplanetary velocities, weighing 60 tons. This defined the basic payload requirement for a heavy rocket.

This minimum weight was chosen by Korolev for rapid development, with maximum use of existing technologies and industrial facilities. The more challenging missions, with landings on other worlds could be achieved by launching several rockets and assembling the interplanetary ships in Earth orbit. Making a more powerful rocket would have greatly increased risk and take more time. There was a particularly favorable launch opportunity at the start of the 1970's – Korolev had 10 years if he was to take advantage of this.

So the preliminary study conducted in 1959 became the basis for the preparation and publication on 23 June 1960 of the document "Decisions of the Central Committee of the CPSU and Ministerial Council of the USSR, № 715-296: About Creation of High-Power Launch Vehicles, Probes, Spaceships and Space Development in 1960-1967." This date is therefore considered the official start of the N-1 project.

Two years later the preliminary design of the N-1 was ready. In July 1962 the expert commission met again to consider the results. On the commission were high-ranking officials from the economic departments and the elite of the space/rocket departments. The commission approved the N-1 rocket project. The documents ran to 29 volumes, plus appendices. In Appendix № 1, by Gleb Maksimov's section, manned flight to Mars was covered using liquid-propellant rocket engines. This was the baseline payload for N-1. Various versions of lunar expedition were also presented to the commission, but not detailed documented studies, so there were no appendices on missions to the Moon in the approved N-1 project. But one thing was added; the Ministry of Defense had not been interested in developing super heavy rockets since interplanetary flights were of no use to them. But without the approval of the military, it would be difficult to get project funding. Therefore Korolev proposed using the N-1 rocket for launching large automatic and piloted military space stations.

After this project was approved, development of the rocket launching complex and the N-1 rocket itself was handled by Sergey Kryukova's division at OKB-1. By December 1962, launch complex construction was underway. But development of the rocket itself was suffering from major complications.

KOROLEV AND GLUSHKO

The only credible way to develop a new high powered engine in the short time available was to use Valentine Glushko's OKB-456 design bureau.[2] Glushko believed the only suitable propellants were nitric acid and UDMH. But these compounds are both highly toxic and highly corrosive. Korolev considered them much too dangerous to use in a piloted vehicle. Glushko point blank refused to consider alternatives, so Korolev turned to the famous designer of jet aircraft engines, Nikolay Kuznetsov, head of the OKB-276 design bureau, who agreed to design the N-1's engines. Glushko was deeply offended and from that day on was an ardent opponent of the N-1 project.

Valentine Glushko

Nikolay Kuznetsov

Historians consider that the conflict between Korolev and Glushko damaged not only the N-1 project, but the further development of the Soviet astronautics as a whole. Kuznetsov's bureau could not create large rocket engines without any experience, so the N-1 was designed to work with a large number of smaller engines. This later led to problems in coordinating the operation of so many units with the KORD control system. But the engines were made and for the time, the quarrel between Korolev and Glushko did not become a hindrance to the TMK and N-1 based projects.

TMK PROJECT EVALUATION (1960-1964)

In the early stages Korolev did not restrict the options and encouraged creative thinking about possible manned flights to Mars. The idea was new, therefore some project concepts for Martian expeditions were truly fantastic and would have required enormous resources for their realization. But such thinking fired great enthusiasm and yielded helpful results.

Just as the nuclear rocket was once considered as the solution to all energy problems, now using nuclear-electric ion engines was viewed as the best method for interplanetary flight (ion engines have very low thrust, but are extremely efficient). However, in comparison with a liquid-propellant rocket engine it was absolutely new technology, with many unknown factors. Therefore Tihonravov's department came up with two versions of the TMK: One using nuclear-electric ion engines and one using liquid-propellant rocket engines.

As has already been said, TMK stood for "Heavy Interplanetary Ship" in Russian. But it was frequently joked that it really stood for "Tihonravov Michael Klavdievich", the Department 9 chief!

TMK, nuclear electric engine version

At an early stage of development, Korolev entrusted the nuclear electric version to Boris Adamovich. The version was promising, but the technology required was too advanced for the state-of-the-art at the time. However, at the start of 1962 a report was prepared on "The Preliminary Design of Heavy Interplanetary Ships with Ion Engines (№3/1033)." The report contained 480 pages of text and schematics. The main task of this study was to estimate the practicality of using nuclear powered ion engines for interplanetary flight, while the design of the ship and its systems were only covered as a general concept. This drawing is from Boris Adamovich's book *Extraterrestrial Life*. The ship was to launch into Earth orbit in a folded state as shown on the left side of picture. Then the nuclear power unit would be unfolded on a rotating boom to a distance about 20 meters from

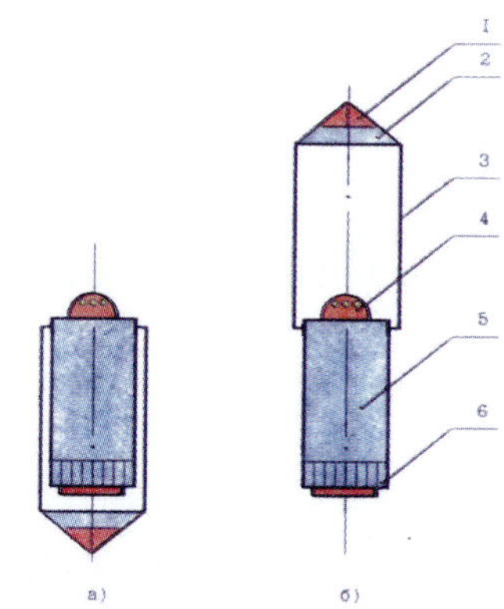

Рис.73. Тяжелый межпланетный корабль ТМК (Тихонравов Михаил Клавдиевич) с электрореактивной двигательной установкой: а) - в стартовом положении; б) - в полете; 1 - ядерная энергетическая установка; 2 - теневая защита от радиации; 3 - поворотная ферма; 4 - спускаемый аппарат (планетная станция); 5 - рабочие и жилые отсеки; 6 - ионные двигатели.

[2] Glusko designed the engines on the R-2, R-3, R-5, R-7 and several other Korolev projects—Ed

the main ship. This was necessary to protect the crew from radiation when nuclear reactor was operating. Apart from the living and service modules, there was a greenhouse with plants. Illumination was supplied by sunlight concentrators, with light directed inside by optical light pipes. Artificial gravity by spinning the ship was not considered in this version. The ship would have a crew of three and flight duration was estimated about three years.

Variant TMK-E or MEK

There was one other version of a Martian ship with nuclear electric engines from 1960. It is known as TMK-E, meaning TMK with an ion engine, (though on the website it has the designation MEK: the Martian Exploration Complex). The mission profile was very ambitious:

The ship would be assembled in Earth orbit from modules launched by the N-1. There would be a crew of six; three of whom would land on the surface with three remaining in orbit. Five unmanned, cone-shaped landing vehicles would descend to the surface of Mars, each containing one module of the wheeled "Mars Train." The three-man surface team would descend in a separate vehicle and assemble the five sections:

1. Crew cabin, manipulator and boring drill.
2. Special plane for researching flights on Mars.
3. Return rocket to bring the crew to bring the crew back to the orbital ship.
4. Back-up crew return rocket.
5. Section with a nuclear power generator.

The "Mars Train" would spend a year[3] on the surface of Mars, going from the South Pole to the North Pole. Its objectives: to carry out surface and atmosphere research and transmit the results to the Mars orbiter. From the Mars orbiter ship the information would be transmitted back to the Earth. When surface activities were complete, the crew, with soil samples and other research results, returned to the Mars orbiter ship. After transferring the surface crew materials to the main ship, it would return to the Earth at the next window of opportunity.

Some sources emphasize that one of the concept authors is Konstantin Feoktistov. But Feoktistov's department was not officially engaged in work on interplanetary vehicles (as mentioned above, his secton was engaged in working on the *Vostok*, *Voskhod* and *Soyuz* vehicles). He was a good engineer, passionate about space exploration and possessed enviable vigor; so most likely we are dealing with one of the early engineering concepts. Those documents were not classified as confidential and were available to many people, unlike TMK documents. The illustration was therefore presented only as a speculative design and that vehicle is not relevant to the TMK projects of Adamovich and Maksimov. But it is an interesting design concept worthy of attention and perhaps use of a planetary train will yet be considered in future projects for studying the Moon and Mars. Feoktistov later became an opponent of piloted expeditions to the red planet, but that's another story.

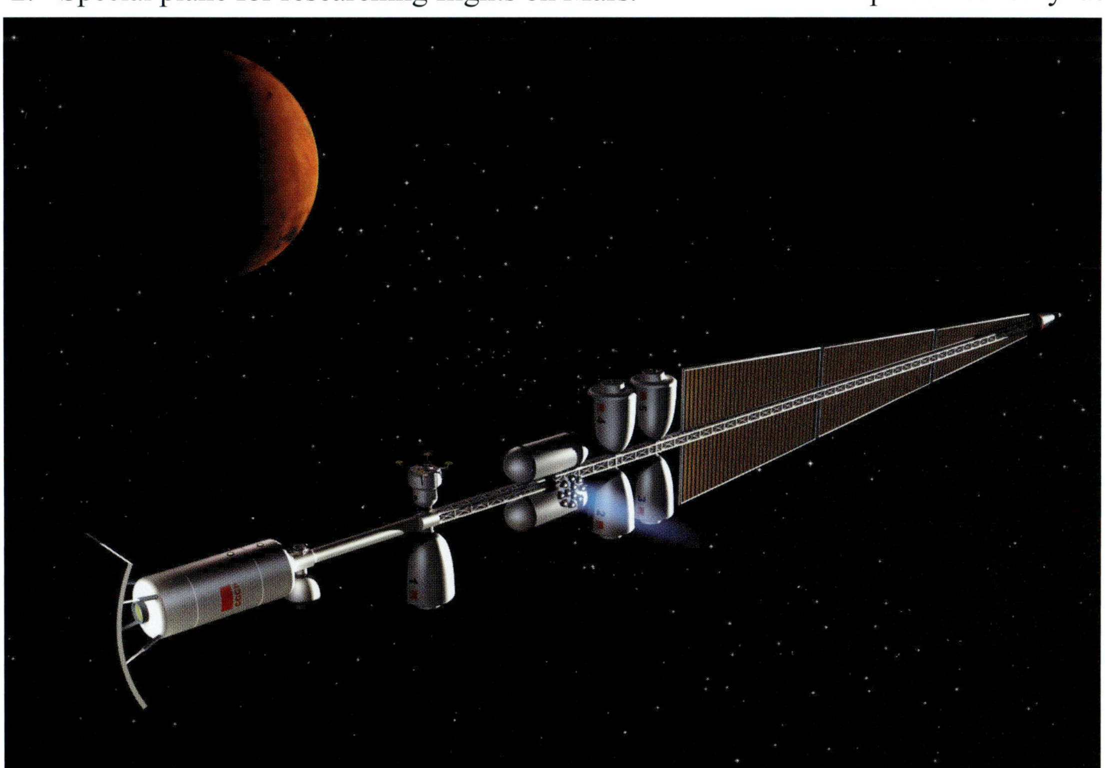

©Igor Bezyaev

TMK-E with Mars Exploration Complex

[3] A Martian year, which is approximately two Earth years—Ed

TMK variant with liquid-propellant engines

Compared with nuclear electric rocket engines, liquid-propellant rocket engines were much more advanced in 1960. Accordingly the study of TMK with liquid-propellant rocket engine was worked on in more detail. The first liquid propelled TMK did not include landing on a planet, it was a flyby version. Here is the design proposal from the study report with the launch configuration to the left and in flight to the right:

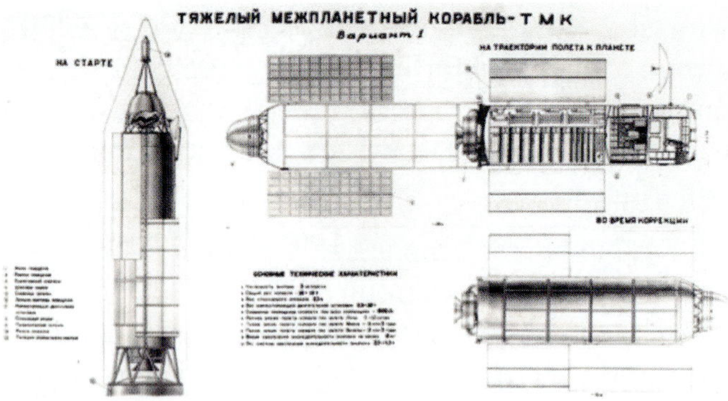

Схема тяжелого межпланетного корабля – ТМК.
РГАНТД. Ф.107 оп.2 д.68.

My CGI rendering of this vehicle is shown below. Like the TMK with ion engines, this shows the ship as deployed for flight. In this version the ship is rotated for artificial gravity and this design placed the living module at the maximum distance from the center of rotation. A greenhouse was part of the ship too. Light for the greenhouse was sent through slot-shaped windows in the body of the ship by the curved reflectors. Electrical power was generated by solar panels near the living module at the front. But the design had some serious problems. For example, the ship rotated in a plane perpendicular to the orbital plane and the mirror reflectors need to be focused on the sun at all times. Thus, during flight, frequent correction of ship's orientation was essential. Calculations have shown that this would use a lot of propellant. The solution was found in changing the rotation axis of the ship. If it rotates in the trajectory plane, the light concentrators always remain pointed at the sun. Here is my computer-rendered image of this version of the TMK:

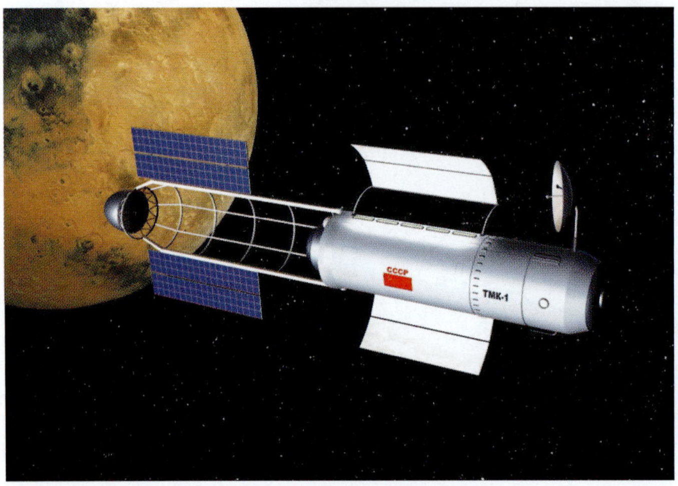

©Igor Bezyaev

Initial TMK design

©Igor Bezyaev

Revised TMK design

So, by the summer of 1962 two versions of TMK, with alternate engines had been designed. Korolev studied both projects carefully and approved the version of TMK with liquid-propellant rocket engine. The one above is the version of the ship presented in appendix № 1 of the preliminary N-1 design document in 1962.

MPRK

Further studies of manned flights to Mars showed that aero-braking using the Martian atmosphere would be very efficient, so starting in 1963 versions of the Mars ship using this technique were studied. Artificial gravity was rejected and a mission variant with a landing was also considered. The ship layout fundamentally changed too. By the end of 1964 the ship became as shown below. This version is known as the MPRK (Mars Piloted Space Complex). The engineers calculated that this vehicle would have a weight of 500 tons when it left Earth orbit. Launch date was planned for 8 June 1971 and successful return to the Earth on 10 June 1974. We have saved the engineers' original conceptual drawing (next page).

Assembling MPRK in Earth orbit would need six or seven N-1 rocket launches and could start within a year. All components of the MPRK would pass a full cycle of checks and the tests similar to Earth based production tests. These activities

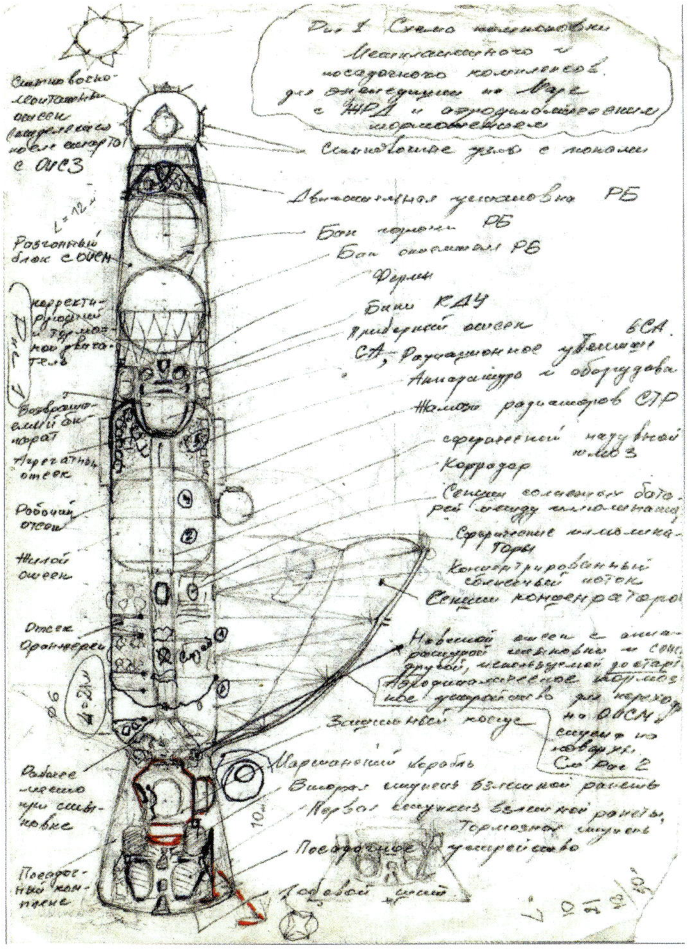

MPRK Concept Sketch

would require special divisions of cosmonauts from among the skilled engineers of OKB-1. They would be flown to the MPRK ship's orbit on *Soyuz* vehicles and they would live in a special orbital living module which was not part of the MPRK ship. After getting everything ready for the MPRK launch, the engineer-cosmonauts would return to Earth, their part in the mission accomplished.

The crew of three would arrive at the TMK in advance, to start the closed biological environment complex and check all the ship systems. Before departure, crew members would enter the reentry vehicle, from which the Mars complex is controlled, including all of the engine burns. Time to leave for Mars! If there was an emergency while accelerating away from Earth, the crew could separate the re-entry vehicle and return safely to the Earth. The empty accelerating rocket stage would separate and be diverted to the side.

During the flight to Mars the ship is oriented with the solar concentrator aligned to the sun. This orientation is maintained by the attitude control system throughout the flight. Communication with Earth would be scheduled rather than continuous and there was provision for course corrections, if required.

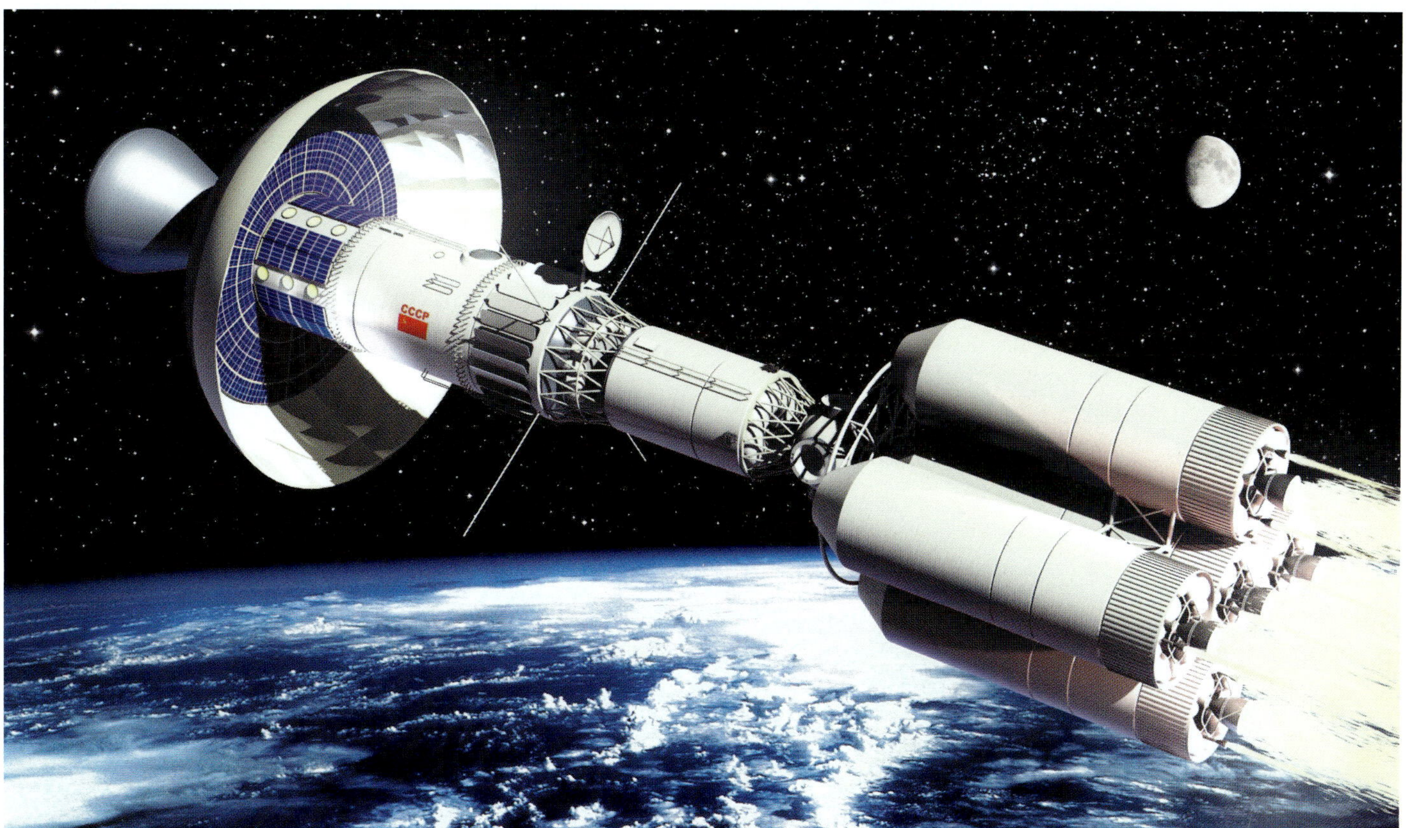

MPRK Departing for Mars!

©Igor Bezyaev

Transition from flight trajectory to orbit around Mars was achieved by aero-braking. During this phase the whole MPRK is repeatedly plunged into Martian atmosphere. After several descents to an altitude of 70-100 km to reduce speed, the MPRK arrives in an extended elliptical orbit. The altitude of apogee[4] is then gradually decreased until after about seven immersions when the apogee altitude becomes equal to the altitude of the future circular orbit. Corrections are made to raise the perigee and the MPRK achieves a circular orbit, with no more dips into the atmosphere.

After preparations and checks of the Martian landing vehicle, two members of crew enter the cabin. The aerobraking shield is jettisoned, then the Martian landing ship separates from the main MPRK, de-orbits and descends into the atmosphere for landing.

Once the surface activities are completed, the ship and cosmonauts launch back to Mars orbit and dock once more with the main MPRK. The main rockets then fire for the return journey to Earth, again keeping the necessary orientation to the Sun for illuminating the greenhouse.

Approaching Earth, the reentry section separates then brakes and enters Earth's atmosphere. Landing is by parachutes. The remaining part of MPRK flies past the Earth and goes into orbit around the Sun. Reuse of the MPRK was not planned.

In 1964 the design began on the Heavy Orbital Station necessary for the development of the MPRK. At the same time, the design of the TMK, (the TMK being the main part of MPRK), was being checked for long flight conditions by the EMBP (Institute of Medical and Biological Problems). This became known as the NEK and in the late 60's and early 70's research was conducted using a simulated crew of three.

Thus Korolev established a firm foundation for a Martian expedition, founded on the decision of 23 June 1960 and systematically developed interplanetary flight options.

But these plans could not be implemented. Further development of Mars ships was disrupted and a new era began: The race to the Moon.

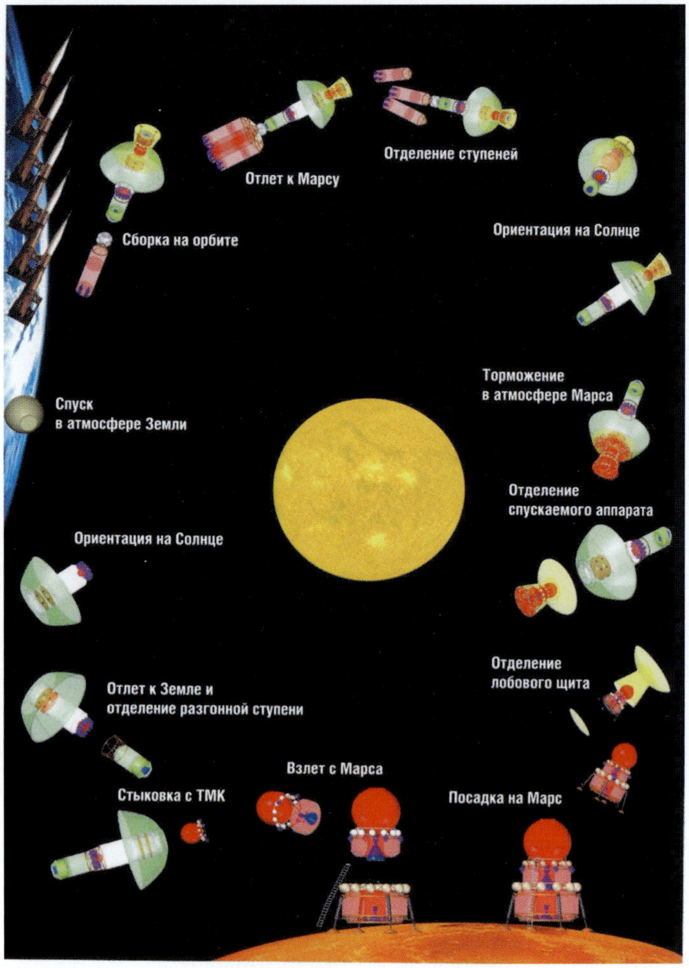

MPRK Mission Profile

Clockwise from upper left next to the five N-1's:

- Construction in Earth orbit
- Departure for Mars
- Jettisoning boosters
- Orienting solar collector
- Aerobraking in Mars' atmosphere
- Separation of landing module
- Front Shield separation (exposing lander)
- Landing on Mars
- Launching from Mars
- Re-docking with TMK
- Departure for Earth and jettison booster
- Orienting solar collector
- Re-entering Earth's atmosphere.

---

[4] Technically "apogee" and "perigee" refer to the highest and lowest points in an orbit around the Earth. The equivalents for Mars are "apoares" and "periares" but that's pretty awkward so we'll leave it at apogee and perigee for clarity—Ed.

LANDING ON THE MOON: THE NEW GOAL

The TMK project received a crushing blow from the government decision of 3 August 1964 when landing a Soviet cosmonaut on the Moon became the main goal. The next month, OKB-1 started the immense task of designing the L3 lunar complex. The Martian expedition was not shut down, but receded into the background as the focus of OKB-1 centered on the lunar program.

Let's review the main events so far.

On 12 April 1961 the first flight of a man in space took place. In May of the same year US president John F. Kennedy declared landing Americans on the Moon as a national goal. Nikita Khrushchev ignored this statement and did not compete with America, because in 1960 Korolev had been given the problem of making a heavy rocket for manned interplanetary flight. But one year later, on 16 April 1962, Khrushchev made a decision to stop work on the N-1 and limited it to a preliminary draft project. He also ordered Mikhail Yangel to create a new heavy rocket project (the R-56). A week later, Khrushchev ordered Vladimir Chelomei to create yet another heavy rocket (the UR-500, which later became the *Proton*). Chelomei was now working with Glushko, who after his quarrel with Korolev, was doing everything in his power to disrupt the N-1 project. What drove the decision to stop work on the N-1, which had already been going at full speed for two years, we cannot say. Most probably it was the result of influence on Khrushchev by political opponents of Korolev. For example, Khrushchev's son, Sergei, was working as department head in a section of Chelomei's bureau. Either way, decisions made in 1962 caused complete chaos in the Soviet space program. Korolev's interplanetary project was under threat and Khrushchev was hearing more and more appeals to land a Soviet cosmonaut on the Moon before the Americans. The result could have been simple - Khrushchev could have ordered Chelomei's main task to be the lunar expedition. In this case, work on N-1 would have been completely shut down and the partially built launch facilities adapted to the new rocket for Chelomei's lunar project.

Mikhail Yangel

Vladimir Chelomei

The next events were almost predictable. The only way to not lose funding for the N-1 (thus continue to work on the Mars project) was to present proposals for a mission to the moon using N-1 rockets…and that's exactly what Korolev did.

In short order the necessary documents were prepared and presented to the leaders of the USSR. Dmitri Ustinov instructed the head of the Rocket and Space Technology Institute (NII-88) to evaluate the three launcher options: N-1, R-56 and UR-500, for the lunar expedition. Of these three boosters, the R-56 came the closest to the American Saturn V since it was designed to use four of the monstrous RD-270 engine which generated 6.1 million N (1.36 million lbs) of thrust, comparable to the 6.7 million N (1.5 million lbs) of the F-1 engine used on the Saturn. These engines wouldn't be ready for several years, though,[5] so the booster that Yangel proposed used 16 of the smaller 11D43 engines, four for each of the RD-270's. (continued page 49)

The RD-270

[5] It was never actually put into production. The project was cancelled in 1972 with some combustion instability problems still to be resolved—Ed

The Zond Missions

While the Zond-L1 program never sent cosmonauts to the moon and it consumed resources that could have gone towards the N1-L3, it was in many ways the most successful Soviet moon program. In place of cosmonauts, they sent turtles, fruit flies, mealworms, seeds and bacteria samples on all the missions to test the life-support systems. They also carried scientific instruments such as micrometeoroid detectors and high-resolution film cameras.

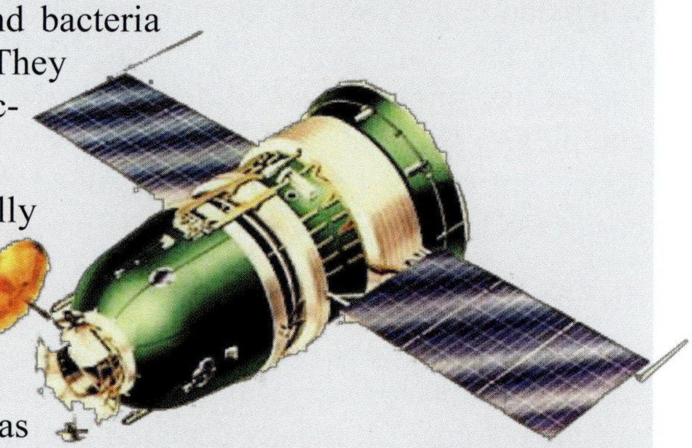

The Zond vehicles, designated Soyuz 7K-L1, were essentially stripped-down versions of the Soyuz 7K-L3 used for the LOK. A lighter vehicle was necessary because of the lower lift capacity of the *Proton* booster. There was no spherical Utility Module at the front, and they used solar panels rather than fuel cells to generate electricity. The Service Module was shorter as there was no Block-I engine needed since the craft never entered lunar orbit. They were boosted from Earth orbit into the free-return trajectory by a Block-D stage developed for the N-1 (see page 148).

There were five lunar missions altogether, *Zond 4* through *Zond 8*, all of which successfully looped around the moon (or a moon-equivalent distance) and returned to Earth. The reentry module followed the same "skip" entry path as the L3 described on page 24. Here is a summary of each mission:

- *Zond 4*—March 1968. Tested out the vehicle in trans-lunar space. It was sent into a looping orbit with a 350,000 km (220,000 mile) apogee, but to avoid the complications of the moon's gravity should something go wrong, it was sent in a different direction from the moon. A guidance system failure prevented it from following the skip-trajectory on reentry, so it was destroyed to prevent it from landing in non-Soviet territory.
- *Zond 5*—September 1968. Actually sent to the moon which it looped around at a closest approach of 1,950 km (1,200 miles). A guidance system failure again prevented the skip reentry, but the trajectory this time allowed for that contingency and it landed in the Indian Ocean where it was recovered by the Soviet Navy. Flying three months before *Apollo 8*, the payload included the first living things to travel to the moon and back.
- *Zond 6*—November 1968. A repeat of *Zond 5*, but this time the guidance system worked perfectly for the skip-trajectory. Unfortunately, a leaking hatch seal during the last stages of the mission depressurized the cabin killing the biological specimens.
- *Zond 7*—August 1969. The first Zond flight post-*Apollo 11*. They had considered sending Cosmonauts, but with the moon race lost, it would have looked like an also-ran. The mission was a complete success with the reentry module landing, as planned, only a few km from the launch site.
- *Zond 8*—October 1970. The last flight of the program took a large number of high resolution photos of the moon and Earth, including their own version of *Apollo 8's* "Earthrise" as it rounded the moon's backside. This time the vehicle did it's skip-entry over the northern hemisphere to allow Soviet tracking stations to monitor reentry directly, leading to a landing (planned, this time) in the Indian Ocean—Ed.

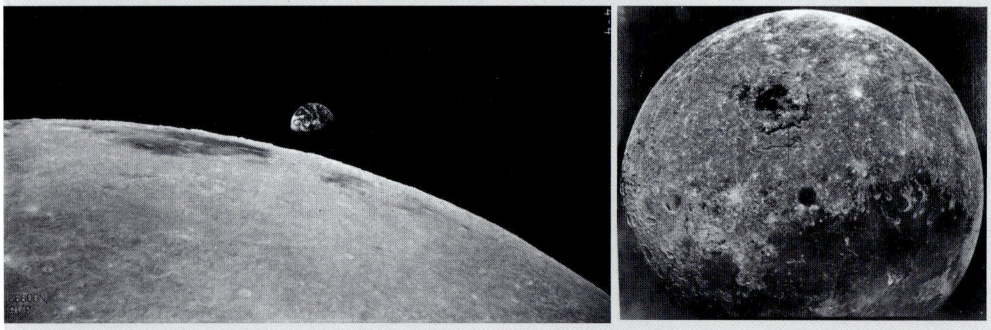

Images from *Zond 8*—Mare Orientale (the Eastern Sea, which is barely visible on the moon's eastern edge as seen from Earth), is prominent in both frames. It is the dark patch to the left of the Earth in the left image and the "bull's-eye" in the upper center of the right.

In June, 1964 they reached the conclusion that the N-1 rocket was most likely to succeed, because it was already under development and some parts have been made, while the rockets of Chelomei and Yangel only existed on paper. It is important to note that all designers at this time were largely bluffing. They represented their projects as best they could, but the question of the time required to complete development was glossed over. It was a struggle for leadership in an industry where it was all-important to win funding.

On 3 August 1964 there was a decision. The prime goal for N-1 was landing Soviet cosmonaut on the surface of the Moon and returning him safely to Earth.

After this, OKB-1 was reorganised. Department 9 was disbanded and Department 93 was created, with the job of designing the LK and LOK. OKB-1's focus was now fully on the lunar expedition.

But that was not all. The same decision entrusted Chelomei with the creation of a new spaceship complex, based on the UR-500/*Proton* booster with the *Zond* ("Probe") spacecraft, for a manned flight around the Moon without landing. This was the simplest manned lunar mission known as "L1." It was a dead end concept, useful only for propaganda, but it gave one more chance for Chelomei and Khrushchev's son to occupy a leading position in space exploration. This flyby program (it just looped once around the moon without going into orbit, see sidebar opposite) was one more blow to the Martian expedition since it meant even fewer resources for the Mars project.

After a while it became clear that Chelomei had no chance of meeting the target dates for a manned lunar flyby mission. With Khrushchev deposed the previous year, national leadership decided in 1965 to transfer the Zond project to Korolev as well. So in addition to the lunar landing, Korolev got the job of developing the complex for the Moon flyby. Korolev's bureau was severely overloaded with work and he made a difficult decision. He transferred all work on automatic interplanetary and lunar probes to the Lavochkin design bureau (OKB-301), retaining work only on main lunar landing project and the "frozen" Martian expeditions. George Babakin thus became chief designer of the "AMC" automatic probes (Automatic Interplanetary Stations).

THE MARS ROCKET BECOMES A MOON ROCKET

The American Saturn rocket was custom designed for the task of landing American astronauts on the Moon and it could lift 130 tons into Earth orbit. This was necessary for a mission with only one launch. The N-1 could launch only 75 tons, which was just sufficient for piloted flyby of Mars. For more complex missions, such as landing on the surface of a planet, the interplanetary vehicle would be assembled in Earth orbit from several N-1 launches. However, multiple launches for each moon mission had been rejected as it was more expensive and could not be done quickly as there were issues with preparing and launching two N-1 rockets at the same time. The decision to use the N-1 for the lunar piloted program, with a one-launch scheme meant the program had a severe deficiency in payload capacity and increasing it caused much difficulty.

Investigations began into ways to increase the N-1's payload. The main options were:

1. Change orbit inclination for launch to get more benefit from the Earth's rotation.
2. Lower orbital altitude from 300 to 220 km.
3. Install cylindrical extensions in the propellant tanks to increase the volume.
4. Decrease temperature of the kerosene fuel from -15° to -20°C and overcool the liquid oxygen to -190°C making both propellants denser thus being able to store more in the existing tanks.
5. Install six additional engines in the central part of Block-A.
6. Upgrade the rocket engines to make them more efficient.

They eventually chose options 4, 5 and 6. The rocket's weight at lift-off increased from 2,200 to 2,800 tons and the payload came close to 95 tons. The N-1 system's capacity could not match the Saturn V's 130 tons, despite the best efforts of engineers and designers.

Korolev probably understood that Soviet Union had most likely lost the Moon race to America as they started 3 years later. But, not having any other way to proceed, he probably hoped that there would come a time when the top management would understand the hopelessness of their lunar program. There would then be a question of what to do next when it might be possible to suggest returning to the initial goal of a flight to Mars with the TMK.

COMPLETE LUNAR EXPEDITION N-1 L3

The real work on the L3 project began in September 1964. On 10 February 1965 the preliminary design was approved. The design, structure of systems and weight requirements of the lunar vehicles were calculated for the first time. The project plan was crude and contained minimum initial data, plus it put tremendous pressure on Korolev. In less than a year's time, the gaps in the plan had to be filled. This project was necessary for Korolev to save funding for the N-1, so that in the future it could launch flights to the red planet.

[See chapter 3 for the detailed plan of the L3 moon mission]

THE LONG TERM LUNAR BASE PROJECT

As I stated earlier, before August 1964 there were no serious studies of piloted expeditions to moon by OKB-1. But it would be wrong to say that Korolev did not have any plans for the moon in the sixties. In 1962 Korolev ordered Vladimir Barmin to develop a project for a lunar base. Barmin was chief designer of the "SPETSMASH" engineering bureau.

They say Barmin replied that he could not design a base, without knowing what rocket would be used for the delivery of its elements to the moon. Korolev answered "You design your base, delivery is my problem. Of course Korolev meant to use his N-1 rocket.

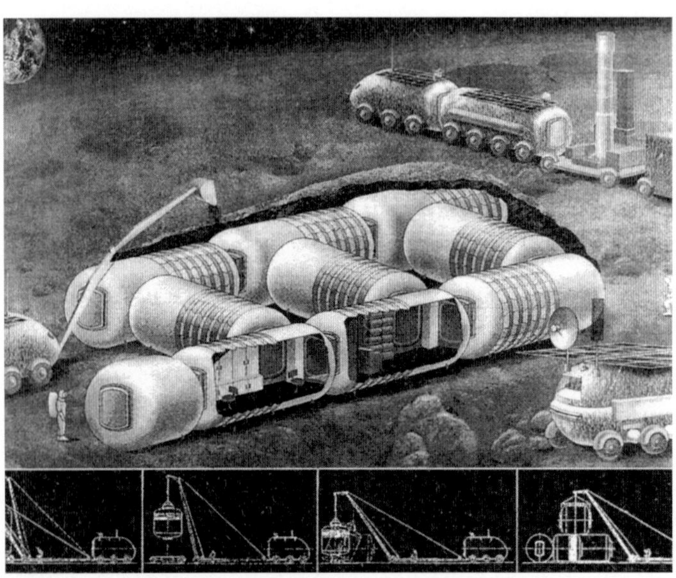

The *Zvezda* Moon base concept

Barmin's bureau took over ten years to design a lunar base and the project was almost completely finished. The project was referred to in documentation as "DLB" (Long-term Lunar Base), but in OKB-1 everybody knew it by the beautiful name *Zvezda* ("Star").

It was assumed that the site of the lunar base would be chosen by automatic survey probes. Next would be a manned expedition with a crew of four and a "Lunar Train", designed for building a temporary small lunar base. At the end of the building phase, it would be used for scientific exploration of the surrounding area. The lunar train consisted of four modules:

1. Main engines.
2. Habitation module.
3. Nuclear power plant, 10 kilowatt capacity.
4. Drilling rig.

The full weight of the Lunar Train was 8 tons.

The basic version of the lunar base for 12 persons consisted of nine standard units with a cylindrical form. At the factory, a standard unit was formed of sections that compressed like an accordion, 4.5 meters long. When pressurized, the unit extended to 8.6 meters and had a diameter of 3.3 meters. The lunar base consisted of these units:

1. Control unit / command post.
2. Scientific laboratory.
3. Storehouse.
4. Workshop.
5. Medical center and gymnasium.
6. Galley and dining area.
7. Three residential units.

In the early seventies the project was almost complete, but was suspended because of the high cost, which was estimated about 80 billion US dollars. And in 1974 the project was left without a carrier - the N-1 rocket. The project died.

1966 – KOROLEV DIES

Korolev did not have time to complete his brainchild, the N-1 rocket. He died on 14 January 1966 about a week after a routine surgical operation. He had been in very poor health for some time suffering several heart attacks and intestinal problems. These conditions can be traced to the severe treatment he received when imprisoned in the Gulag (1938 to 1944) as the result of one of

Stalin's purges, but they were made worse by his enormous work load that he took on voluntarily.

Vasily Mishin, Korolev's assistant since 1946, became his successor. The heavy burden of achieving all the great programs fell to Mishin, including the military projects. Mishin was an exceptional engineer and a good organizer and coped with these tasks. He continued the work begun Korolev many years previously. But the political arena of the space program had other strong players and eventually they won the struggle.

Vasily Mishin

N1-L3M

On 20 July 1969 the lunar module *Eagle* touched down in the Sea of Tranquility and six hours later Neil Armstrong stepped onto the Moon. The goal of the Apollo program, initiated by President John Kennedy in 1961, had been reached. In the Soviet Union during this period there had been two launches of the N-1, both ending in failure. Khrushchev's challenge had not been met. What next?

For a time the N1-L3 project continued. The Zond/Soyuz spaceship was showing good reliability. In October 1969 three standard Soyuz craft flew simultaneously, carrying a total of seven people (a record for the greatest number of people in space at once). Eight months later Soyuz 9 set another record by keeping two cosmonauts aloft for nearly 18 days. The LK made three successful unmanned test flights in Earth orbit in November 1970, February and August 1971. The engine was fired repeatedly, simulating hovering over the lunar surface and flight from the moon's surface to a rendezvous with the LOK. Following these flights, the moon ship was ready to carry cosmonauts. But the need for a project to land a man on the moon before the USA was gone and the project made no sense. The N1-L3 lacked a serious purpose. Vasily Mishin made a decision to update the lunar program and in 1970 started work on a new improved variant, the N1-L3M.

For this project it was planned to improve the N -1 and to increase its payload capacity to 100 tons, by replacing the kerosene with synthine ($C_{10}H_{16}$). The plan was for an expedition to moon using two launches of an N-1 (i.e. a return to the multi-launch scheme for the N-1). In this plan the lunar orbital unit and the lunar lander are launched from Earth orbit by separate N-1 rockets. Then, by their own engines, the lander and orbiter are injected into a trans-lunar trajectory. They would dock in low lunar orbit. If docking fails, the lunar orbiter can return using its own engines. If there's a successful docking, the booster unit is used to start the lunar lander down from orbit towards landing. The final landing is provided by the lander's propulsion system and landing gear. So far this is similar to the L3 mission described previously (Chapter 3). However, take-off from moon and return to the Earth were to be direct, with no lunar orbit rendezvous. Such a plan provided:

1. Two cosmonauts to the lunar surface.
2. A wide choice of areas or the lander to pick from, (strong cross range capability)
3. A longer stay on the surface of the Moon.
4. Better electronic and radio equipment in the ship, due to an increased weight allowance.
5. Heavier scientific equipment and more of it.

According to engineer's calculations, it seemed that within the N1-L3 current annual budget, by 1978-1980 the USSR would have the capability for gradual development of the necessary infrastructure to create a lunar base and launch lunar expeditions of three months duration.

But this manned lunar program did not receive approval from senior management and later was completely buried, along with the N-1.

DEVELOPMENT OF TMK TRANSFERRED FROM OKB-1 TO NII-88

So, the race to the moon was lost. What next? One logical step is to return to the idea of men landing on Mars since there was ten years work already done on the subject. This was real groundwork and the project looked achievable.

In 1966, after Korolev's death, the manned Mars expedition project was transferred to NII-88 (Research Institute-88, today known as "TSNIIMASH").

MAVR

Before concentrating on a landing on Mars, The engineers at NII-88 created a flyby mission idea that was quite remarkable.

They started with the Maksimov "TMK" project and developed the concept of a manned Mars ship for a crew of six, with a flyby of both Mars and Venus. The project had the code name MAVR (Mars and Venus in One).

The technical report on MAVR was presented in July, 1968. The ship was cylindrically shaped, 6 meters in diameter and divided into 7 sections:

1. Service section.
2. Gymnasium section.
3. Living section.
4. Laboratory.
5. Airlock.
6. Section for servicing automatic probes.
7. Greenhouse.

Sections were connected by a central passage with a diameter of 2 meters. The exterior of the ship was shaped like a bottle, similar to the TMK concept from 1962.

There was a centrifuge in the gymnasium section to keep the crew's vascular systems healthy. The greenhouse area was 14 square meters with three ring-shaped planting beds. The parabolic concentrator focused sunlight into the greenhouse through a central window. The light was then spread out using Fresnel lenses. The parabolic mirror was made from aluminum alloy and also served as a reflector for the ship's radio telescope (to conduct studies while enroute) and as the antenna for communication with the Earth (similar to the TMK concept of 1962).

At the aft end was the maneuvering system for course correction. In the centre was the reentry vehicle for 6 people. It is similar to a Soyuz reentry module, but larger. Fuel for the course correction system was in elongated tanks placed on an external surface to give additional protection from cosmic radiation. During interplanetary flight six solar panels would unroll, to provide electrical power.

Finally there was an astronomy module, approximately 7 meters long.

The plan was to launch MAVR in 1975 and fly on a trajectory for 480-600 days, using a gravitational assist at Venus. Research probes would be sent down to the planet's surface as MAVR flew past.

AELITA

On 30 July 1969 there was an order from the ministry of general engineering №232 concerning the design of a rocket complex that could reach Mars. The project had the code name "AELITA".

Various versions of the Martian expedition were studied. The most well known concept on the theme of AELITA is in the TCHIIMASH museum and on the official web site of RKK ENERGIA.

This Martian vehicle was shaped like a long needle, with nuclear powered ion engines at one end and the orbital complex with the landing craft at the other end. For cooling the reactor, a conical telescopic radiator was installed. The ship would be put into Earth orbit by two launches of the N1-M and assembled into one piece in orbit. Ion engines are low powered, but very efficient and the ship would then slowly accelerate while unmanned, spiraling slowly up and away from Earth. After the ship had passed through the Earth's radiation belts, the crew would join the ship by a modified LOK from the lunar program. There were four crew members and the ship weight was estimated 150 tons. Launch was planned for 1985.

MAVR Complex

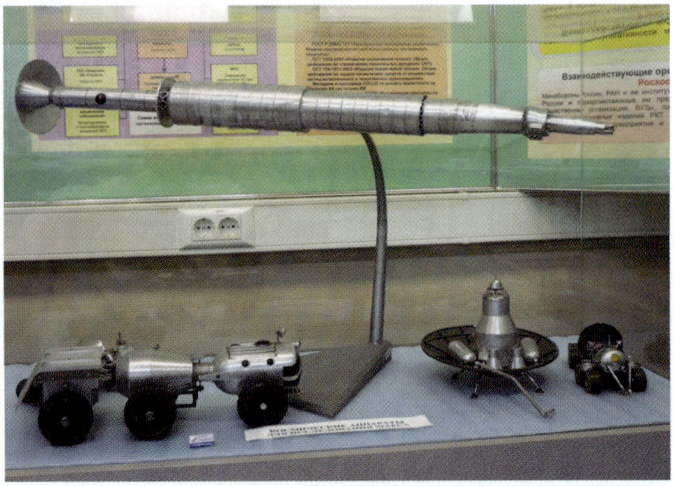

AELITA Complex

AELITA "Exploration Train" Model

One version of the expedition envisioned the use of a six-wheel Martian Train, with a crew of three. It was designed for 30 days independent travel on the Martian surface, with an average speed about 12 km/h. The front module was the control room and laboratory plus the crew living cabins and the airlock for exit to the surface. The second module contained the rocket for returning to orbit and a system to erect it in a vertical position for launching. As one of the project engineers joked, "You must carry your rocket behind you; otherwise the evil Martians will steal it!" The third module is the power unit, with a nuclear reactor, capacity 100 kW, radiation protection and a heater. It was estimated that such train could stay for a month on Mars and travel a preplanned route of 1500 km, collecting a great deal of scientific information.

The report on AELITA comprised two thick volumes. Next would have been a decision on which was the best design for the Mars expedition.

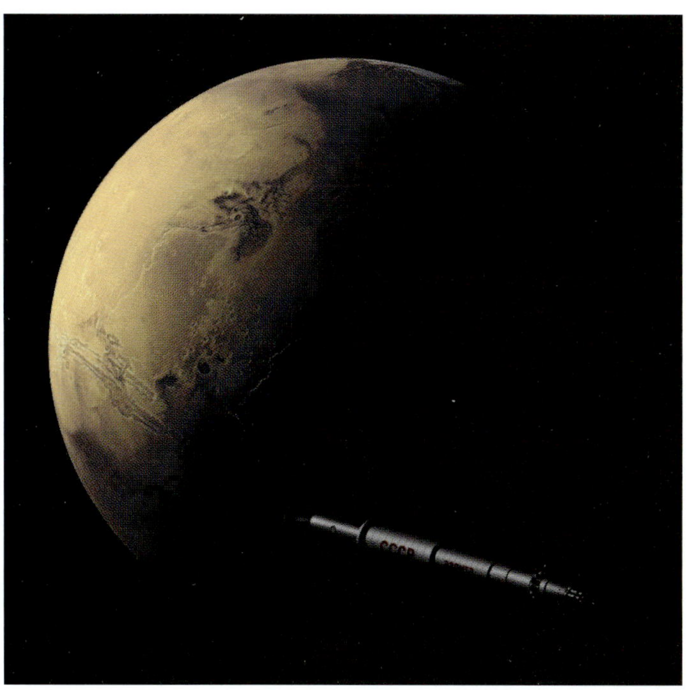

©Nick Stevens

AELITA in Mars Orbit

But AELITA was doomed. It was buried together with the lunar plans in 1974 when all work on the N-1 was cancelled.

End of the N-1

In 1974 the Central Committee decided to remove Mishin as head of the OKB-1 bureau and combine it with Glushko's OKB-456. Ironically, this put Glushko in charge of the N-1 project which he had opposed from the beginning. Even before the merger was complete, Glushko declared his intention to stop all work on the N-1 and to begin on a new family of rockets. His basic argument was that the engines were not reliable. Today this looks ridiculous as the NK-33 engines are considered one of the great achievements in aerospace history. Roscosmos has restarted production of the NK-33A for use in an upgraded Soyuz rocket (photo "D" on page 10).[6] the decision to close the N-1 project was a personal tragedy for many workers. The work of millions of people over 15 years (1959-1974) was wiped out and destroyed. The next N-1 rocket had many significant improvements over the original and chances of a successful launch were considered very high. The fourth launch had come very close to second stage ignition and the N1-8L had greatly improved components.

Glushko immediately terminated all activity on the hated N-1 rocket[7] and thereby also killed all the other projects which depended on it. Moreover, he gave the order to destroy all documentation, all hardware and did everything he could to ensure his orders were carried out. Now everything had to be started again. The N-1 era was over and the era of a new rocket, *Energia* ("Energy"), began.

Were such actions justified? Most likely it was just the result of personal ambition and vindictiveness of Valentine Glushko.

Many of Korolev's colleagues consider that if not for his death, men would have walked on Mars by the end of the 20th century. There was only one essential reason that the Soviet Union did not landed cosmonauts on the Moon or on Mars: the death of Korolev. All the remaining tasks and problems would have been solved by him. There

---

[6] As we go to press, *Samara Today* reports that the NK-33A was successfully hot-fired in a Soyuz-1 stage on 15 January 2013 – Ed.

[7] N1-8L was so close to being launched it had even been moved to the launch pad and was undergoing fueling tests.

is no doubt in their minds; such a man was Sergey Korolev. He was not only a brilliant engineer, but many who worked with him considered him a truly inspirational leader, taking time to encourage and inspire others despite his heavy workload. He was effective at operating within the political constraints of the Soviet Union and had also managed to maintain effective quality control of his space projects - something which caused many of the problems on the N-1 program after his death.

Legacy

Ironically, the greatest lasting impact of the Soviet moon program is in the commercial US space program. In the 1990's aerospace companies in the west were offered the "forest" of completed NK-33 engines (photo "A" on page 10) that had been secretly stored in defiance of Glushko's orders to destroy them. Engineers from the American company Aerojet could not believe the specifications for the efficiency of the engine until they tested one. After it was proven, many advanced concepts, such as the completely reusable Kistler K-1, were designed using the NK-33. Aerojet sells a version of the engine, called the AJ26, that it modifies in the US. It is used in launch vehicles such as the Orbital Sciences *Antares*. The RD-180, a two-chambered derivative of the NK-33 developed for the *Zenit* booster, is used on the Lockheed-Martin *Atlas V*.

Aerojet testing an NK-33

Epilog

So, now we can answer many of our questions. As the sealed records are opened, we learn more about the events of those years. We will probably again revise the history of the N-1 rocket and will understand it better. But we can never go back and correct the mistakes of the past. And it is important to remember remarkable people like Sergey Korolev and his project for interplanetary travel on the N-1 rocket. Mankind was one step from realizing one of his most treasured dreams, a flight to the red planet Mars!

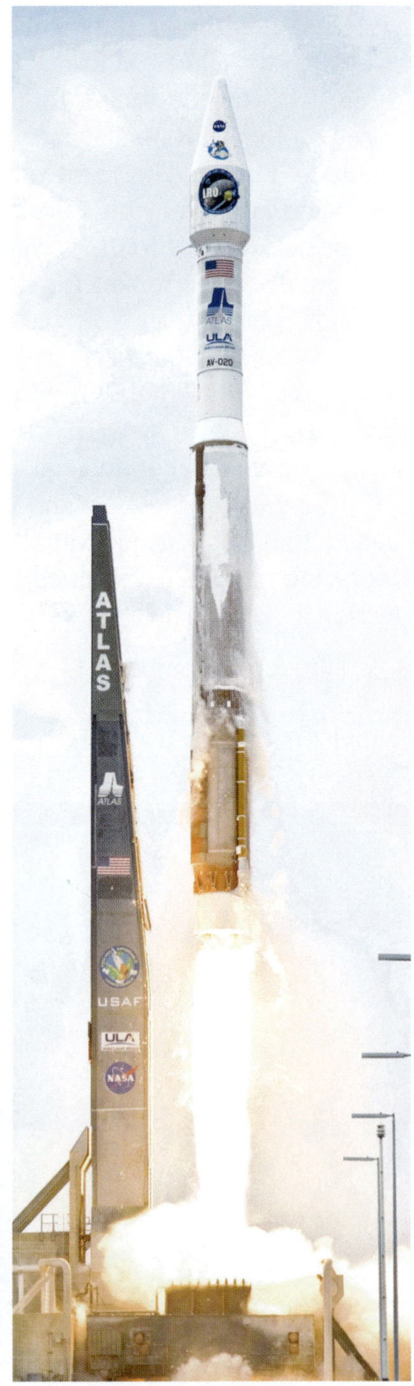

An *Atlas V* Lifts-off with the Lunar Reconnaissance Orbiter

Kistler K-1 with three NK-33 engines

A Pair of AJ26 Engines on the Orbital Sciences *Antares*

Part II

Detailed Descriptions and Renderings

5: History of the N-1 Diagram (Part 1)

5: История схемы Н-1 (часть 1)

Developing the Diagram of the N-1 Rocket

By Alexander Shliadinsky

Before beginning the story about the creation of the diagram of the N-1, it is necessary to understand three facts:

1. The development of the N-1 rocket was a top secret affair. Information stored by official sources was not published publicly, nor was there any information leakage at the time. Frequently the engineers developing separate parts of the rocket did not know their true assignment; workers did not know the details or purpose for which they worked.
2. After an American stepped onto the Moon, the management of the Soviet space program decided to close down the N1-L3 program. The decision made was not only to close the program, but to destroy all equipment and documentation created during the N-1 program. This meant that all N-1 drawings, equipment, remaining unused rockets, data collected during rocket tests, photos and videos were subject to destruction.
3. As they said in the USSR, laws can be very strict, but their severity is often compensated for by how they are enforced. In a twist of fate, after closing of the Soviet lunar program, the OKB-1 bureau was headed by V.P. Glushko. He was one of the main opponents of the N-1. The quarrel between Korolev and Glushko is the one of the main reasons for the failure of Soviet lunar program based on the N-1 rocket (the heart of the quarrel was centered on the type of rocket propellant to use). I don't argue against the role of Glushko in the implementation of the Soviet lunar program. But I do point out that Glushko implemented the order to destroy the memory of Korolev's N-1 in his subordinate departments, and that that order was to be executed exactly. Despite the order to destroy all of the materials related the N-1, much of the material would remain static in the debris field, carefully hidden or would be used in the workings of the Soviet space program in the form of Energia-Buran.

The Earliest Diagram (1985)

The true history of the creation of the N-1 diagram cannot begin without mentioning the name of an American researcher of Soviet rocket techniques, Charles Vick. He possessed satellite photos of the N-1 rocket on its launch pad with its projecting shadow. Through these photos, the size of the rocket was estimated and the assumption made that the rocket had a conical form. From other

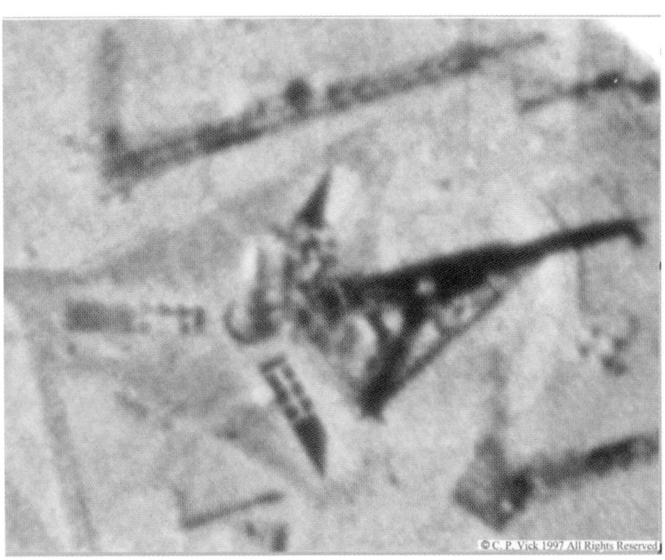

Both © C. P. Vick

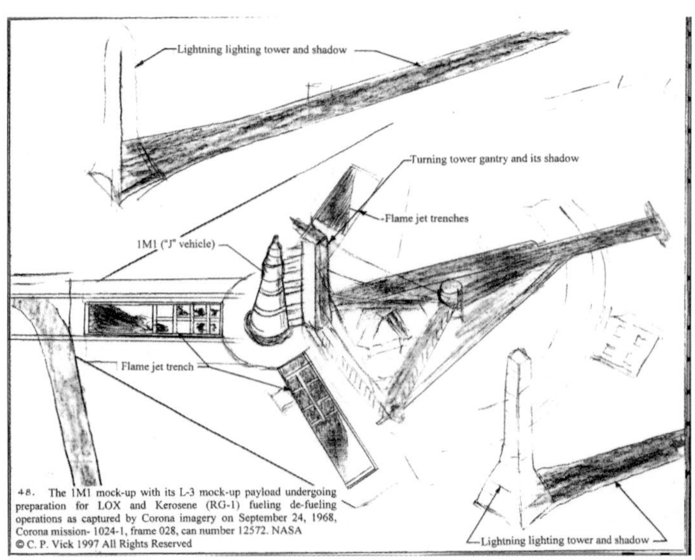

CORONA satellite image of N-1 M1 on the pad and Charles Vick's drawing of it

sources the information about the number of stages and engines in the rocket was determined. With this information in hand, Vick tried to reconstruct the rocket, which received the designation G-1 by the CIA.[1]

This reconstruction of the rocket appeared in the Journal of the British Interplanetary Society, JBIS №1, 1985. The rocket had a conical first stage weighing 4,200 tons. A cylindrical interstage connected to a second conical stage weighing 850 tons. For the third stage, he placed the third stage from the *Proton* rocket which was of his own reconstruction. He imagined that it had a diameter of 4 meters (13 ft), a length of about 18 meters (59 ft) and a weight of 140 tons. At the top he placed a space ship based on the Soyuz with landing and take-off stages. With a booster starting weight at nearly 5,200 tons, the weight of the payload was estimated at nearly 100 tons.

Instead of believing the reconstruction of such a monstrous rocket, the West relied on the perceived extreme backwardness of the Soviet engineers in designing rocket engines and the level of Soviet rocket technology in general at the time. Their analysis of the Vick rocket reconstruction and a poor understanding of Soviet space technology left the West with the conclusion that the G-1 simply would not have flown.

I spent much time studying Vick's drawing. Thanks to my degree from VOENMECH (Technical University in Leningrad), I knew about the real status of the Soviet space techniques and could not agree with the West's conclusion. But, other data which could help me to see the real picture was not available at this time, until…

The most unexpected things found in the most unexpected places

To understand a discovery, it is not enough to just have preliminary information. For true understanding, it is necessary to search for the primary sources. The primary data about the N-1 was found unexpectedly—in ordinary textbooks, which I could easily buy in the technical section of our local bookshops! That situation existed at a time when all my colleagues and friends, who were experts in

Alex holds his gorgeous model of Thor-Able 127, the first US attempt to reach the moon (*Pioneer 1*). Perhaps fittingly, the real mission exploded just over 1 minute after launch.

moon rockets, swore that the all of the N-1 documentation had been destroyed and that to search for them was useless.

The textbooks were, *Basics of Designing Flying Machines* (1985) and *The Design and Designing of Space Flying Machines* (1986) by academician V.P. Mishin. Illustration № 2.17, on page 44 of *Basics* shows a schematic view of rocket stages with weight bearing and suspended propellant tanks. The stage had a conical form with the spherical tanks suspended in it. Vick's drawing made it obvious that this was the N-1 first stage. The draw-

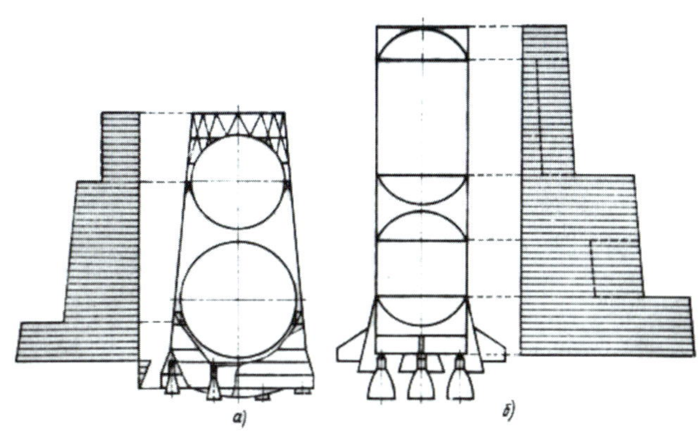

Рис. 2.17. Схема ракетного блока с моноблочным топливным отсеком и хавая

[1] Note that some documents, such as Charles Vick's drawing, refer to the N-1 as the "J-Vehicle" since the intelligence community had named its launch site "Launch Complex J" (a sequential naming scheme starting with "Complex A" for the first Sputnik launch site) – Ed.

ing also showed the first stage of a Saturn V alongside it in scale. That made it possible to estimate its weight.

In another page of that book there is one more schematic view of a conical stage. Unlike the previous stage, it had no conical tail-end skirt. Unfortunately, the scheme did not allow estimating the sizes of this stage, but already it made possible to tell with confidence, that at least two stages were conical with suspended spherical propellant tanks.

With Vick's reconstruction and my new knowledge, I made my own reconstruction of the N-1 rocket. That rocket had two conical oxygen-kerosene stages with suspended spherical tanks. On top of that I placed the second and third stages from a *Proton* rocket. For the payload I placed a *Soyuz* space ship. Calculations showed that such a rocket booster should have a starting weight of nearly 2,860 tons. That estimated value turned out to be correct. I made the necessary approximate calculations and prepared a report which was presented at the "Conference of the Historical Section of the Leningrad Department of Astronautics Federation" in 1987.

Publicity – The Engine of Progress

My report was a success and began to place me in a position to receive additional information. Despite the fact that "everybody knew" all photographic materials had been destroyed, soon, a photo of the N-1 on the launch pad came into my hands. The photo was of poor quality. But it made it possible to surmise that the rocket had three conical stages and a huge streamlined nose cone. At the same time, one of my colleagues, a rocket-modeler named Sergey Andreev, gave me a photo of a public bulletin board taken in one of the rocket agencies by a relation of his. The left side of bulletin board contained the picture of the N-1 made by some artist. Of course the picture was greatly distorted, but it also gave some information about elements of the rocket. For example, the picture showed external pipelines. Moreover, that information gave me the opportunity to estimate the number of rocket engines in the stages! The second stage contained eight engines and third stage contained four engines. Also, I could see the third stage was conical. The data obtained in this way made it possible to estimate more exactly the dimensions and weight of the stages.

Unfortunately, in estimating these technical specifications, I was using data that was obtained illegally, so I could not present my findings without being questioned about the source of the information. I was afraid for the safety of my colleagues who had given me secret data and I kept it all under the table. I could not present it so I waited until an event would come that would change the situation. I believed it would take a long time.

That Which Is Not Permitted For the Engineer, Is Permitted For the Modeler

It was 1989 and I made my first drawing of N-1 based on the obtained data. After that the idea came to me to make a flying model of the N-1. By this time there was a regular model rocket demonstration launch in Leningrad (St. Petersburg) on the Sunday before the 12th of April (the anniversary of Yuri Gagarin's space flight—called Cosmonautics Day in Russia). The tradition of this annual model rocket launching began at that time and continues to the present day. It is possible to see the show if you come on that day to Peter and Paul Fortress in St. Petersburg. At 12 o'clock in the afternoon, just after the noon gunshot, the show begins. (The noon gunshot is also a tradition for St. Petersburg).

I decided to make my N-1 rocket model for the 1990 show. On one hand, I wanted show my ability as a modeler. On the other, I hoped that some people who would have knowledge about the N-1 would come to the show, and I could meet them.

I planned to make a large model in 1:50 scale. The model had a height of more than two meters (six feet) and a diameter at the rear of 34 centimeters (13"). Launch weight was nearly 1 kg! For flying that huge of a model rocket I used six rocket motors with 20 Newtons of thrust for each one.

In April of 1990 this model was ready and I presented it to the people at the show. The weather was fine and it was good conditions for testing such a huge flying model. Even the preparation of the model was interesting for the viewers since the process resembled the preparation a real space rocket. At first I inspected the recovery parachute system. Next, I checked the electric circuits and installed the rocket motors. I mounted the model in a horizontal position on the launch pad and then erected to a vertical position. The next step was the insertion of ignitors into the rocket motors. After that I again checked the electric circuits. Finally the model was ready to fly!

COUNTDOWN! Everybody was standing ready.

LAUNCH! The model slowly lifted off on its railing and fell down on its side! It was an accident like the real N-1 rocket had! The reason for the fall was a faulty motor ignition system. Actually, only two motors fired and the thrust was not enough to lift it successfully. Fortunately, the model was not broken. During the next half hour I replaced two motors and reloaded the parachute system. Again the model was ready to launch. This launch was successful. The model left the launch pad and when it reached height about 30 meters (100 ft) the parachute system was activated. Unfortunately, the model fired only 4 motors. However the flight was successful.

This launch justified all my hopes. The flight was enjoyable for the viewers and it was seen by engineers from "VOENMEH" - Baltic Government Technical University of D.F. Ustinov. In the archives of the university there was information about the flights of the N-1. After some time, I met with people who helped me in the further continuation of my work. The next year I received information which allowed me to improve the model. For example, on top of model I placed the mockup of the Launch Escape System. Earlier photos of the model N-1 appeared without the LES.

I launched the model again at the next Astronautics Day in April, 1991. But the rocket again showed its temperament. On this day, cold weather prevented all the motors from firing. As a result, during launch, only one rocket motor fired. The thrust was too low to lift it and the model just stood on the launch pad. The flame from motor's nozzle reflected up to the model and the bottom began to catch fire! At the risk of get under an igniting motor, my companions and I were successful in putting out the fire and my model was saved. There was much time left until the end of the show and I decided to try to fly the model again. In the cold weather this was not so easy. But, after forty minutes, the model was ready.

LAUNCH came at last! The model confidently soared up into the sky. This time all the motors work very well and the rocket made a big arch to a height of about fifty meters. The model passed through apogee and began to come down. The seconds ran on but the parachute system did not activate. After some time the model fell straight to the ground. Most likely, bad weather dampened the ejection charge and the system did not fire. My model crashed, but she flew very nicely.[2]

Now I smiled. The question: how many times was the N-1flown? I answer – eight.

[2] This is a universal sentiment among model rocketeers. It is also one of my favorites post flight comments! –MJ

New Acquaintances

The best result of these public launchings of my N-1 model was that I made some very important acquaintances. The first of these was Dmitry Pajson. He was the editor of the newspaper *Apogee* which was published by the Moscow Aviation Institute. Dmitry told me that he needed articles on space techniques for the newspaper so I wrote a series of articles which students read to this day. The most popular first two articles: *The Goal–Moon Flyby* (1991) and *Touch the Moon* (1992).

My next new acquaintance was Igor Afanasiev. He lives in Moscow and is still the historian and editor of the Russian magazine *Novosti Cosmonautica* (*Astronautic News*). Igor gave me details about the engines of the N-1 rocket and some data on the propellant weights in the different stages which confirmed my calculations. In addition, Igor had some material on the tests of the N-1, dates of launches and the reasons for failures. We agreed to work together to reconstruct an accurate picture of the rocket.

My next important meeting was with Vladimir Antipov. He is a military man who had served at Baikonur as a launch equipment engineer. At this time he also organized a children's team, the "Group for studying the N-1 Rocket." Through this group he had gained permission to visit the lunar program debris fields at the Baikonur Cosmodrome, shoot photos and interview veterans of the Cosmodrome. In this way, he collected a large library of photos and other interesting information about the N-1 and the L3 lunar ship. I met with Vladimir in St. Petersburg and received important materials regarding the rocket design from him. For example, I received, for the first time, information about the hypersonic lattice fins. And most importantly, I received the invitation to visit Baikonur where I could look at the remaining rocket fragments. From that point on, I began the transition from imagining the external and internal elements of the N-1 rocket to creating its accurate diagram.

Vladimir took me to various parts of the Cosmodrome where parts of the N-1 have been used in such capacities as tanks for water, canopies for gazebos, garages and warehouses. But the most interesting trip was to the crater which was created after the explosion of the first and second stages of the N-1 in the crash of 27 June 1971 (N1-6L).

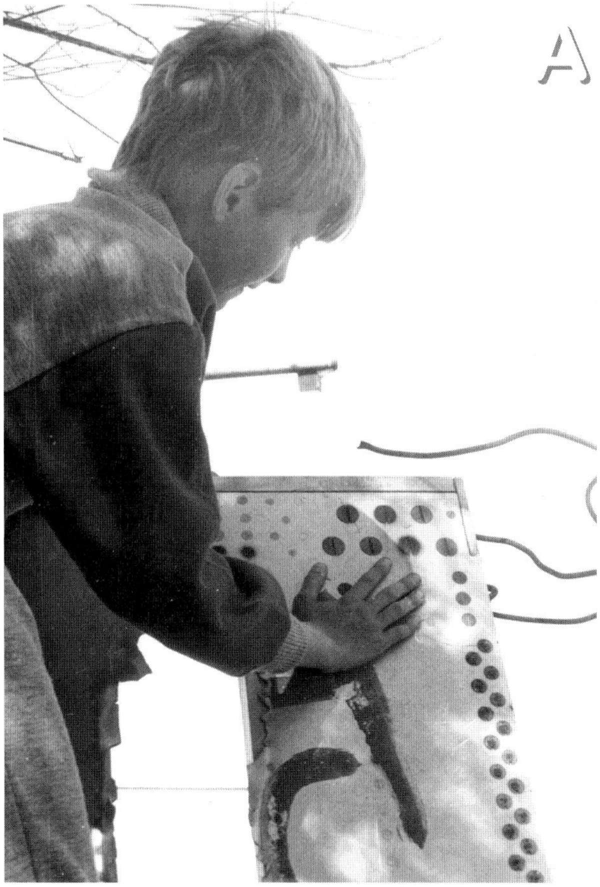

Vladimir Antipov's children's team
Group for Studying the N-1 Rocket

A - Touching History
B - First Stage Lower Skirt
C - Piece of the First Stage Skin
D - Block-A Dome Blast Deflector
E - Block-A Central Engine Heat Shield

There I soon found a large fragment of the rocket. It was a part of a transition between a rear skirt and a load-bearing frame of first stage. In the same place, there was a part of the load-bearing frame itself. I made measurements and wrote the data down in my notebook. Next, I found one of the triangular antennas for the transmitting of telemetry. Next, I found the load-bearing frame of the second stage. And the next part and the next part and the next part... So, in this way I made a detailed reconstruction of many of the elements of the rocket which I could not see on the photos, and I understood the total design of rocket better. But unfortunately, many of the parts were missing or in bad condition after the explosion, which did not allow me to make a good drawing. Additional data were still necessary.

The next stage of studying the N-1 began in February 1992. This is when I started my intensive exchange of letters with Igor Afanasiev. I sent to Igor draft variants of rocket sections. He found the people that designed the N-1 and they kept some information in their personal files. In this way I received pictures or drafts showing the location of rocket systems and elements of it. Later, Igor organized a meeting with a designer of the N-1, Renald Dolgopyatov. In that meeting he explained in detail all modifications made to the rocket as new ones were built – modifications that were constantly improving the design. Unfortunately, he died in 1995 soon after our meeting.

Alex visiting the debris fields
A - Interstage truss
B - Block-A debris
C - L3 Lower fairing being used as shed
D - Block-V antenna

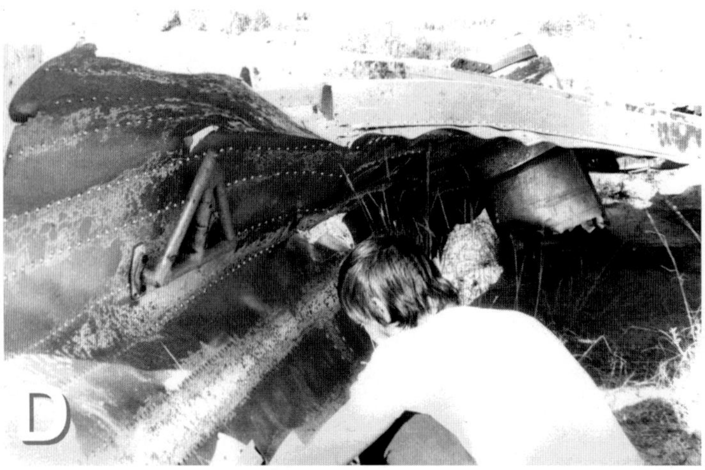

Crater left by impact of N1-6L

The second person that helped to restore the measurements of the rocket was Valentine Bobkov (1931-1999). He was an engineer and participated in the design of the reentry module of the *Soyuz* spacecraft. He gave to me the basic sizes of the stages of the N-1 rocket.

The 1993 Diagram

In 1993 I had enough photos, drafts, descriptions and sizes for making a detailed drawing of the N-1 and proceeded to create it. At that time Igor Afanasiev prepared an article describing the development, production and testing of the N-1 rocket. He proposed to use my drawings in that article. This diagram proved interesting for modelers. This article was printed in *Krylja Rodiny* (*Wings of the Motherland*) magazine (http://kr-magazine.ru/) № 9-11 of 1993 titled "N-1: Top Secret."

The drawing was very complete and possessed so much detail that it has been used not only by individual modelers, but also for the creation of display models for the international exhibition "S.P. Korolev Rocket and Space Corporation Energia."

The 2010 Diagram[3]

I knew that the 1993 drawing was neither perfect nor exact. For some years I continued searching for new data and materials about the N-1. In 2010 I finished a new, more exact, AutoCAD version. I am very grateful to my American friend, Mike Dorffler, designer at Estes Rockets in America. He pushed me to complete its updating. He asked me to make such a drawing for Estes which he hoped could set up manufacture of a flying N-1 model. Unfortunately, with Mike's untimely death in 2010 this project has stalled.[4] But I continue to improve the drawing, depending on the availability of new information.

There are rumors that not all drawings of N-1 were destroyed in 1974 and that some of them remain in confidential archives and personal collections. I do not know if this is true or not, but if it is true, they will be published some day. But that is in the future. Today, to let us see the detailed drawing of the epitome of S.P. Korolev's design skill, I have passed through a long and difficult path. It is not possible to repeat this path as the majority of fragments of the rocket are now gone, and many of the participants of the program have already died, taking their memories with them.

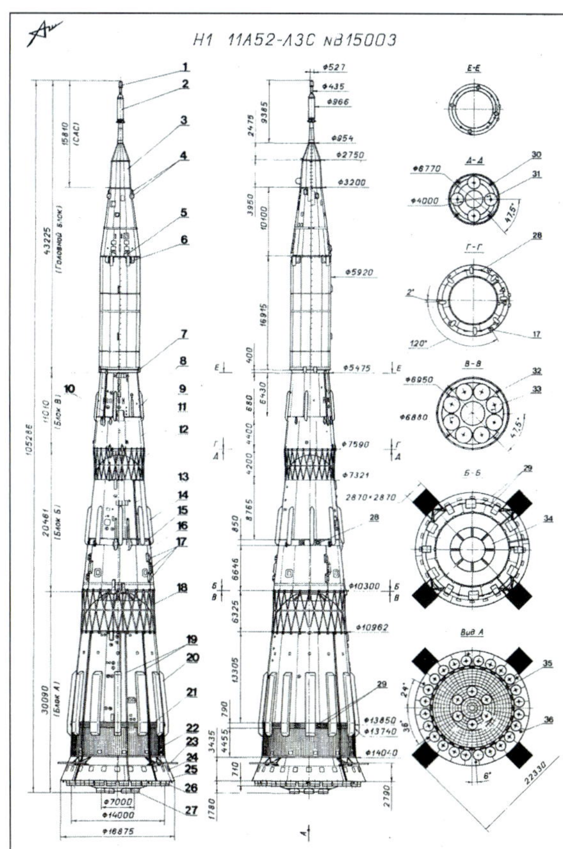

1993 Diagram

[3] The 2010 Diagram is provided as a foldout in the next chapter–Ed.

[4] Mike's successor at Estes, John Boren, has expressed an interest in continuing the project using the information in this book as a basis–Ed.

Debris Field Gallery

More of Vladimir's images of the debris from the N-1 flights as found by the researchers

Top of Block-G Fuel Tank

Bottom of Block-G Oxygen Tank

Tanks and Engines

LES Support Tower

NK-15V Engine

NK-15 Pump and Nozzle

NK-15 Pump Shaft

NK-15 Engine Nozzle

Trusses and Fairings

Block A-B Interstage Truss

Block A-B Interstage Truss

Block A-B Interstage Truss

Block-B Engine Fairing w/ Separation Motors

Block-V Engine Fairing w/ Separation Motors

Lattice Fins

Partial Fin w/ one mounting strut

Mounting Truss Detail (Rotated)

Fin w/ Mounting Truss (End View)

Fin w/ Mounting Truss (Side View)

Pivot Detail

General Debris From Flight N1-3L

6: History of the N-1 Diagram (Part 2)

6: История схемы Н-1 (часть 2)

Differences in Design Between N1-3L thru N1-7L

By ALEXANDER SHLIADINSKY

For an understanding of all the changes made to the N-1s as a result of testing, it is necessary to first review the rocket's basic design.

The N-1 was a three-stage rocket with consecutive stages. All three stages contained suspended spherical propellant tanks for the fuel (kerosene) and oxidizer (liquid oxygen). The volume of the oxidizer tanks was approximately 1.8 times greater than the volume of the fuel tanks (meaning they were 1.2 times larger in diameter). This ratio defined the conical form of the stages. All three stages had an identical design.

The numbers in the following description refer to the foldout drawing on page 69.

The design of all three stages started with a main load-bearing frame (№ 11; 25; 38) at the bottom. On top of this frame there is the spherical oxidizer tank (№ 12; 26; 39) held by thermal-bridges (insulating supports). The main frame of Block-A (№ 11) has a transition section (№ 10) that transfers the thrust of the twenty-four outer rocket engines to the rest of the structure. The conical load-bearing frame (№ 8) is also attached to the Block-A main frame (№ 11) and transfers the thrust of six central rocket engines.

At the bottom end of Blocks B and V there are conical load-bearing frames (№ 20, 34) to take the thrust from the eight rocket engines in the second stage and four rocket engines in the third stage respectively. Fairings (№ 23, 37) cover the rocket engines until staging when they are jettisoned.

The bottoms of all three stages are covered by thermal shields (№ 3, 21, 35) to protect the structure and oxidizer tanks from the heat of the rocket engines. Above the engine frames there are conical transition sections (№ 14; 28, 41) that carry the thrust load between the tanks. Above these are the conical top frames (№ 16, 30, 43) that support the kerosene tanks. The kerosene tanks of the first and second stages have hemispherical shields (№ 17, 31) to protect them when the stage above ignites.

Attached to the top of the third stage by a Payload Adapter Ring(№ 44) is the L3 lunar spacecraft. At the top of the L3 is the Launch Escape System (LES, № 46).

The separation of the stages in flight was accomplished by having the engines of the next stage ignite when engines of the previous stage were still burning (hence the need for the shields). Between Block-A and Block-B is an open interstage truss made of round tubing formed into twenty-four rhombuses (diamond shapes). A ring of the same tubing joins the mid-point of these rhombi forming 96 triangles, which results in a very strong but light structure. A similar truss made of smaller tubing connects Block-B and Block-V. The trusses are open to the outside so that the start-up exhaust from the engines from the stage above, can escape and not push the stages apart prematurely.

One of the new aspects in the design was to control the pitch and yaw of the rocket by changing the thrust of rocket engines rather than steering all of the engine assemblies. This is possible because the engines are located in a circle, placed as far as possible from the central axis of the rocket. This has the advantage of great weight savings as there are no gimbal assemblies or steering mechanisms

Block-V

33. Rocket engines NK-19 (4 units)
34. Engine load-bearing frame
35. Bottom thermal shield
36. Roll control nozzles (4 units)
37. Block-V Fairing (3 sections)
38. Main load-bearing frame
39. Oxidizer tank (liquid oxygen)
40. Fuel line cover (4 units)
41. Upper transition section
42. Fuel tank (kerosene)
43. Upper load-bearing frame
44. Payload Adapter Ring

Spacecraft section

45. Block-G & -D Fairing
46. Launch Escape System (LES)
47. Block-G load-bearing frame
48. Block-G
49. Block-D
50. Lunar Lander (LK)
51. Lunar Orbiter Vehicle (LOK)

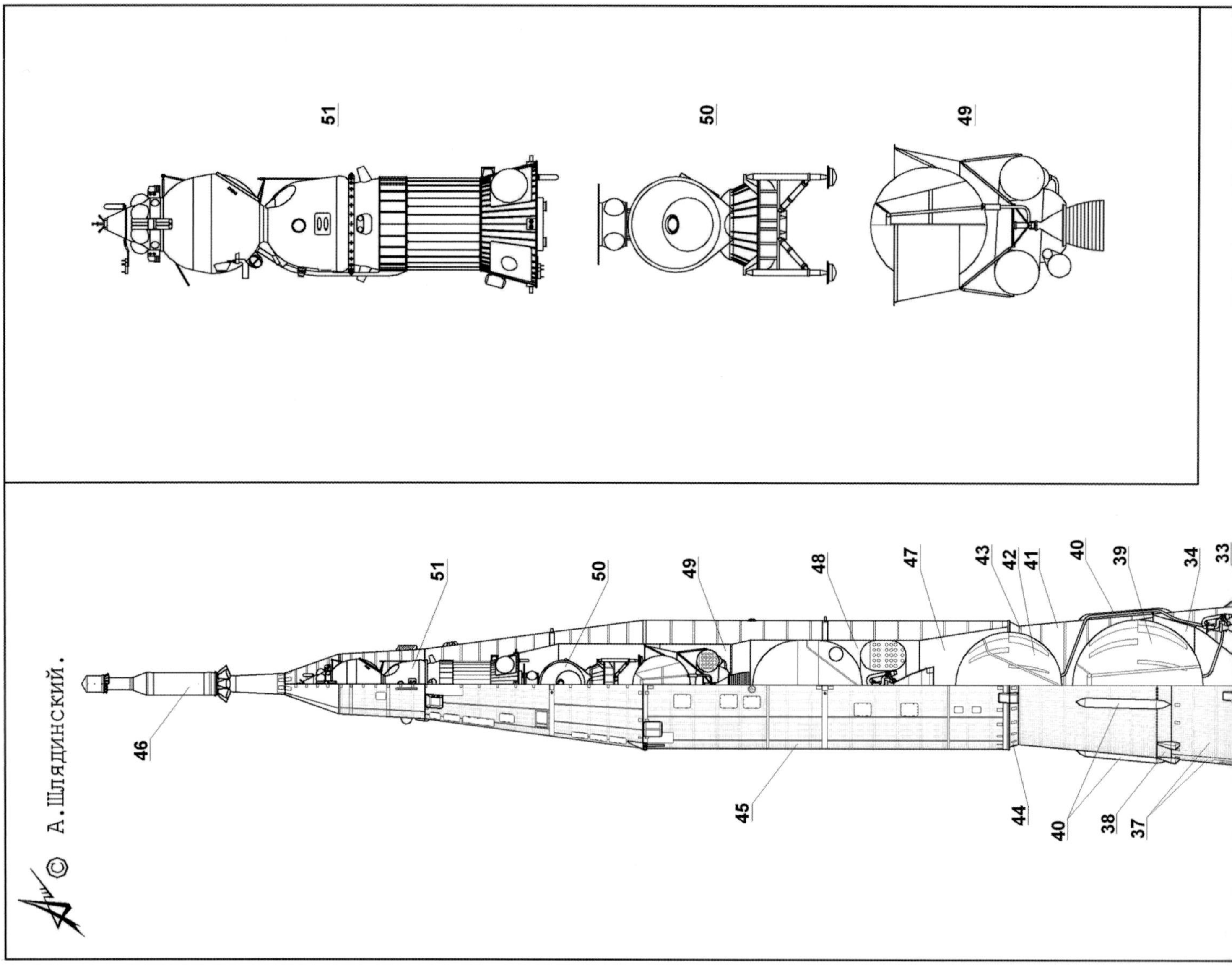

THE N-1 DIAGRAM

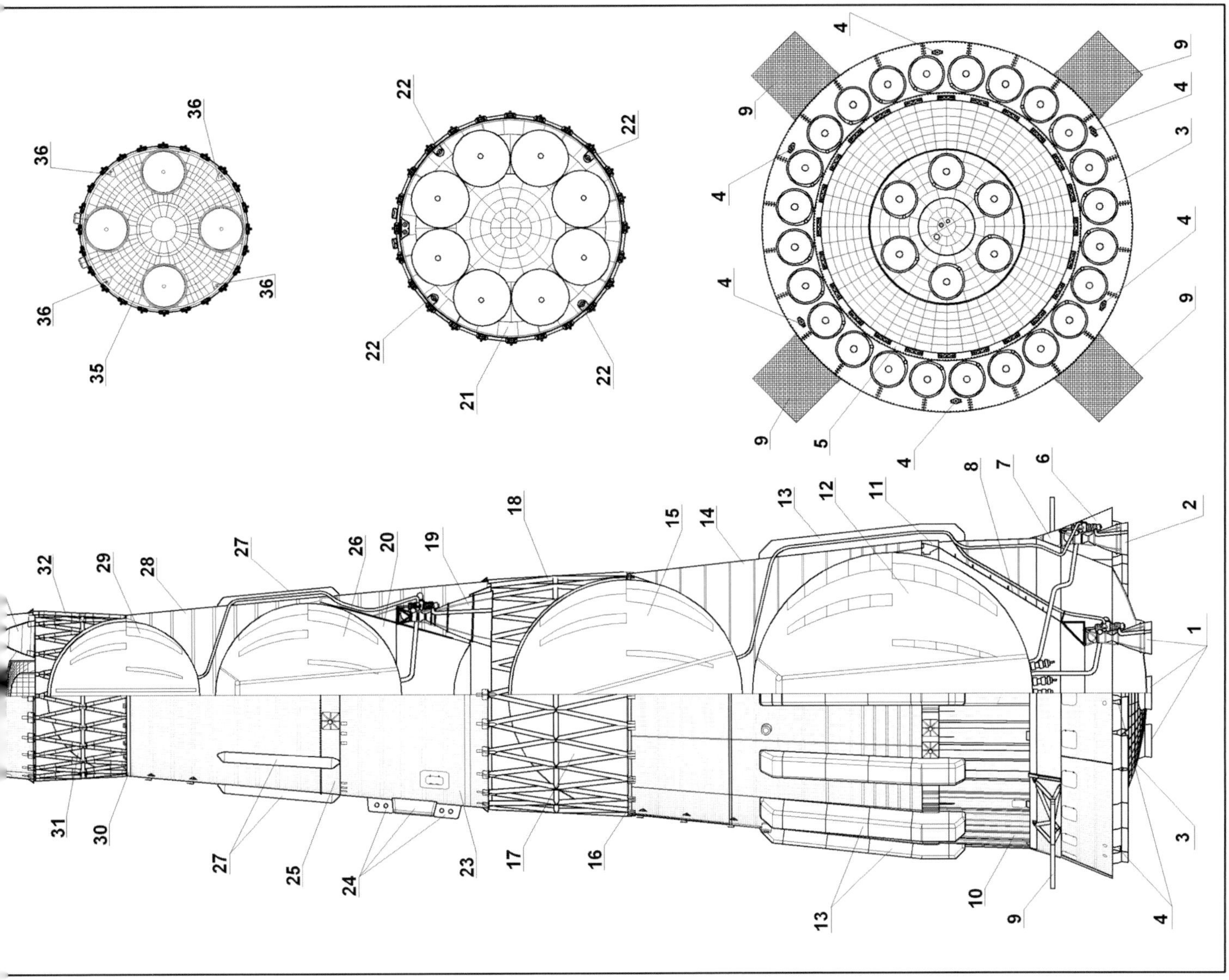

(2010 version)

THE N-1 DIAGRAM

Block-A

CRG - Central Ring Group

ERG - External Ring Group

1. CRG Rocket engines (6 units NK-15)
2. ERG Rocket engines (24 nits NK-15)
3. Bottom thermal shield
4. Roll control nozzles (6 units)
5. Load-bearing ring with thermal bridge
6. ERG Rear skirt
7. ERG engine load-bearing frame
8. CRG engine load-bearing frame
9. Lattice fin stabilizer (4 units)
10. Transition section
11. Main load-bearing frame
12. Oxidizer tank (liquid oxygen)
13. Fuel line cover (12 units)
14. Upper transition section
15. Fuel tank (kerosene)
16. Upper load-bearing frame
17. Hemispherical fuel tank shield
18. Interstage truss

Block-B

19. Rocket engines NK-15V (8 units)
20. Engine load-bearing frame
21. Bottom thermal shield
22. Roll control nozzles (4 units)
23. Block-B Fairing (3 sections)
24. Fairing jettison motors
25. Main load-bearing frame
26. Oxidizer tank (liquid oxygen)
27. Fuel line cover (8 units)
28. Upper transition section
29. Fuel tank (kerosene)
30. Upper load-bearing frame
31. Hemispherical fuel tank shield
32. Interstage truss

Block-B Roll Control Nozzle

that would have to be duplicated on each engine. The difficulty is that it requires very fine control of each rocket engine's thrust level. Roll control was by small steerable nozzles that vented gases from turbine pump units of the rocket engines. These are much smaller than the main engines and only have to steer around a single axis.

The first stage has thirty NK-15 rocket engines: 24 in the External Ring Group (ERG) and the other six in the Central Ring Group (CRG). The original design of the N-1 first stage contained only the 24 ERG engines. The central group was added when the payload for the N-1 was increased to 95 tons. This required many changes in the first stage design and the other two stages as well. Since the liquid oxygen tanks are at the bottom of each stage, the oxidizer pipelines can pass directly to the rocket engines inside the rocket body. But since this tank takes up the entire diameter of the body, the fuel pipelines must bypass the oxidizer tanks on the external surface of the stage and are enclosed by covers. For the twenty-four rocket engines of the ERG in Block-A, there are twenty-four outside pipelines which are combined in twelve pairs. Each pair of lines is has a single cover for a total of twelve covers around the body.

When the decision was made to increase the payload to 95 tons, components of the rocket were already in production. So the six pipelines for the CRG rocket engines were added to six of the existing pairs of pipelines resulting in every other cover being wider since it now enclosed three pipes. This is plainly visible in photos of the rocket. The second and third stages have the covers placed in regular intervals around the body. Covering only one fuel line each, their number corresponds to the number of rocket engines: eight for the second stage and four for the third stage.

Before making a flight version of the N-1 rocket, a full size mockup was created for testing launch pad contraction and the ground support systems. The main visible difference between the flight rockets and the ground test version was that on the top of the nose there was no Launch Escape System (LES). During the test of the Grasshopper raising the rocket to the vertical position, cracking was detected on the transition section (№ 10) and the main load-bearing frame (№ 11). The decision was made to make those elements stronger by installing stringers on the transition section. This change consisted of metal straps with thickness of 10 mm and a width of 50 mm. Some of them passed along all of the transition section and some of them stretched from the main load-bearing frame down to about 720 mm. On the main load-bearing frame (inside the skin) there were installed metal L-bars 70x70x750 mm. The presence of stringers is easily seen in photographs to help distinguish the mockup from the first three launched rockets (see pages 19 and 20).

I want to note one more peculiarity of the first N-1 flight. The short schedule for developing and construction of the rocket was driving development of the NK-15V rocket engine. The NK-15V engine is a version of the NK-15 engine with an extended nozzle for high-altitude flight used in

NK-15 (above) and NK-15V (right) with extended nozzle

Block-B. But the first flight was planned to be a non-landing round trip to the moon by the 7K-L1A *Zond* spaceship that had no lander. This spaceship was lighter than the combined LOK and LK and without a lander there was no need for Block-D either. So the decision was made to use the non-modified NK-15 engine on the first N-1 flight. All existing photos of the N1-3L show Block-B with the NK-15. I have never seen photos of the second stage with the NK-15V high-altitude nozzles.

The test vehicle, 11A52 №3L, was launched on 12 February 1969. The first two stages were grey in color. The third stage was split equally between white and grey (the part facing the launch tower was white and opposite side was grey). The entire L3 spacecraft section was white. The rocket successfully ignited but large vibrations caused the failure of one of the rocket engines. At 55 seconds, a fire started due to a short circuit in one of the electric cables in the KORD engine control system. The shorted cable caused the KORD system to issue the wrong command and it turned off all of the rocket engines. The LES was activated and saved the mockup of space ship 7K-L1A and the rocket came crashing down on the local steppe while on fire.

After a failure analysis of the first flight, there were modifications in the design. The main focus was reducing the heating of the aft section of the first stage by the rocket exhaust. A shield was installed, beginning with rocket № 5L, to prevent this (see page 20). During flight, the plastic shield would burn and carry the heat away, thus not allowing the transfer of heat to the inside of the aft section. The plastic shield was fixed by bolts, the heads of which were protected from high temperature by a special plastic cover. In the photos these bolts can be seen as a lot of dark points spaced uniformly around the bottom shield.

It was also believed that the volume of aft section was big enough that the air would heat slowly and prevent damage; but it was found that there were some areas with very high temperatures. To eliminate the heat buildup, forced ventilation was provided by opening the bottom end of the six 2-pipeline fuel line covers (№ 13). These openings were covered by a grid for protection from kick-back debris. During the flight, the openings provided an extraction effect drawing cold outside air in to mix with the inside air. Similar openings were added to the covers on the second and third stages. Also, the KORD system, which had provided the wrong command to turn off all rocket engines on the first flight, was moved from the aft section to the upper transition section (№ 14).

Next, during the firing of the second stage rocket engines during the 3L flight, the engine covers shot from the nozzles at high speed (the covers allow the rocket motors to start in a vacuum). Analysis showed that the high speed of the covers was the reason for the damage to the top side of the first stage fuel tank (№ 15). This situation was dangerous and could cause an in-flight accident. The decision made to install extra shielding on top of the fuel tank opposite the rocket nozzles.

The engineers also worried about the strength of the aft section of the second stage. Rocket № 3L had L-shaped frames on the bottom side of the jettisoned rear skirts (№ 23) and the interstage truss (№ 18) was fixed to that L-shaped frame. The frame changed to a triangular cross section. These triangular frames look like small conical projections on the base of second stage.

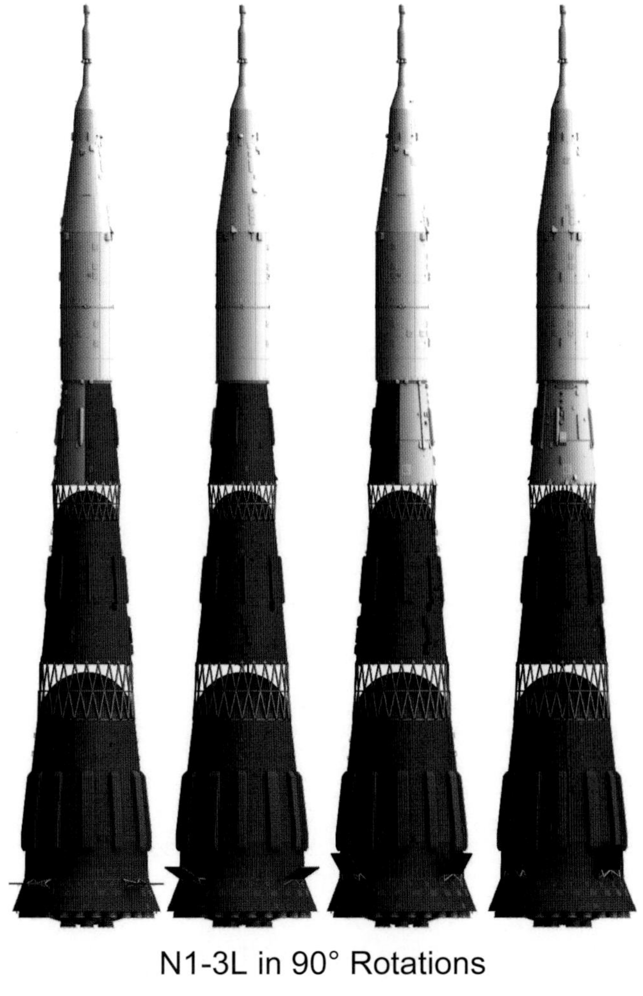

N1-3L in 90° Rotations

patterns. During tests of the full size mockup, the staff complained about the high temperature inside rocket because of the sunlight shining on it. Sometimes the temperature of the interior spaces was as high as 70°C (160°F)! Rocket № 3L was launched in winter so the rocket's color did not affect preparation. But in the summer it was a problem. The decision was made to change the rocket color to nearly all white. Only the side opposite the launch tower remained a grey color for all three stages. These grey parts take up less than half of the rocket circumference. Rocket № 5L had ¼ of the circumference of Block-A grey while Rocket № 6L had ⅓ of the circumference of Block-A grey.

The launch of a second test vehicle, 11A52 № 5L, was on 3 July 1969. It repeated the same flight program of rocket № 3L. Only 0.4 seconds after ignition, rocket motor № 8 exploded and a fire began in the aft section. The KORD-system turned off all motors and after 18 seconds the rocket fell back to Earth, destroying the launch pad.

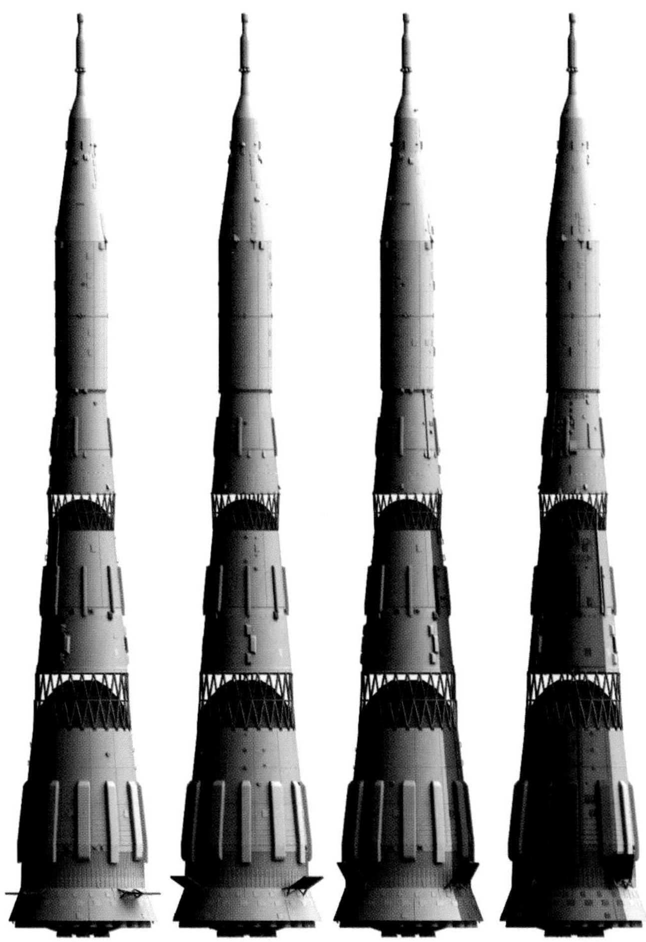

N1-5L in 90° Rotations

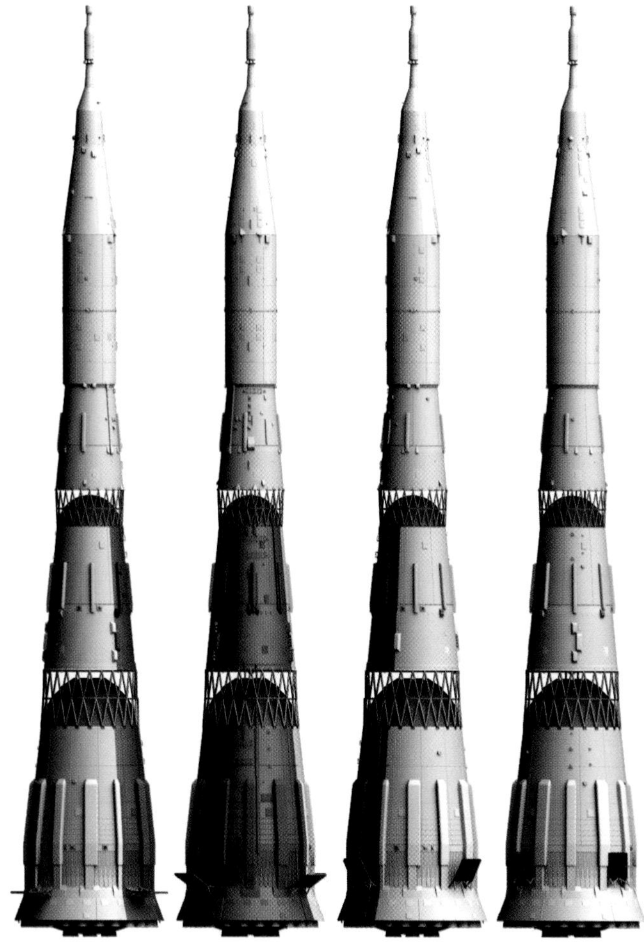

N1-6L in 90° Rotations

The next launch of an N-1 wasn't until 1971 and engineers made fire-prevention the focus of the improvements. A fire extinguishing system was added to the first stage of the № 6L rocket. The ends of the fuel line covers (№ 13) were opened to admit the pipelines of the fire extinguishing system as can be seen in the photo of N1-6L on page19. The pipes passed along under the cover (two pipelines through each cover), but still on the outside of the transition section (№ 10), until they reached the rocket engine section. Here the pipelines go inside and are distributed to their respective engines. If there was a fire, foam would be sprayed into the rocket motor's section through these pipelines.

The № 6L carried a mockup of the LOK with a mockup of the LK lander meaning that for the first time, Block-D was used.

N1-6L was launched on 27 June 1971. To avoid the previous launch pad destruction just after ignition, the rocket was placed into a special maneuver. All motors worked correctly, but the rocket began to slowly rotate around its axis. The rotation of the rocket made it impossible to control. The angle of attack became overcritical and the rocket began to tumble down and break into pieces. The first break was in the area near the upper interstage truss (№ 32) so the third stage with the payload still attached broke away and fell to the ground. The first and second stages continued to fly and fell near the *Soyuz* launch complex.

The most probable reason for that failure is the effects of unknown aerodynamic flow associated with rocket the rocket's shape. But for the next test, № 7L, it was not important because the big changes had been made in the external design.

First, the shape of the first stage tail skirt was modified. Initially, rocket engineers thought that the conical shape of the rear engine skirt (№ 6) would work as a stabilizer. Actually, the opposite was true. In addition, the large diameter of the stage trapped air inside the shock wave, and this extra pressure reduced the thrust of the rocket motors. To fix both of these conditions, the bottom part of the rear skirt was changed from a conical to a cylindrical form. This reduced the diameter of the rocket from 16.9 to 15.7 meters. Also, the upper transition section (№ 14) was changed to give it better aerodynamics and strength. The fuel line covers (№'s 13, 27 and 40) were modified to give better aerodynamics form by tapering the top to form a point, and tapering the rear edge inward towards the body.

A second major modification was to the roll control system. Instead of having six rotating nozzles on the base between the main engines, four much larger gimbaled rocket motors were installed in special housings on the cylindrical part of the rear skirt (№ 6).

There was also a big change in appearance. There were no more grey sections, rocket № 7L was completely matte white, including the interstage trusses and tank dome covers.

Test vehicle № 7L had real LOK and LK modules which were planned to execute the lunar mission by automatic mode. The rocket was launched 22 November 1972 and ignition was successful. All motors of the first stage worked well until 107 seconds into the mission. Seven seconds before the ignition of the second stage an explosion began in one of the fist stage engines and the rocket suffered the same fate as its predecessors.

Test vehicle № 8L was also planned to complete a lunar mission in automatic mode with standard LOK and LK modules. The chances of this N-1 succeeding were quite good because the new, more reliable engines were installed:

- Block-A replaced the NK-15 with the improved NK-33
- Block-B had the NK-43 instead of the NK-15V
- Block-V replaced the NK-19 with the NK-39

All of these new motors were better that the originals. Also, while the № 7L rocket was matte white, from № 8L onward, the rockets were to be gloss white.

Alas, rocket № 8L was completed, transported to the launch pad and was undergoing fueling tests in May 1974 when the program was cancelled by Valentine Glushko as soon as he had been made head of the OKB-1 Design bureau.

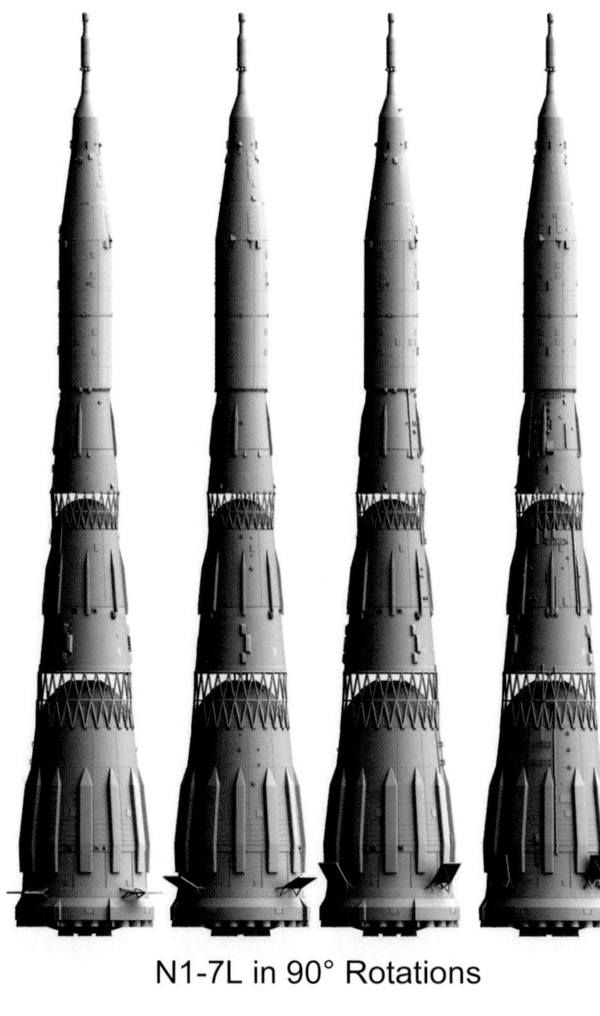

N1-7L in 90° Rotations

7: N-1 Reference Guide - Detailed Documentation

7: Справочный раздел - Подробная документация

CAD Drawings by ALEXANDER SHLIADINSKY
CGI Renderings by NICK STEVENS

Editor's note: This is the section of the book for modelers, aerospace historians and folks who like to dive deep. It is where we reveal the finest level of detail that the researchers have been able to extract from the fragmented remains of the program. In the following chapters we will break down the N-1 one Block at a time. Each section will be a combination of Alex Shliadinsky's drawings (such as on the left below) with dimensions, Nick Stevens' photorealistic CGI renderings (like on the right) interspersed with actual photos of the hardware where available. The drawings make it easier to identify and dimension all of the components, but the fully rendered and shaded CGI gives a powerful intuitive sense of what the N-1 looked like sitting on the launch pad.

The drawings are based on the N1-3L version, the first one flown. The end of each chapter will show the variations in the later three variants, if any. Finally, we have a chapter on the launch complex itself; as much a technological accomplishment at the rocket.

Before getting into the details, though, this chapter gives you an overall comparative look at the four flight rounds. The illustrations below show all four as seen from the tower side (considered Rotation 0°). The next few pages show each round individually at 90° rotations so you can see all around them—JH

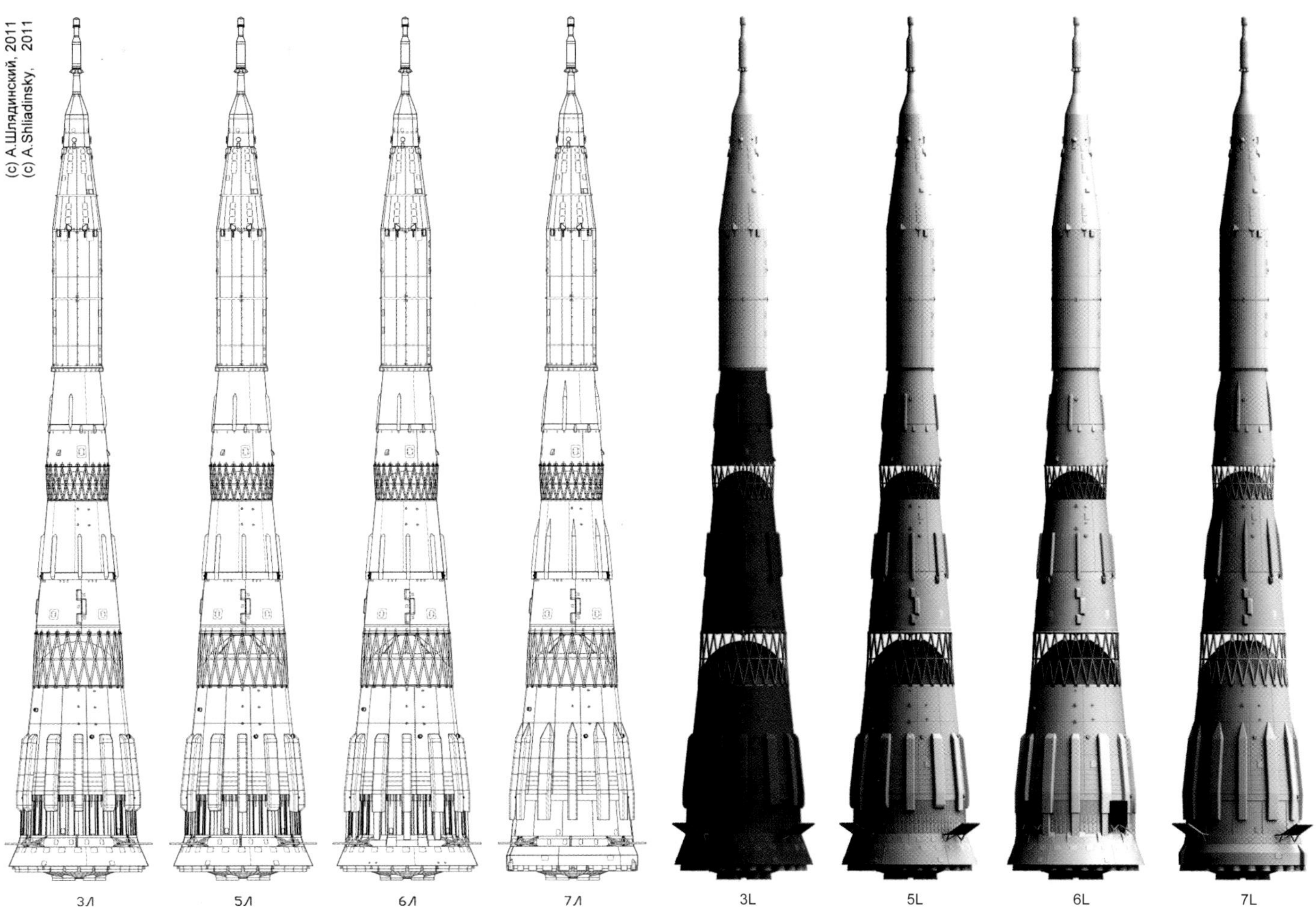

Vehicle N1-3L
Rotational Views
Scale 500:1

Top View

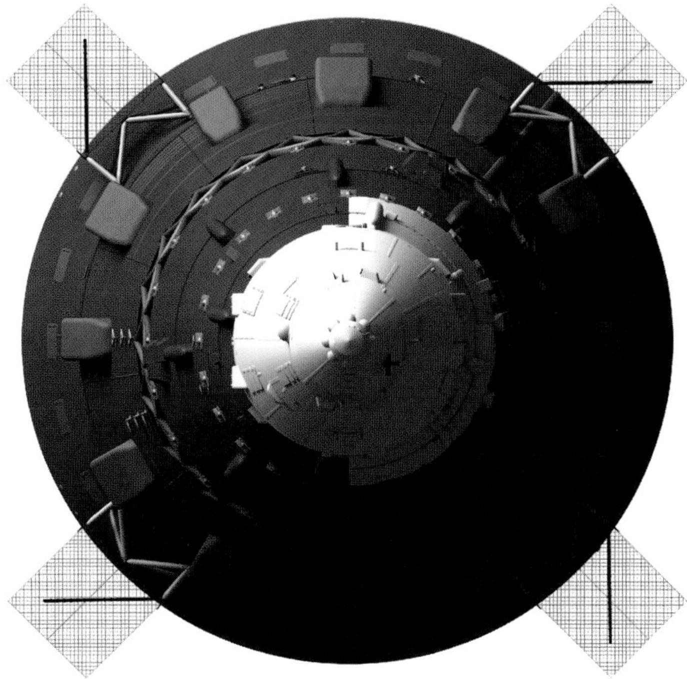

Bottom View

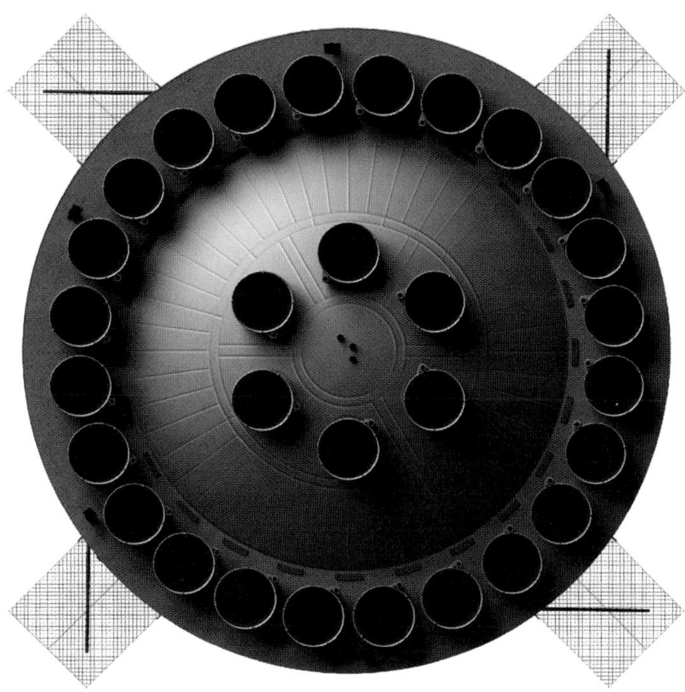

Vehicle N1-3L
End Views
Scale 200:1

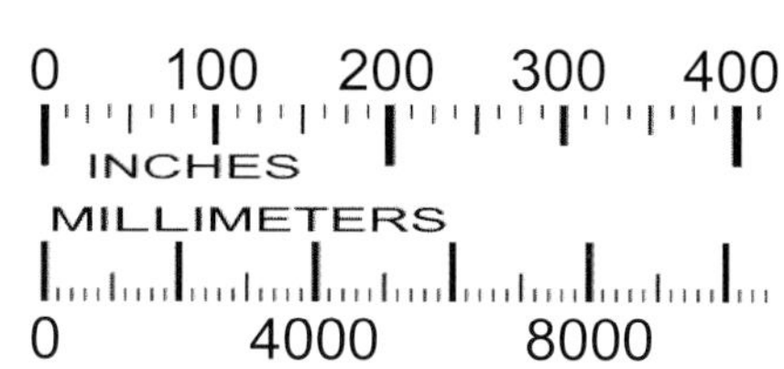

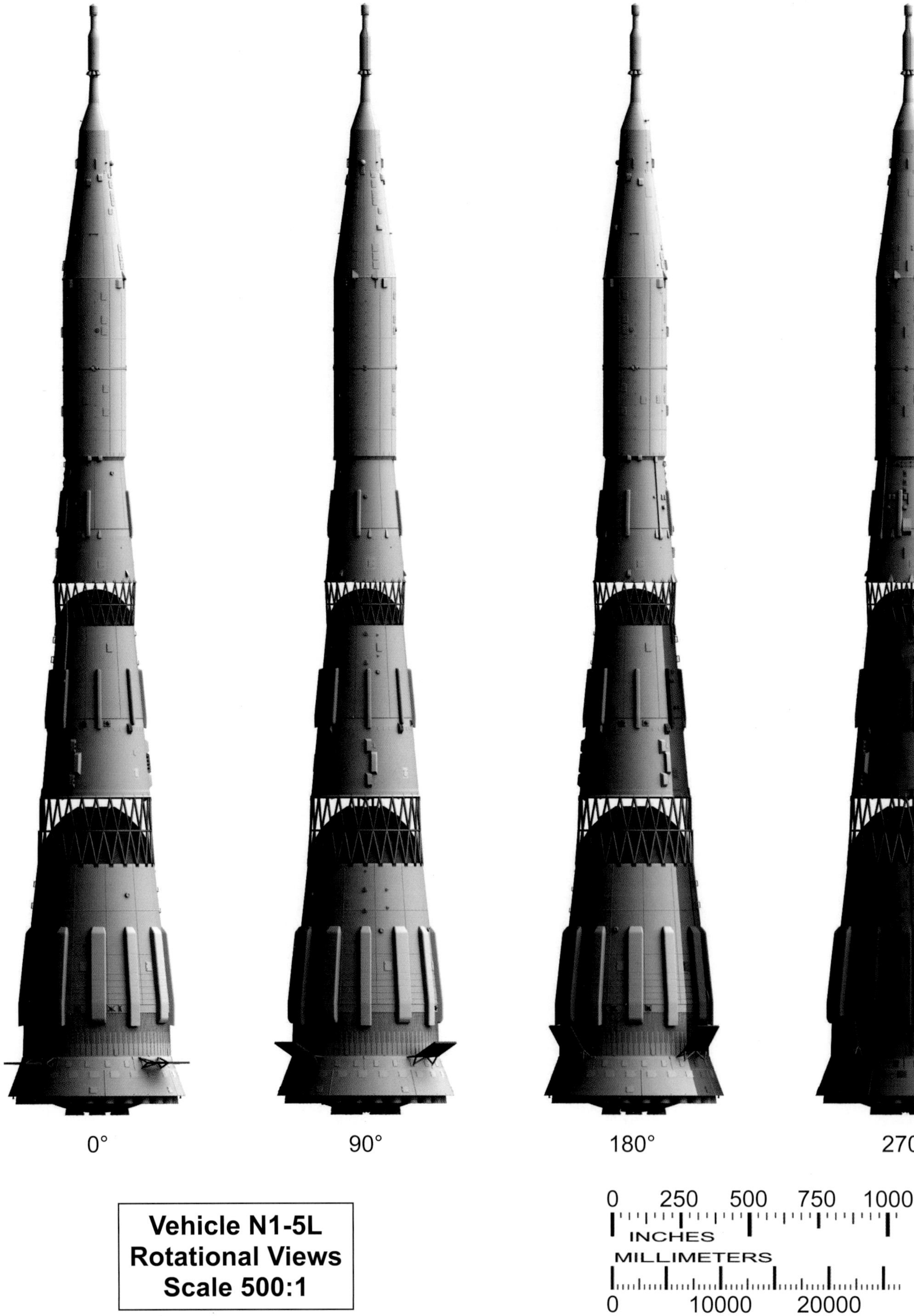

Vehicle N1-5L
Rotational Views
Scale 500:1

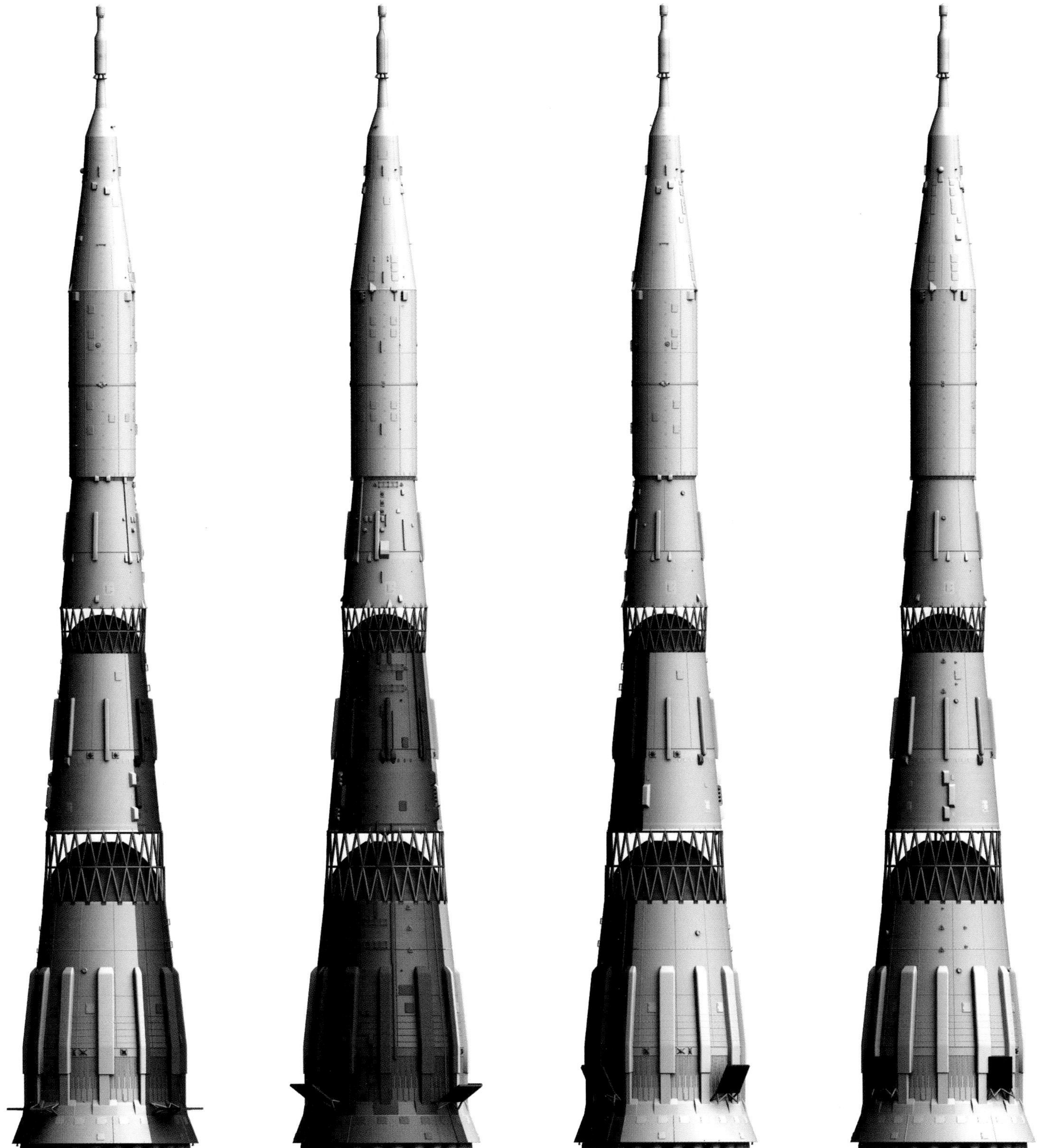

0° 90° 180° 270°

Vehicle N1-6L
Rotational Views
Scale 500:1

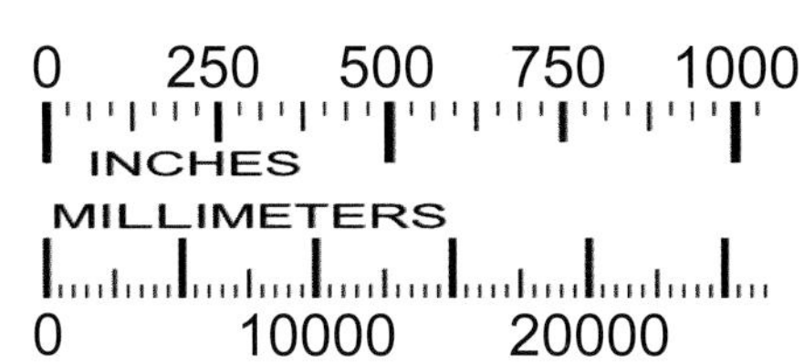

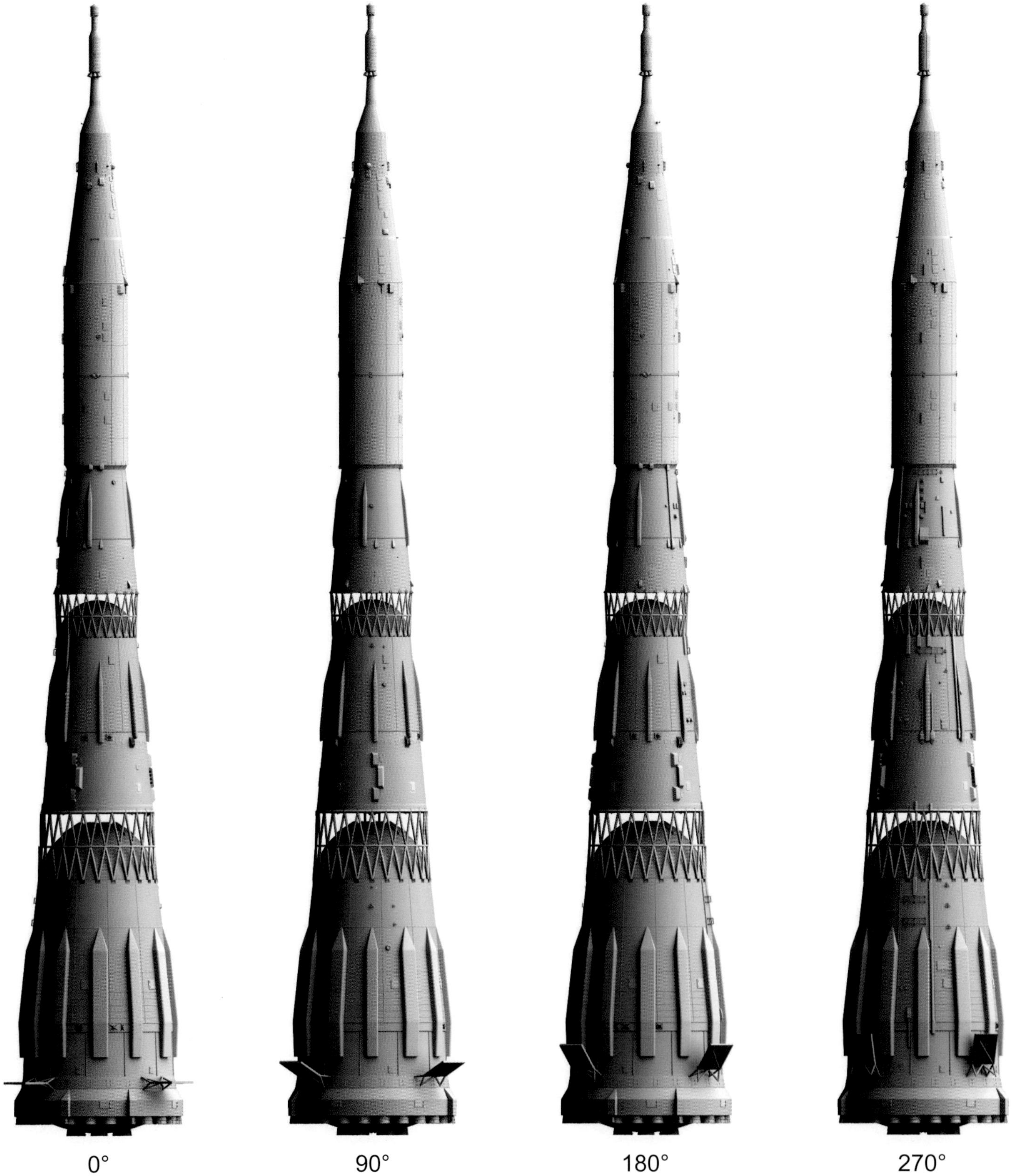

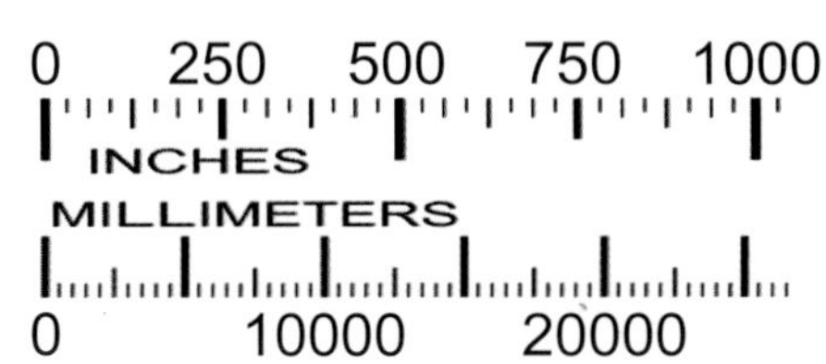

Vehicle N1-7L
Rotational Views
Scale 500:1

8: N-1 Block-A Reference Guide

8: Справочный раздел: блок А

CAD Drawings by ALEXANDER SHLIADINSKY
CGI Renderings by NICK STEVENS
Dimensions Applied by MATT JOHNSON

The Block-A is the most powerful rocket stage ever flown. The 41 million Newtons (9.2 million lbs.) of liftoff thrust from its 30 NK-15 engines is 23% greater than the 33 million Newtons (7.5 million lbs.) of the American Saturn V, still the largest successful booster.

Block-A Engine Configurations

		Thrust[1]					
Flight No.	Engines	Sea Level (kN)	Vacuum (kN)	Propellant (tons)	Burn time[2] (seconds)	Flow rate (tons/sec)	Total Impulse (megaN-s)
N1-3L	NK-15 (30)	41,400	46,310	1,744	25+113+10	14.85	5,437
N1-5L	NK-15 (30)	41,400	46,310	1,744	25+113+10	14.85	5,437
N1-6L	NK-15 (30)	41,400	46,310	1,744	25+113+10	14.85	5,437
N1-7L	NK-15 (30)	41,400	46,310	1,744	25+113+10	14.85	5,437
N1-8L[3]	NK-33 (30)	45,174	50,345	1,759	25+108+10	15.51	5,709

[1] The thrust values, and all the other values in the table, are for the stage as a whole.

[2] First value is the burn time for all 30 engines, the second value is for the 24 engines in the ERG and then for the 6 engines of the CRG. This last burn is to keep the acceleration positive (and the propellants at the bottoms of the tanks) while the ERG engines are shutting down and the Block-B engines are spooling up.

[3] Not flown.

Left: Block-A under construction. As there are no stringers, this is either N1-M1 or N1-L3 before they were added. Below: The unpainted Adapter Ring stands in sharp contrast to the rest of the N-1's base. Also, nothing implies scale like human figures in an image.

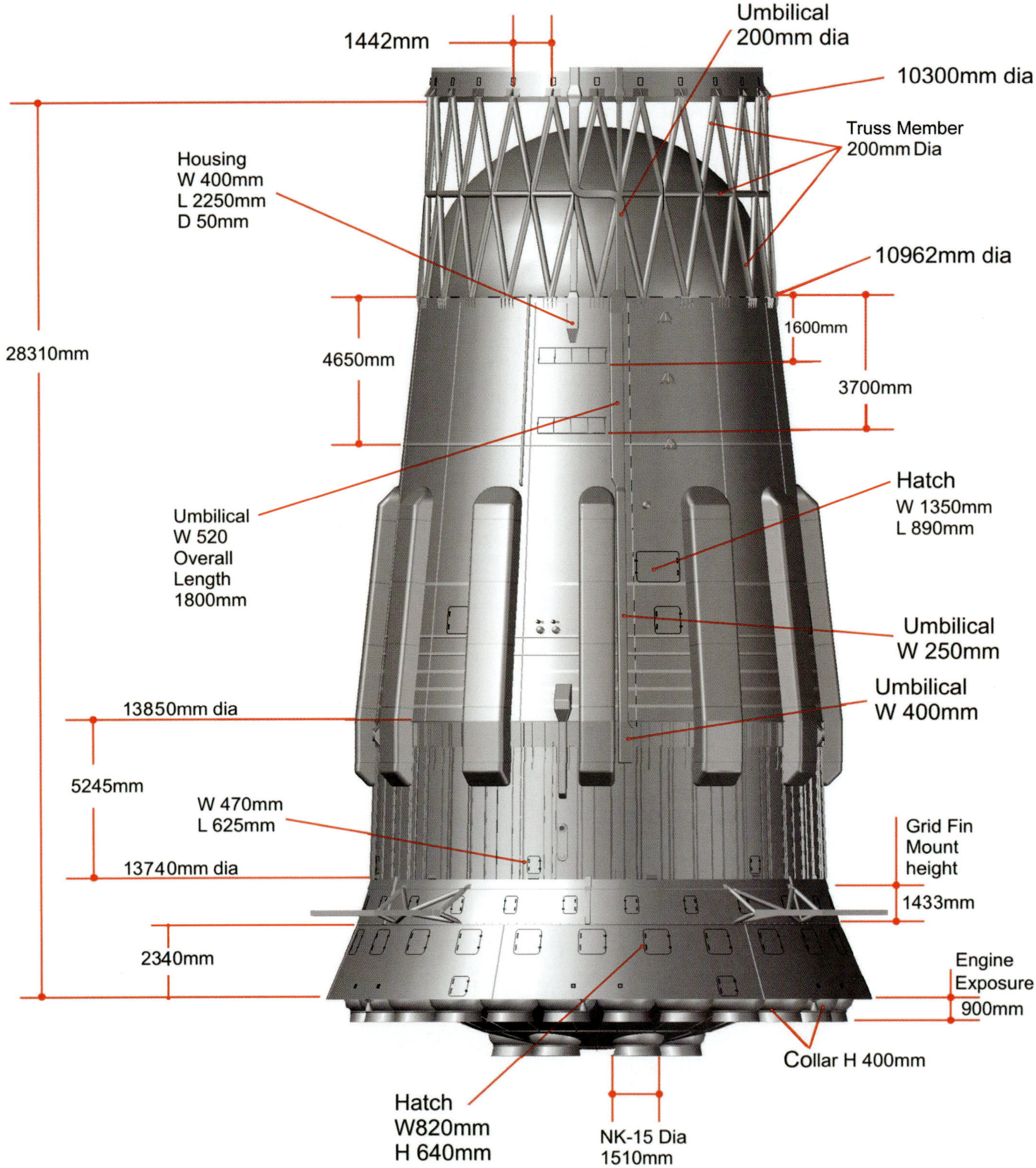

Vehicle N1-3L, Block-A
0° Rotation, Scale 180:1

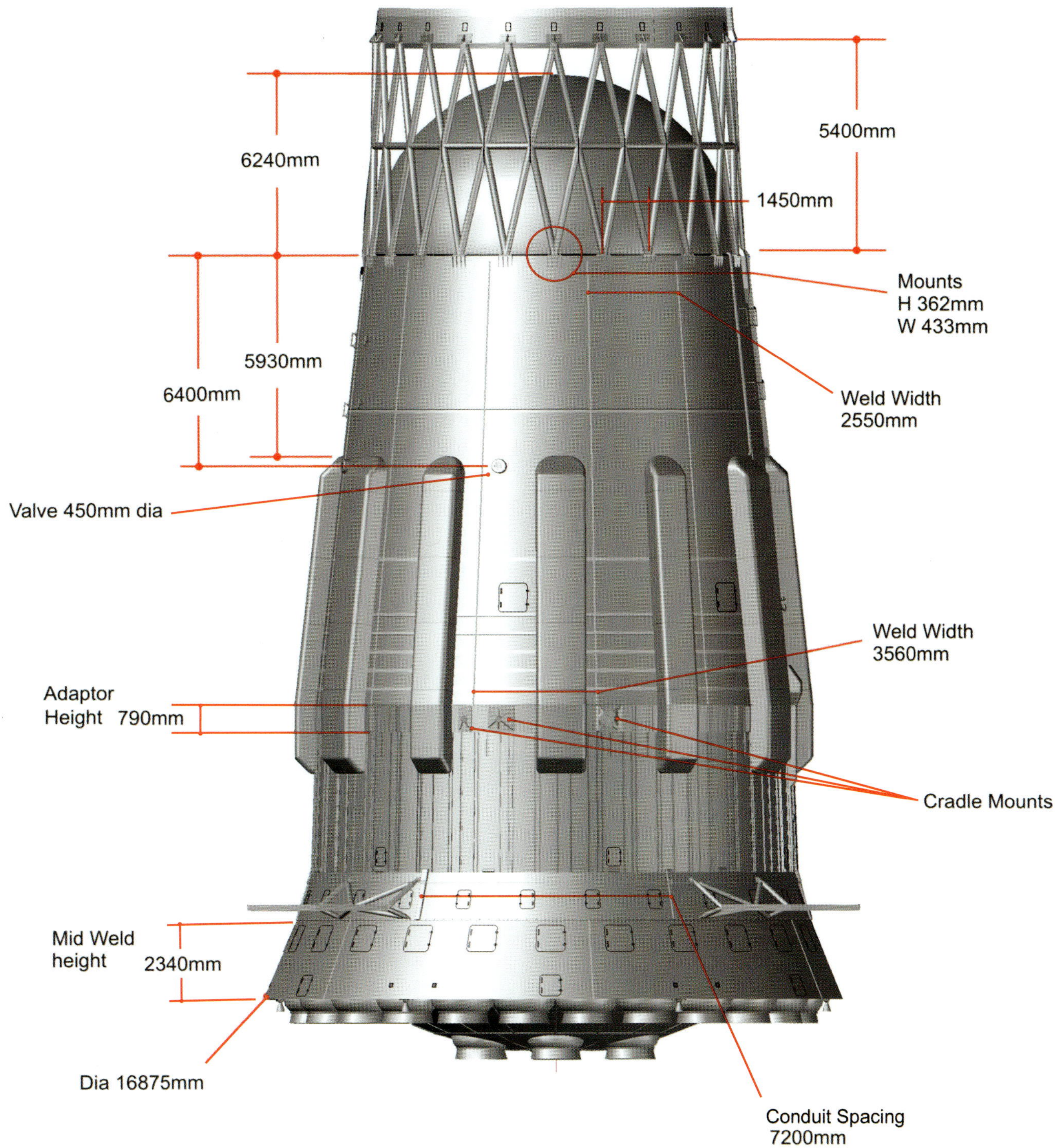

Vehicle N1-3L, Block-A
90° Rotation, Scale 180:1

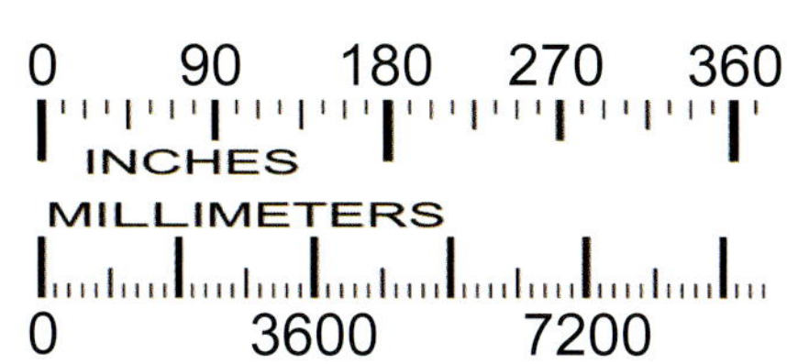

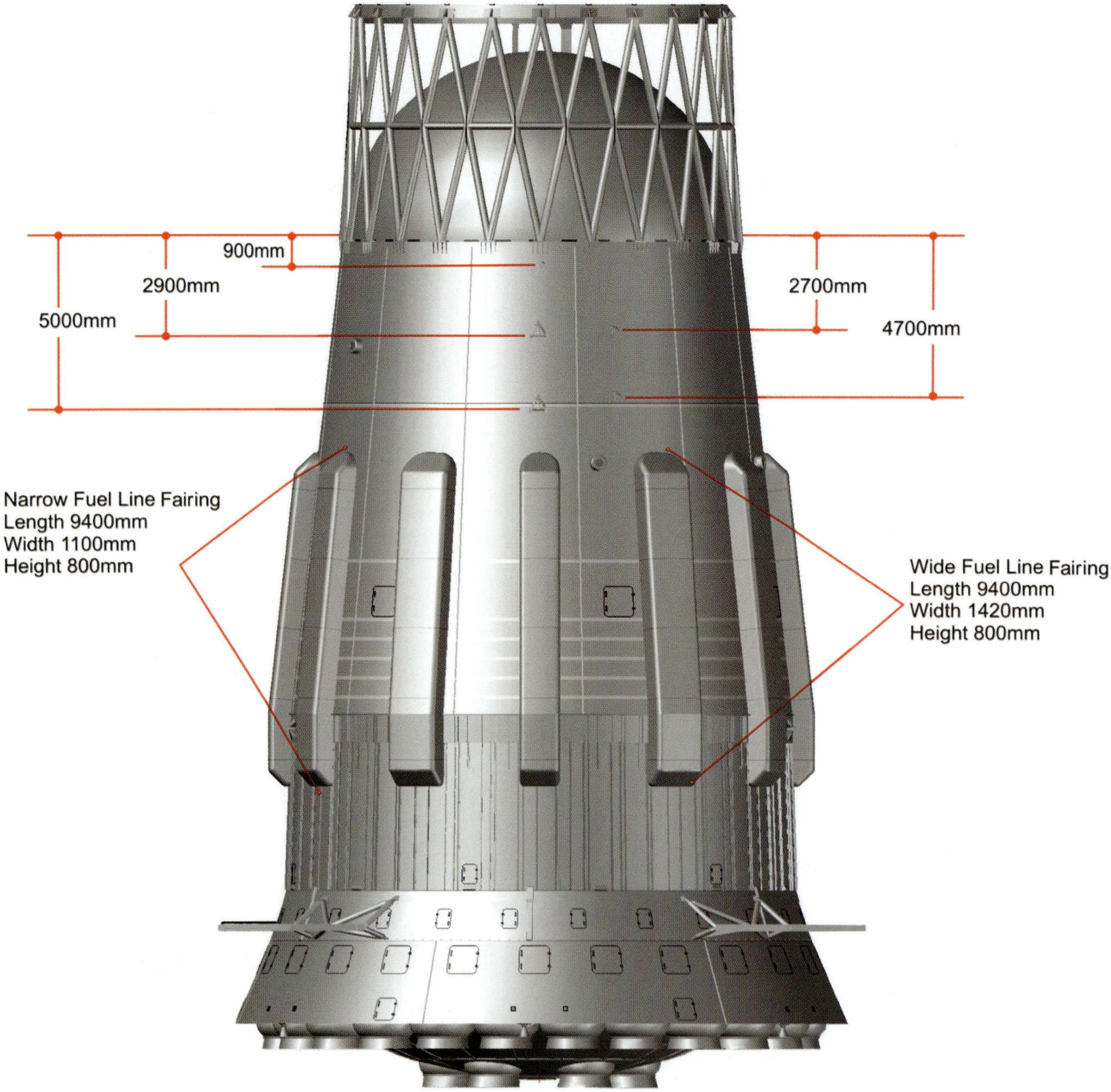

Vehicle N1-3L, Block-A
180° Rotation, Scale 180:1

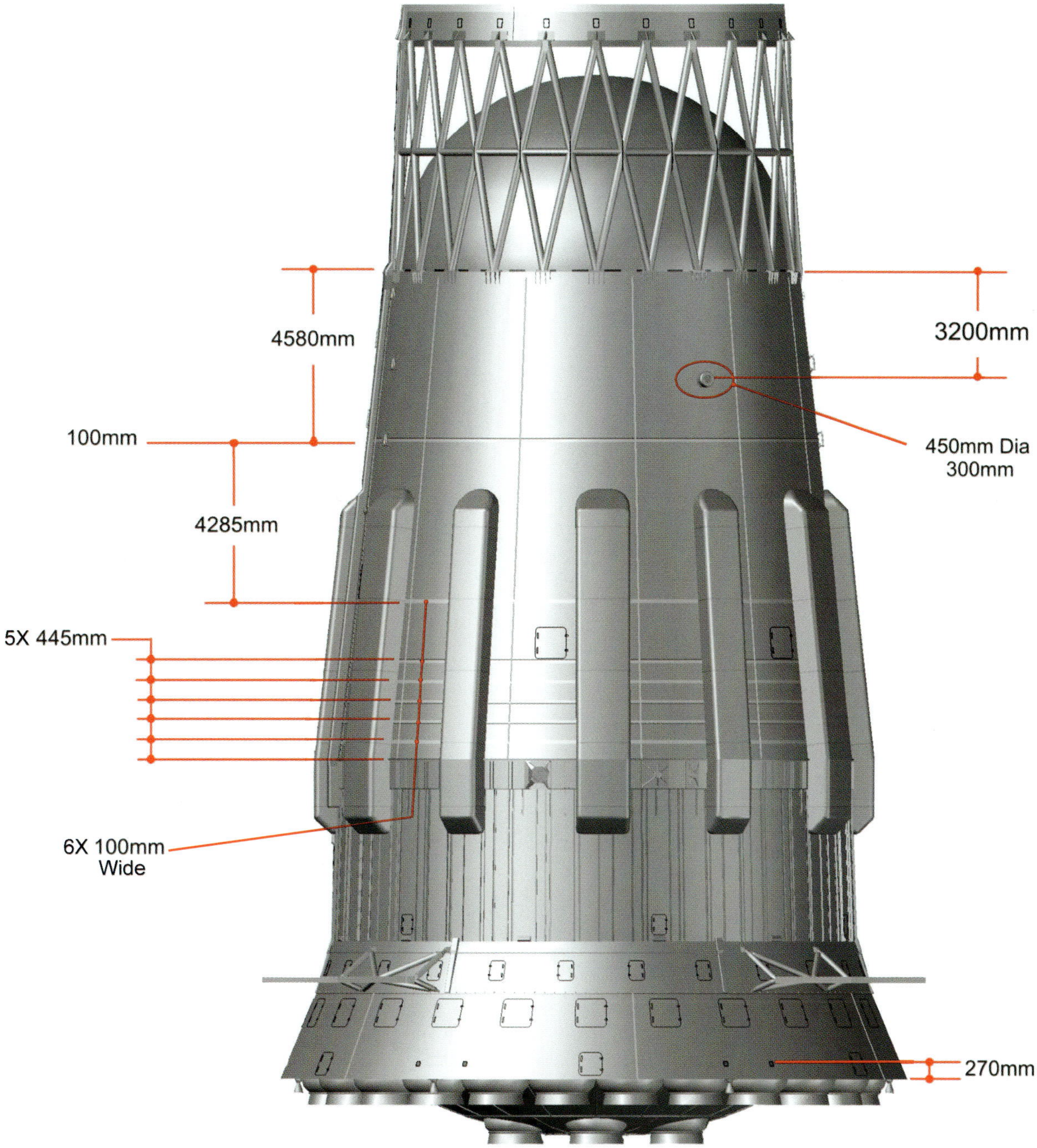

Vehicle N1-3L, Block-A
270° Rotation, Scale 180:1

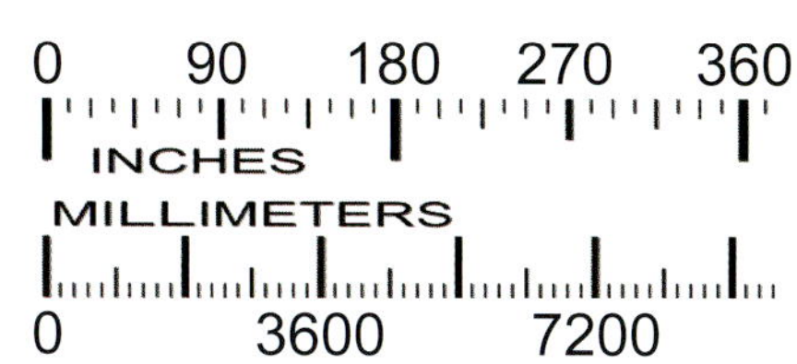

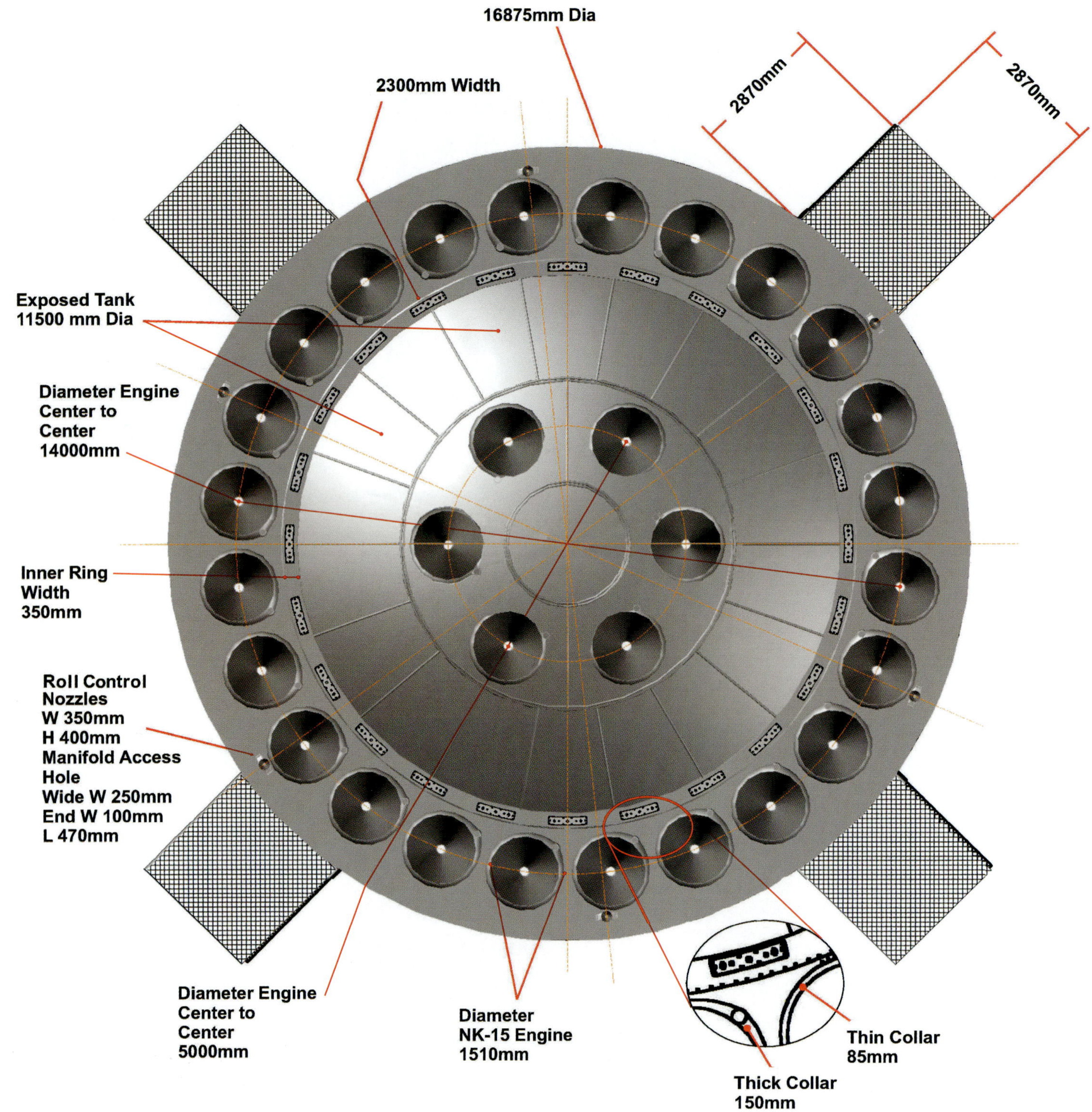

Vehicle N1-3L, Block-A
Bottom View, Scale 120:1

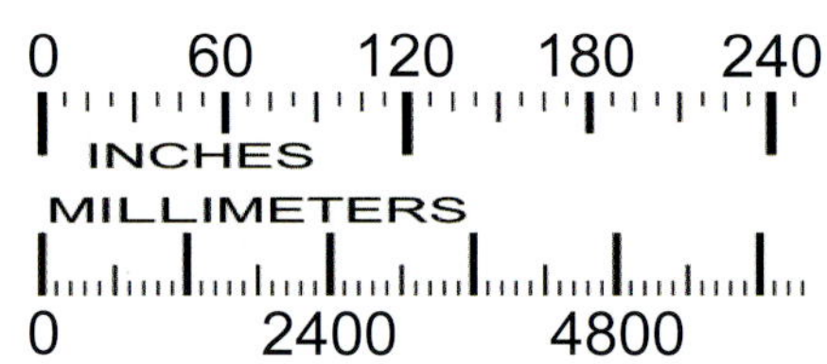

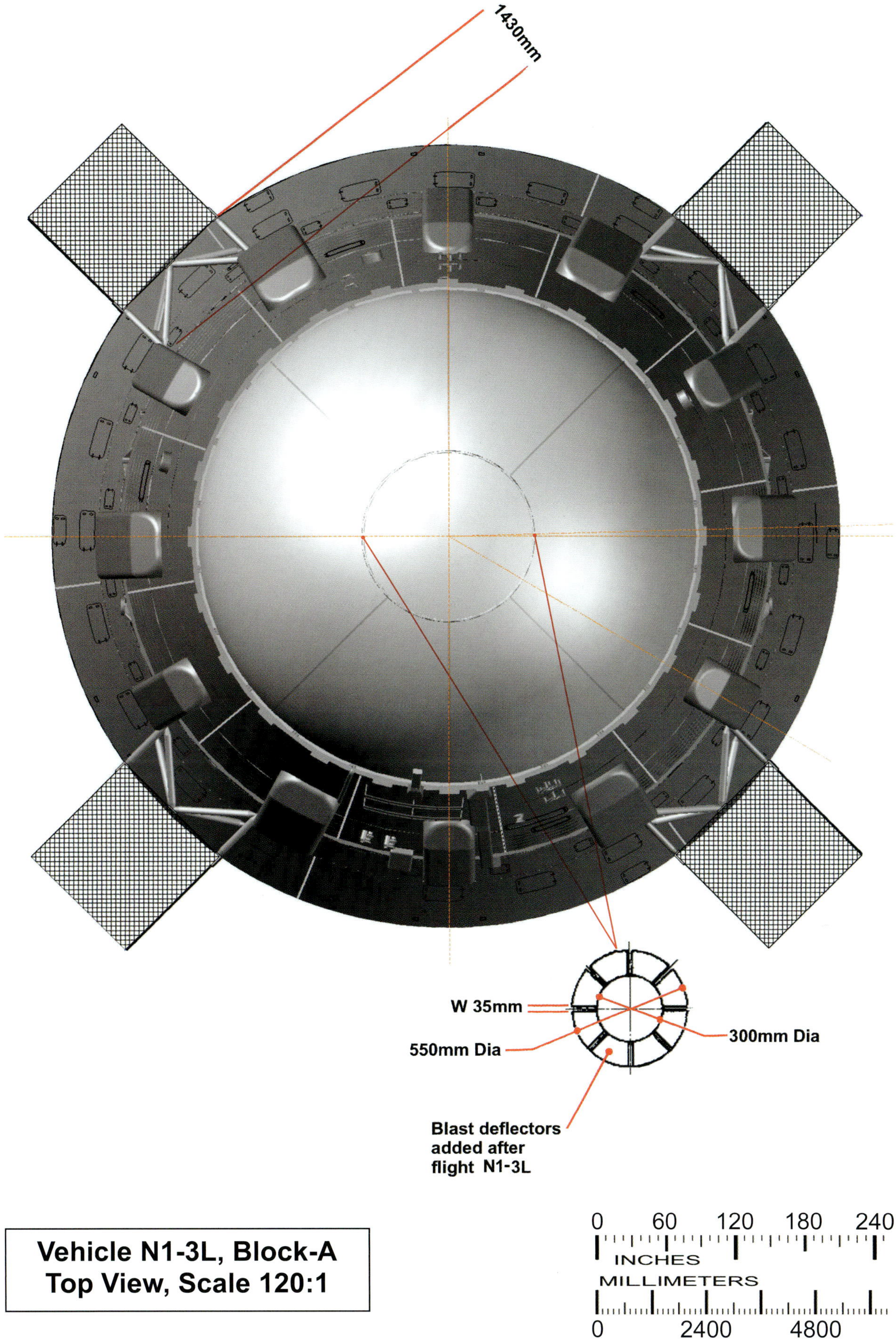

Vehicle N1-3L, Block-A
Top View, Scale 120:1

Launch Pad Supports

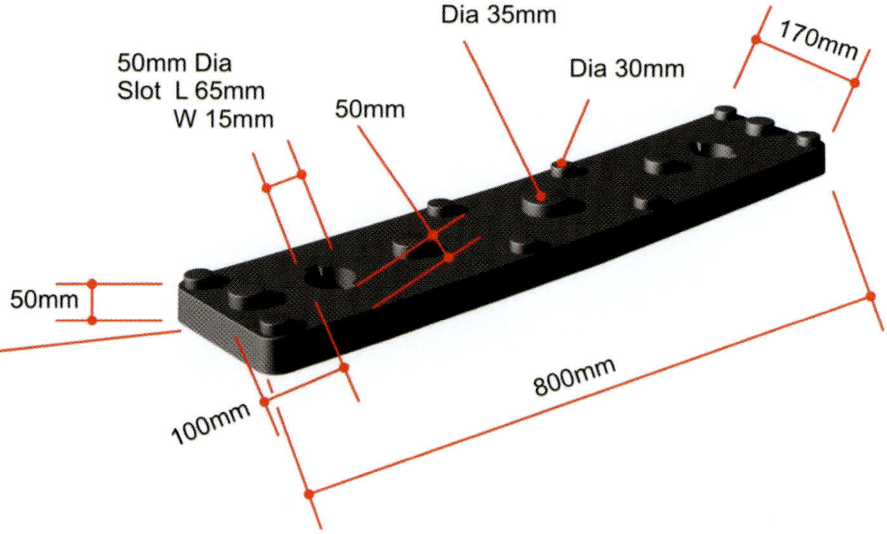

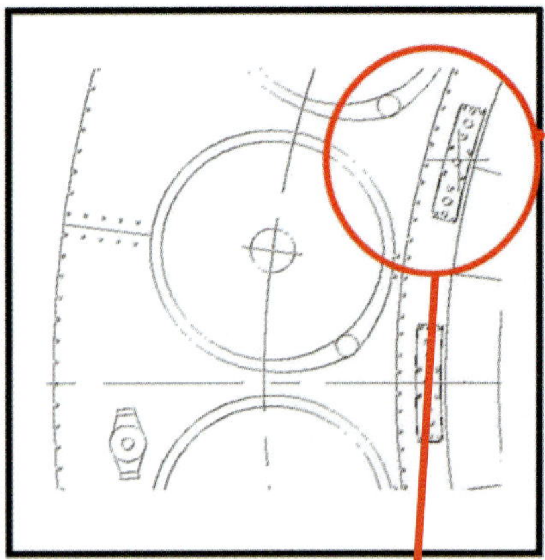

These Pad Supports are the only place where the N-1 contacted the launch pad. They formed the point of attachment to the Adaptor Ring (which was left behind at liftoff) and had to support the enormous weight of the entire booster stack. They were attached to the bottom of the Block-A mainframe between the ERG and CRG engine groups halfway between each of the ERG engines.

Transport Cradle Mount

At the bottom is the Block-A tail skirt being fabricated. The detail below left shows the main erector "hard point" attached to the main frame for transportation and erection onto the pad (see pp 188-192). To the right is the same mount, looking a little worse for wear, in the debris field.

Note the end caps of the fuel line covers already installed at the top of the skirt.

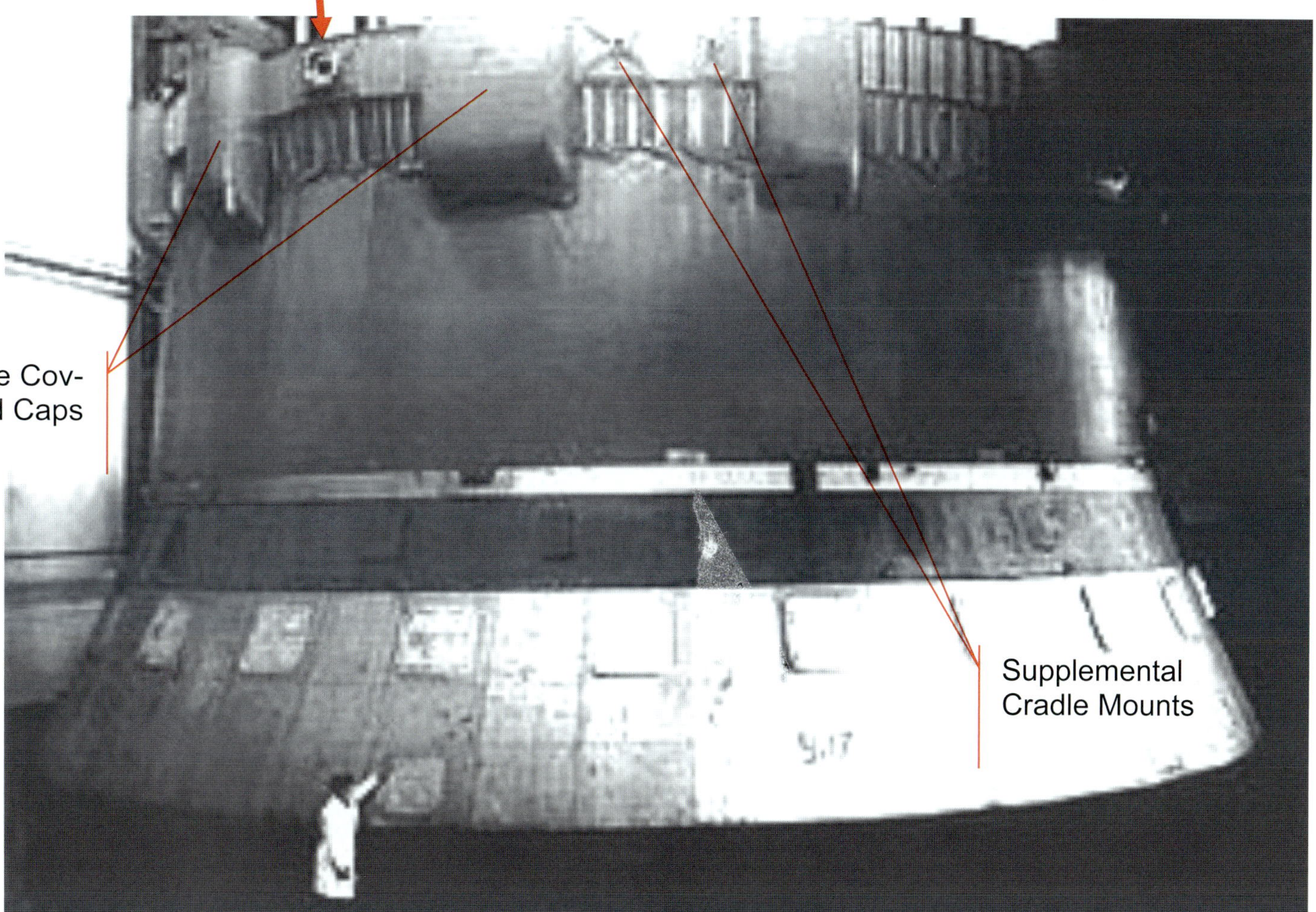

Thermal Bridges

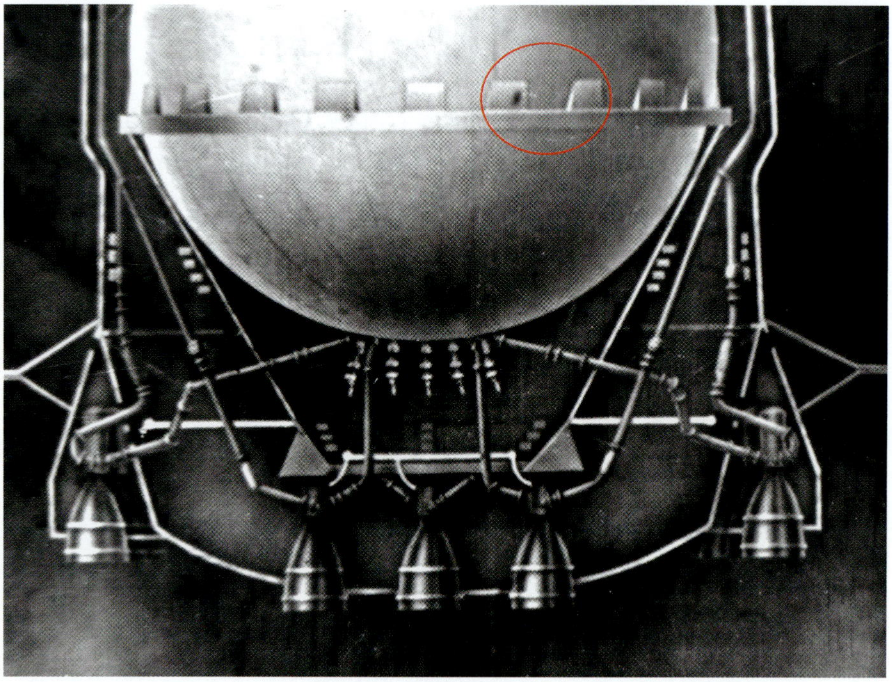

The liquid oxygen tank was attached to the main frame by 20 insulated mounts called "Thermal Bridges." To the right is a drawing showing how the Bridge Assemblies (circled) formed a cradle supporting the tank. These helped reduce the flow of heat from the structure into the tank which caused propellant boil-off before launch However, even with these mounts the rocket lost over a metric tonne of liquid oxygen between the time the LOX filling hose was detached and engine start. Still, this was about $^{1}/_{10}$ as much as would have evaporated otherwise. Each Bridge consisted of three parts: A bracket welded to the tank, the support bracket riveted to the Main Frame and between them the Thermal Bridge element.

Below left is a bridge assembly from Block-V (Block-A was the same, just larger) identifying these pieces. Below right is an actual Thermal Bridge element which consists of a swiveling metal socket to receive the Tank Bracket and an insulating material (possibly Bakelite or an epoxy) that slides into the Frame Mounting Bracket. At the bottom is a section of the Main Frame in the debris field with the mounting brackets an a partial thermal bridge still visible

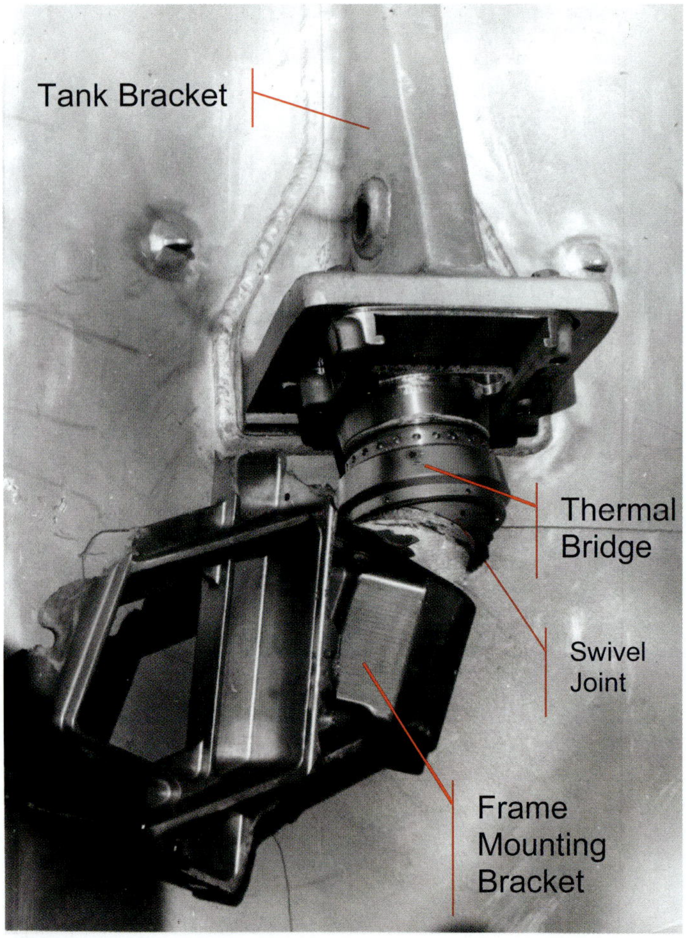

Block-A Lattice Fin Side View Scale 50:1

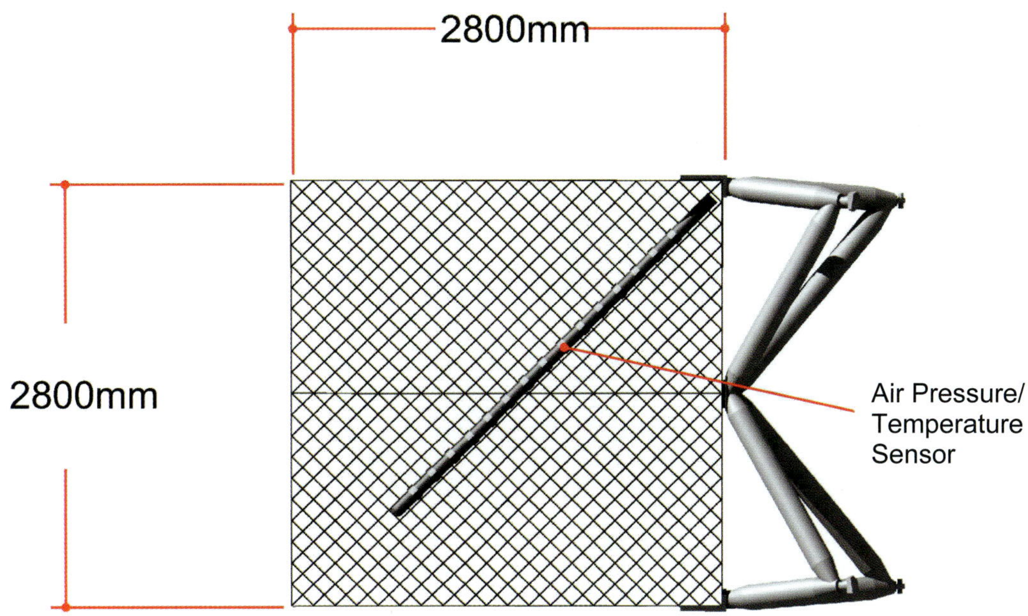

Block-A Lattice Fin Bottom View Scale 50:1

Lattice Fin Mounting Truss
Scale 50:1

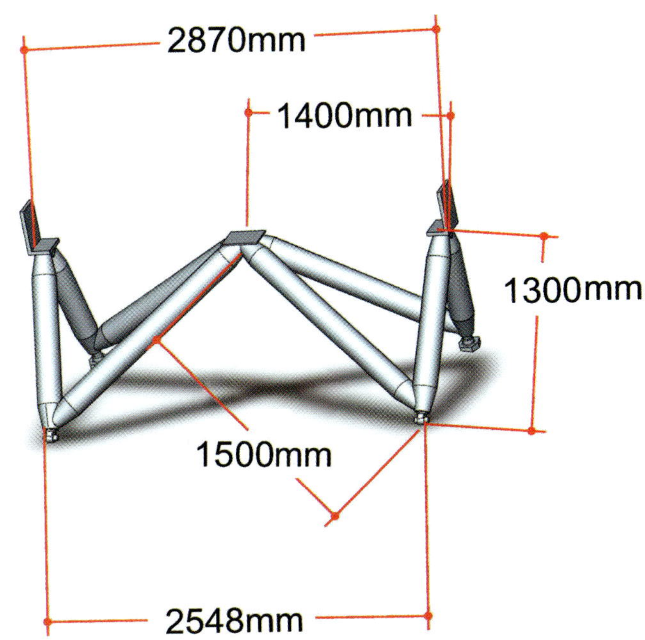

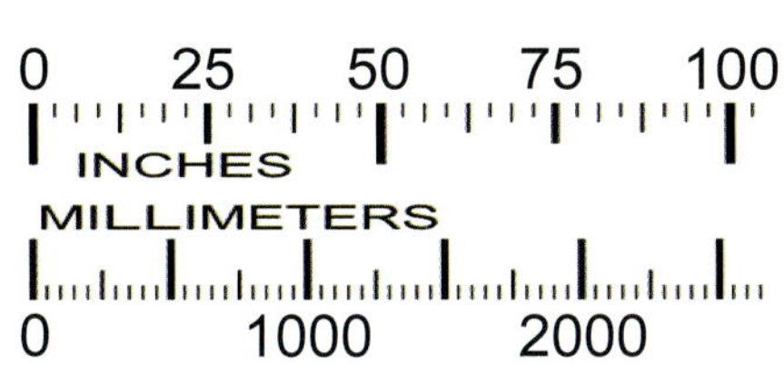

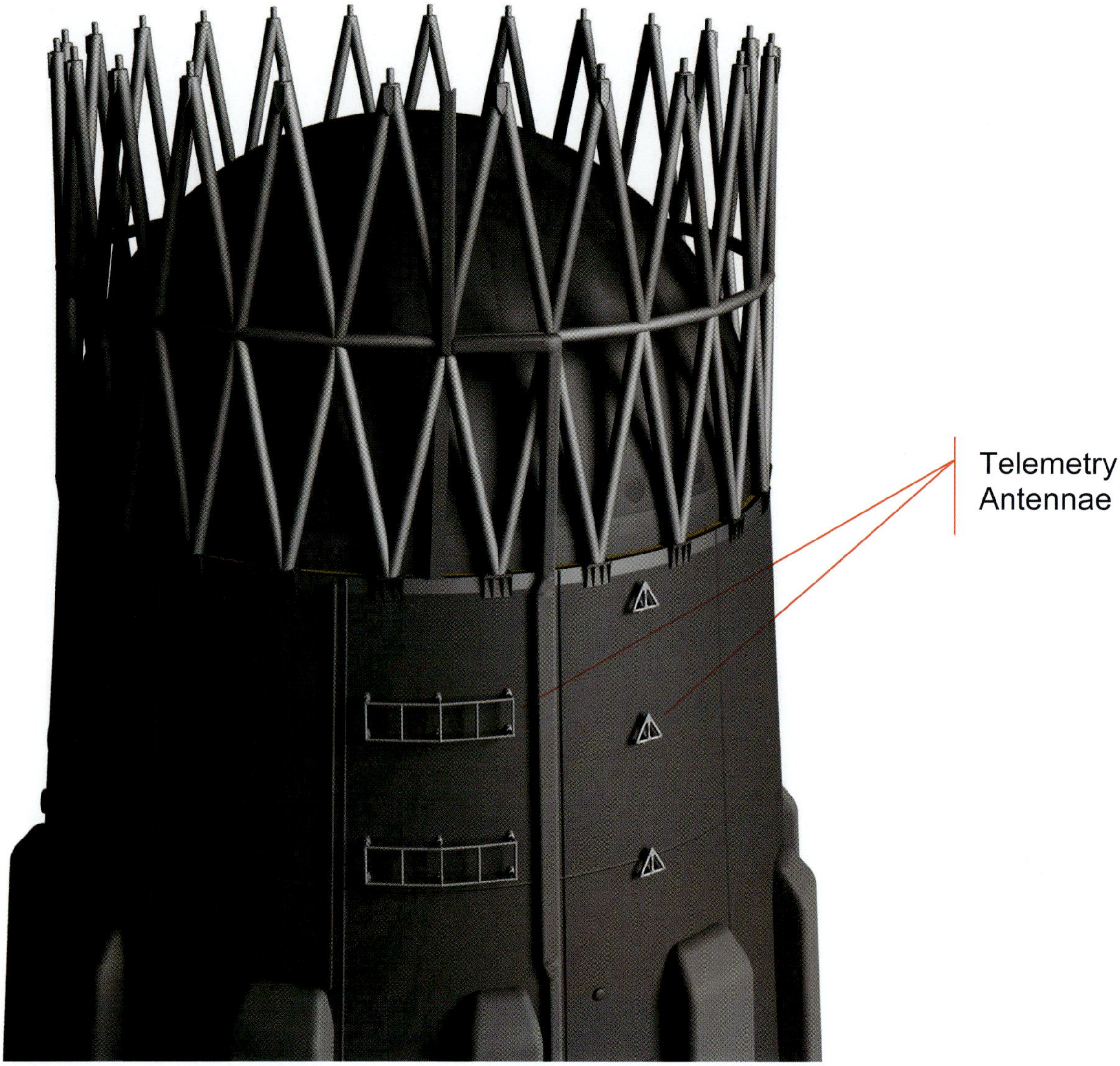

Completely Stowed

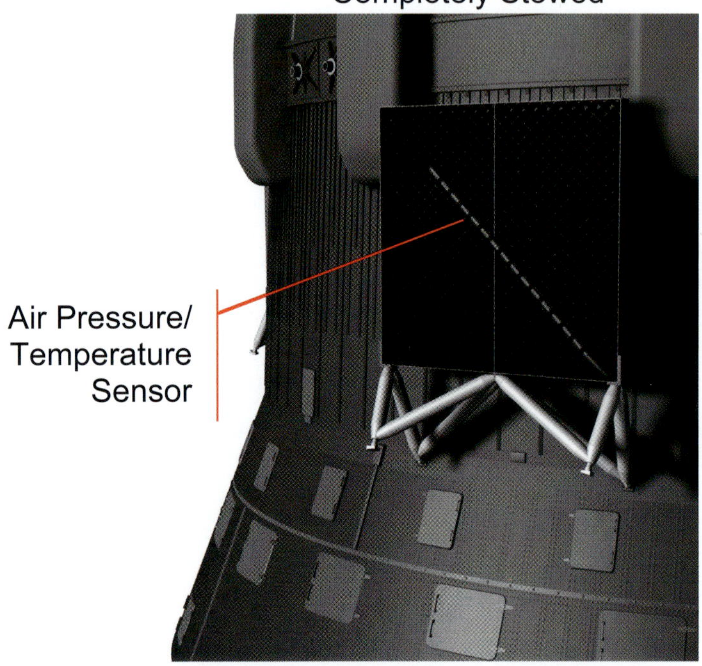

1/3 Deployed

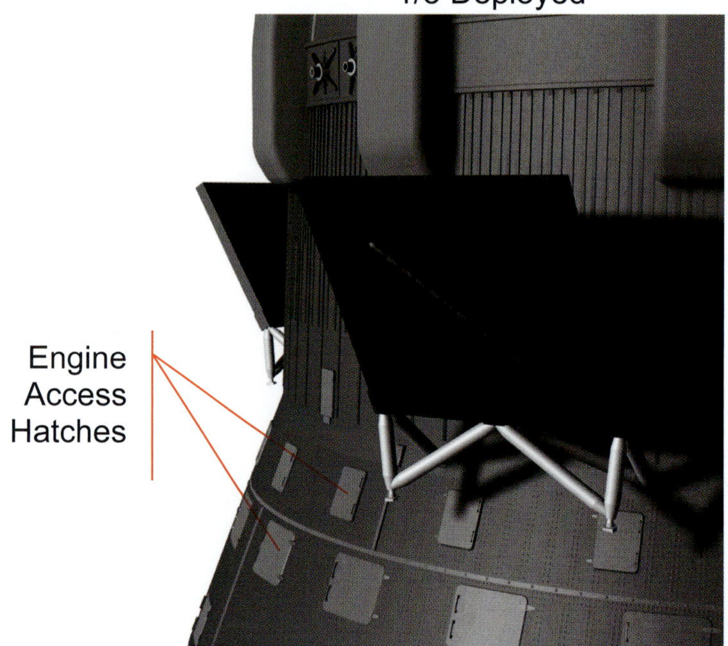

Across above: Perspective views of the top (left) and bottom (right) of Block-A

Across below: Lattice Fin Deployment

Cradle “hard points” for horizontal transporting

Hard Point in Debris Field

2/3 Deployed

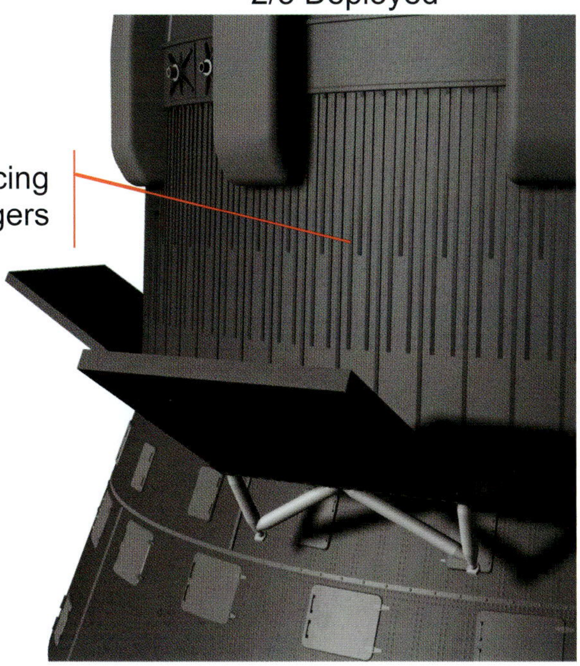

Fully Deployed

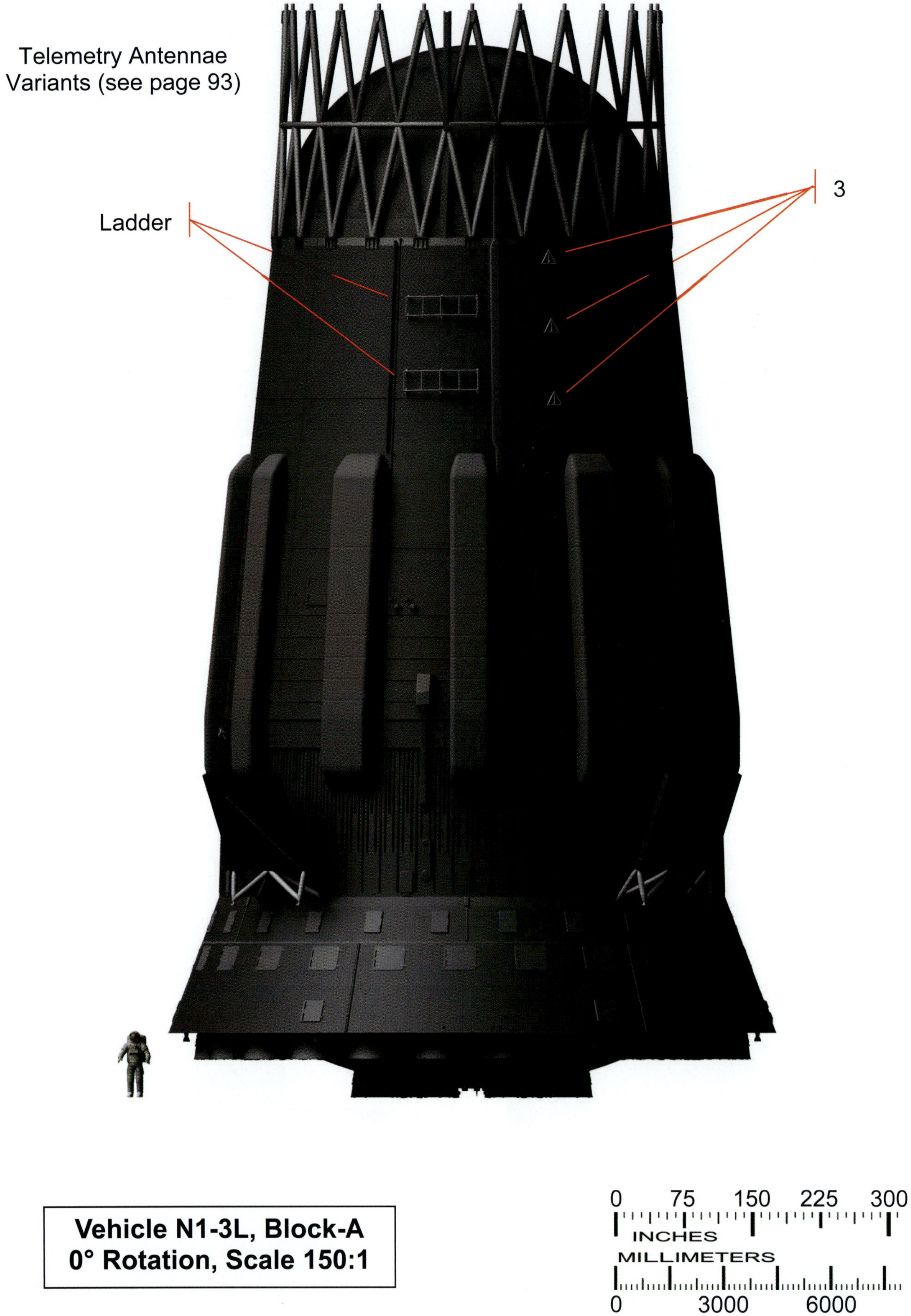

Vehicle N1-3L, Block-A
0° Rotation, Scale 150:1

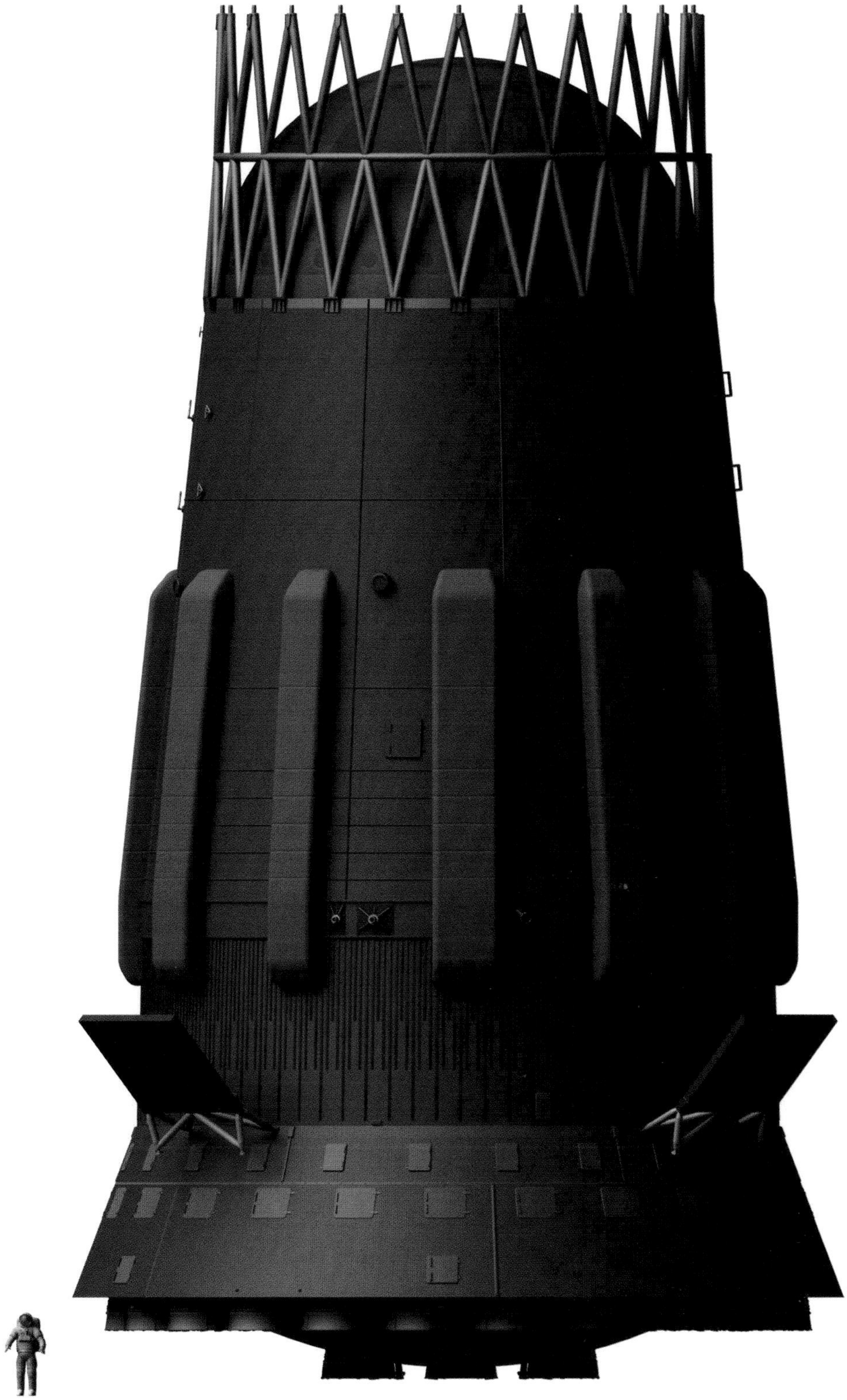

Vehicle N1-3L, Block-A
90° Rotation, Scale 150:1

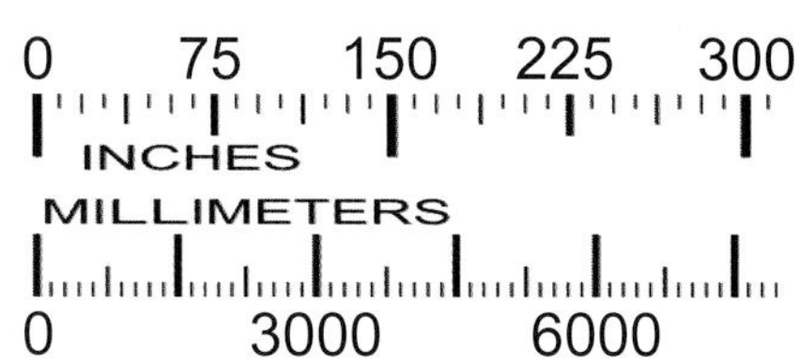

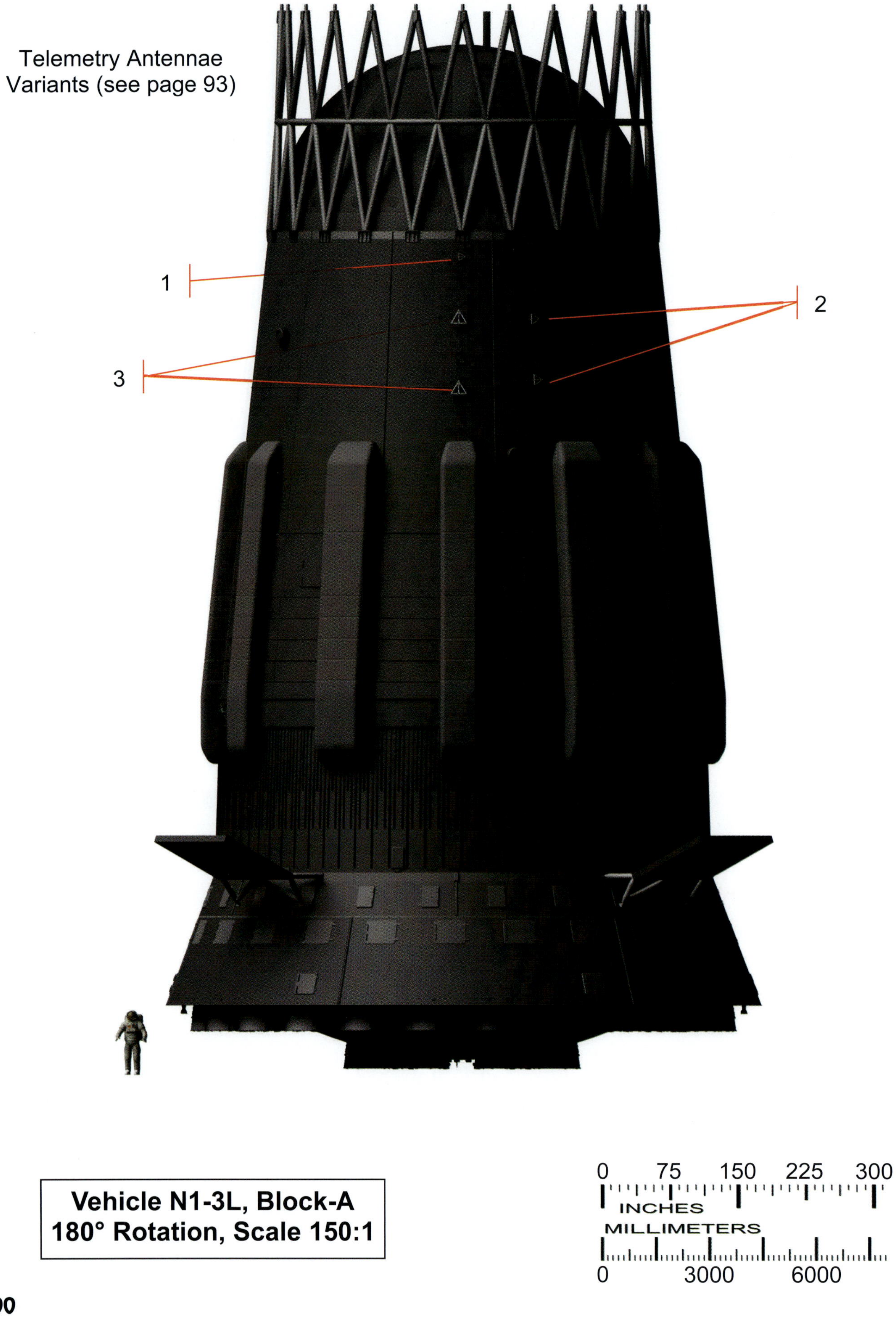

Vehicle N1-3L, Block-A
180° Rotation, Scale 150:1

Vehicle N1-3L, Block-A
270° Rotation, Scale 150:1

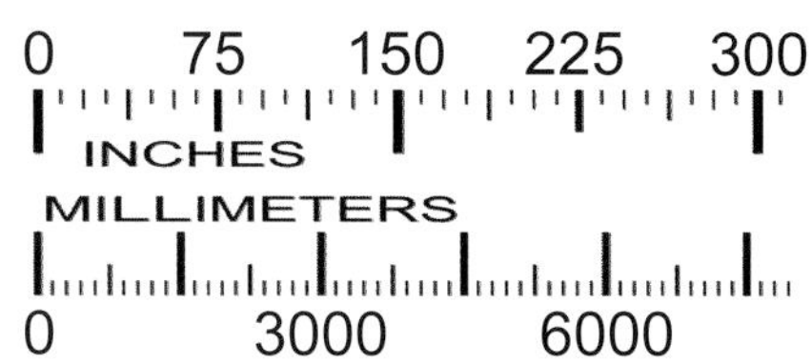

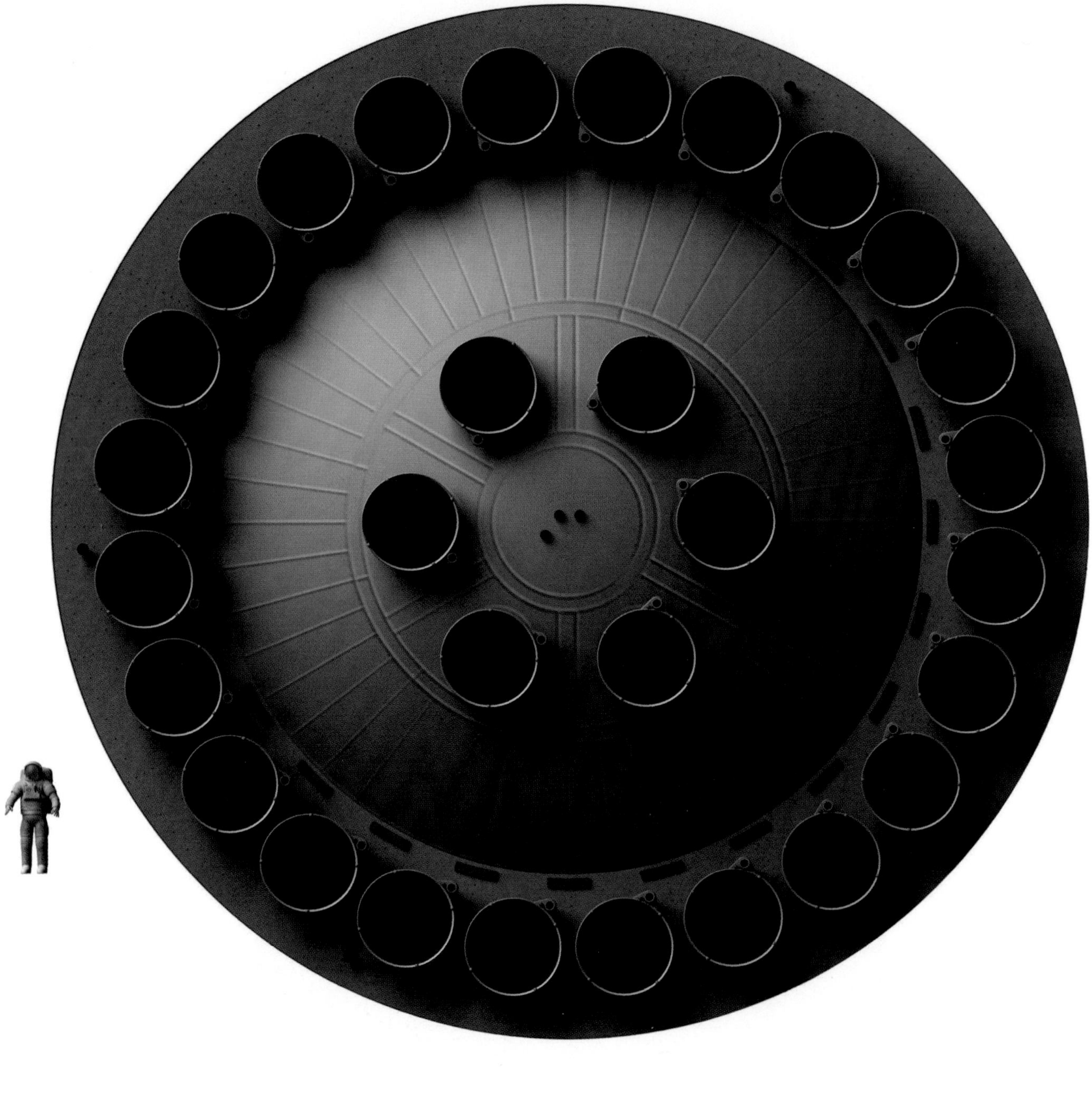

Vehicle N1-3L, Block-A
Bottom View, Scale 100:1

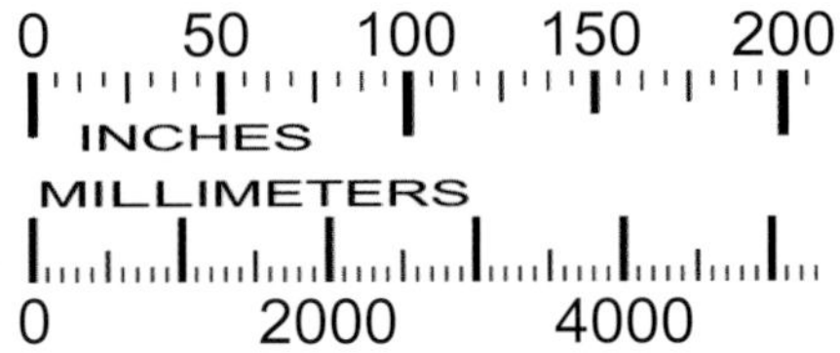

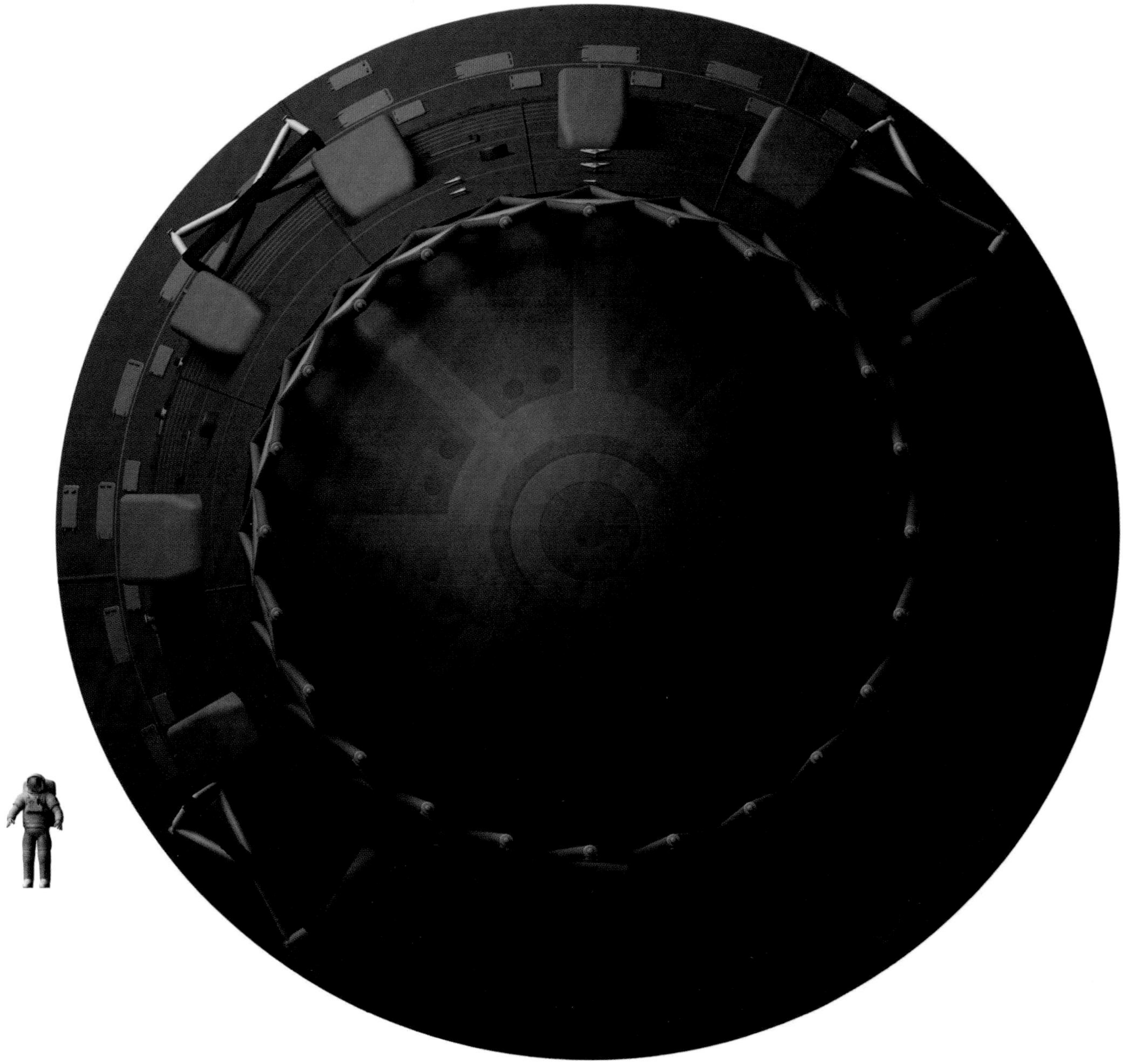

Vehicle N1-3L, Block-A
Top View, Scale 100:1

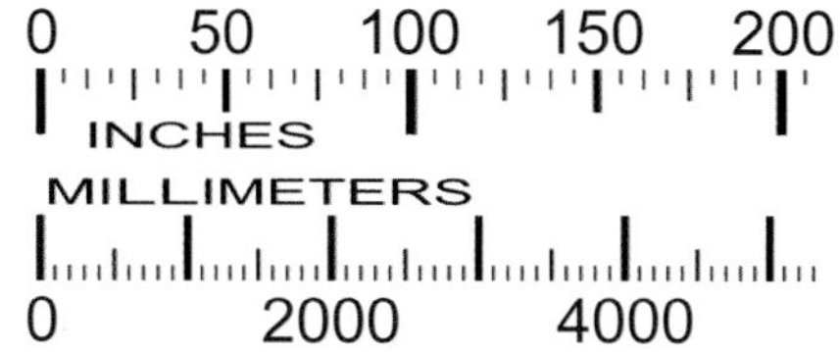

Vehicle N1-5L, Block-A Changes from N1-3L

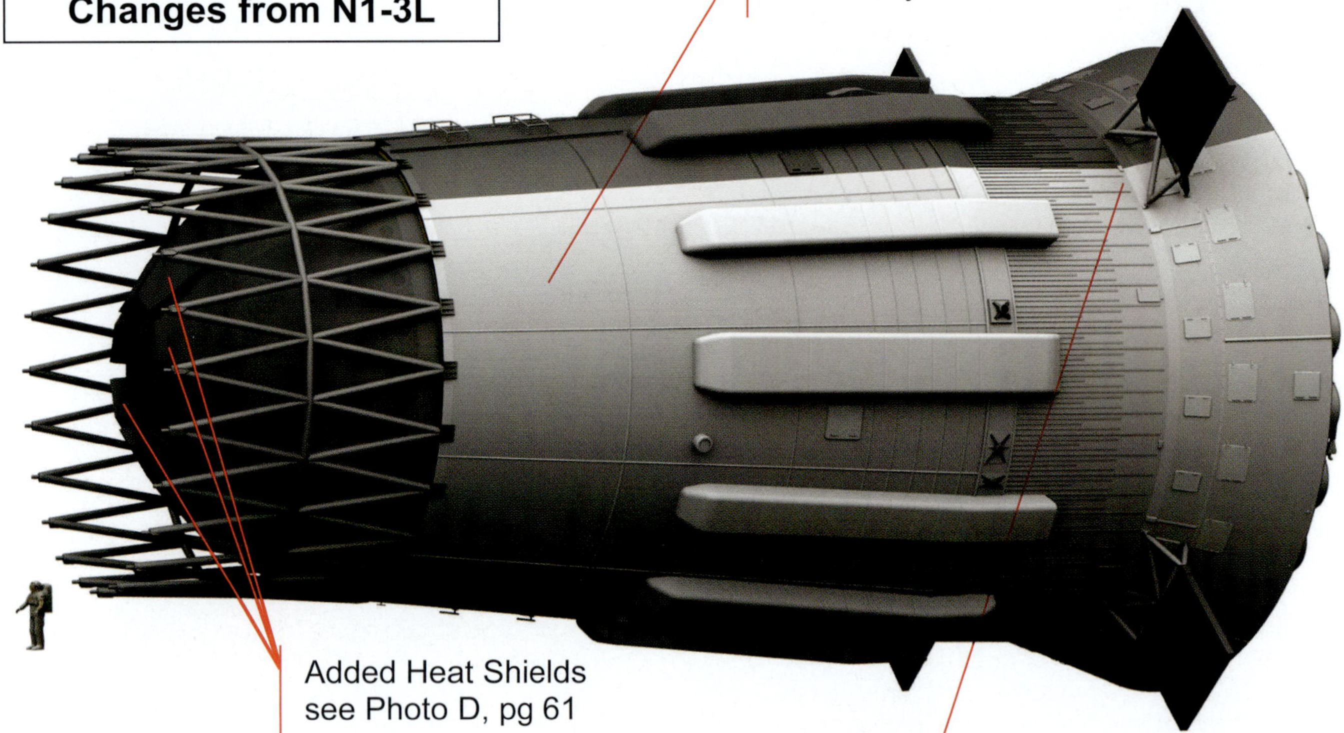

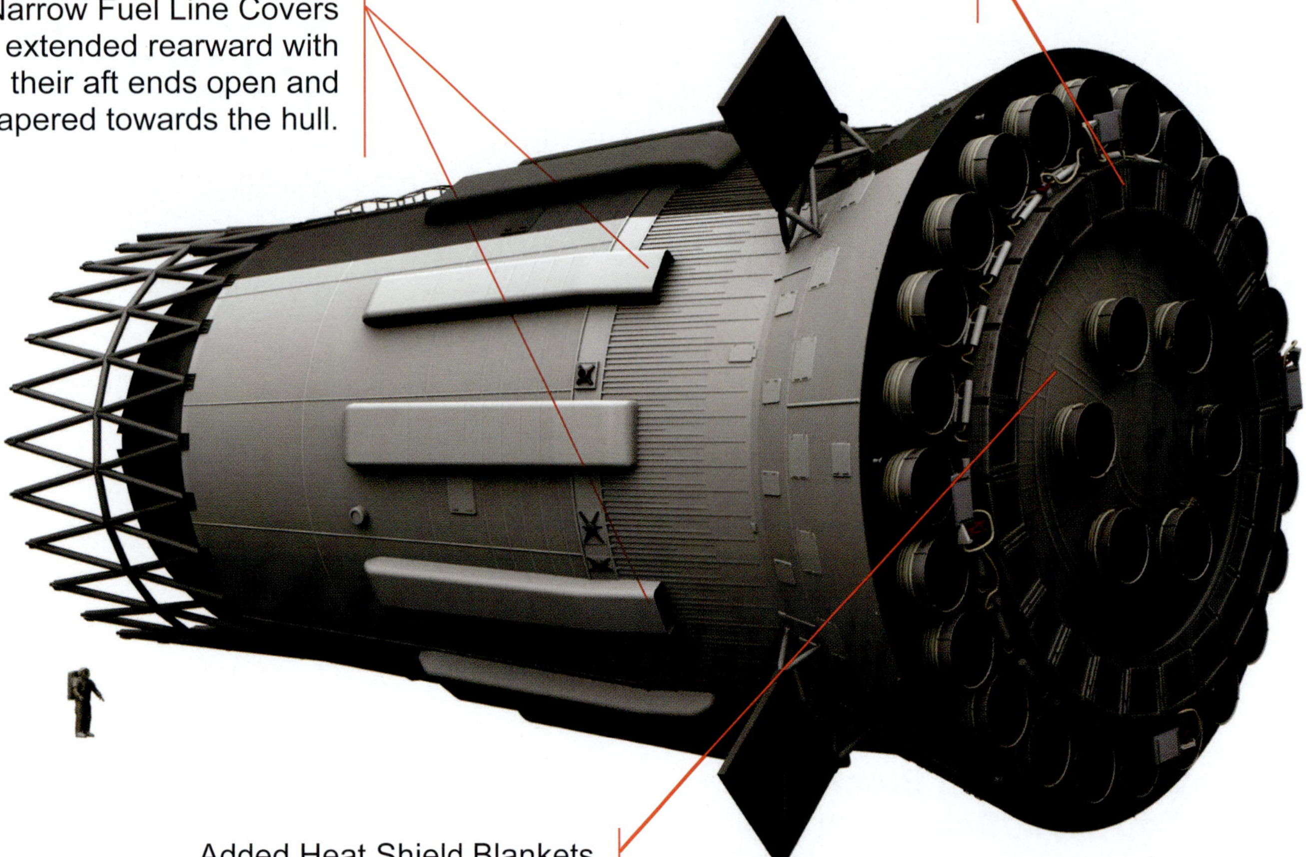

Vehicle N1-6L, Block-A Changes from N1-5L

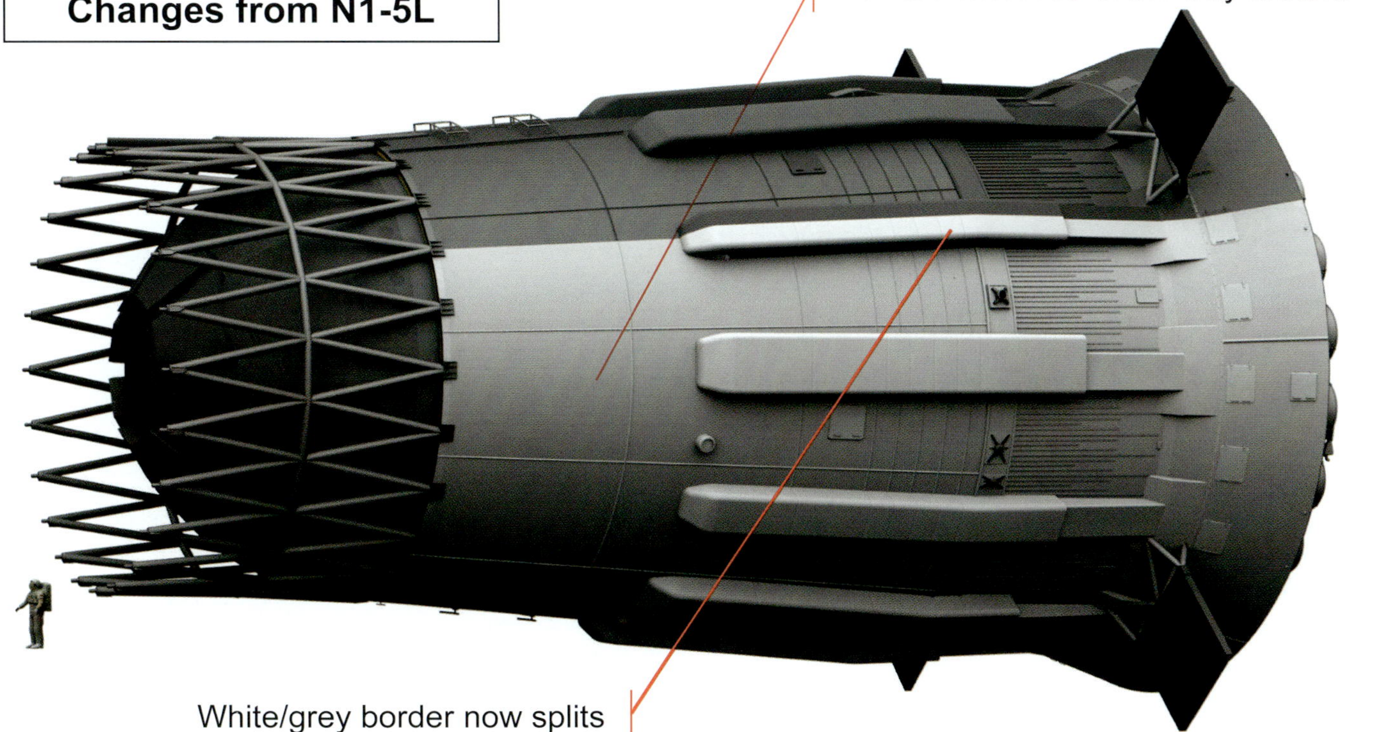

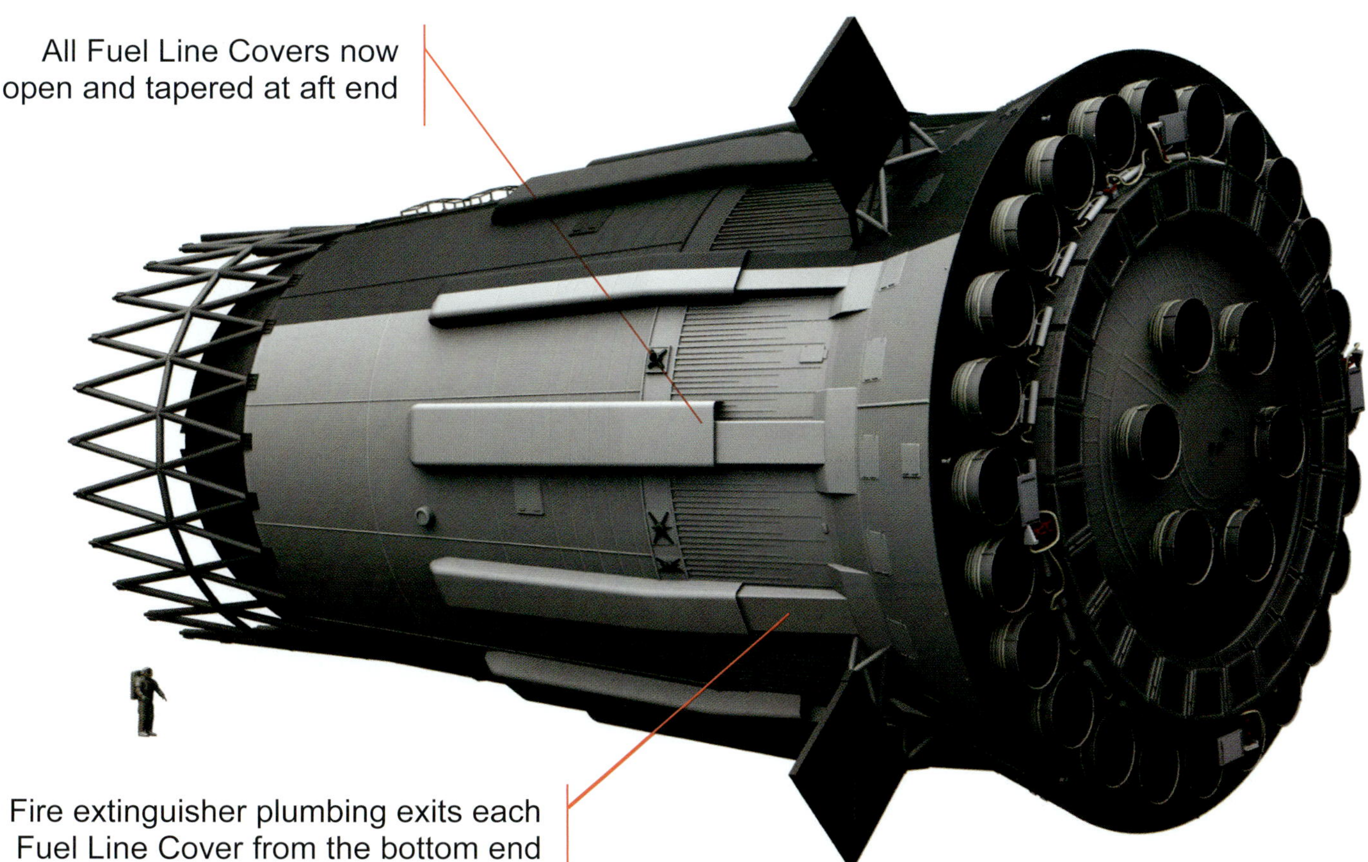

Vehicle N1-7L, Block-A
Changes from N1-6L

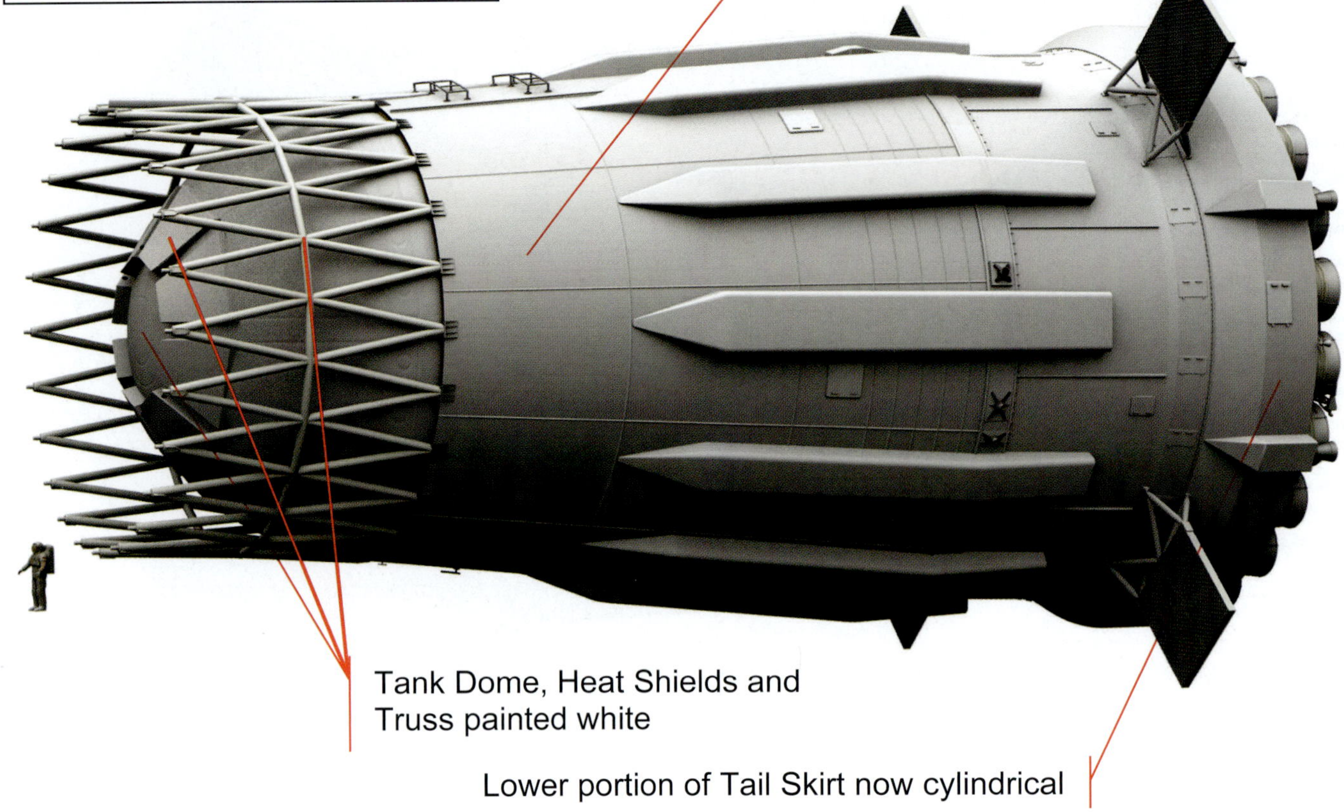

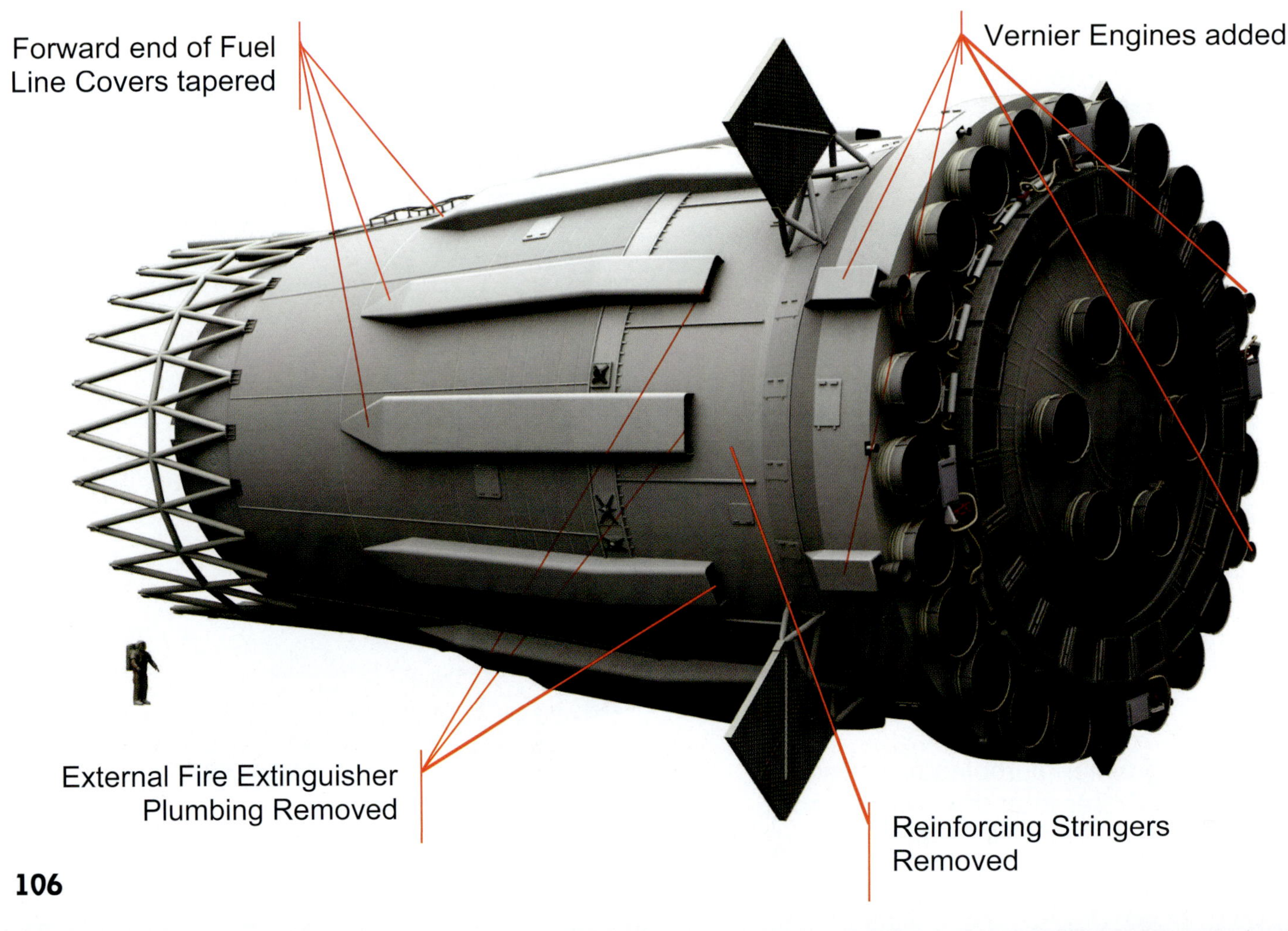

9: N-1 Block-B Reference Guide

9: Справочный раздел: блок Б

CAD Drawings by ALEXANDER SHLIADINSKY
CGI Renderings by NICK STEVENS
Dimensions Applied by MATT JOHNSON

The Block-B used the same basic engines as Block-A. Since it ignited at high altitudes it used a version with an extended bell called the NK-15V (for "Vacuum") that improves efficiency when the outside pressure is very low (or, even better, zero). The -15V wasn't ready in time for the N1-3L test flight so the standard NK-15 was used. The lower efficiency at high altitudes was acceptable since the payload wasn't the full L3 Complex, but a mockup of only the L-1 Zond spacecraft without the LK lander or a Block-D stage.

The NK-15V was eventually to be replaced with the improved NK-43 (vacuum version of the NK-33 used in Block-A). N1-8L was built this way, but the program was cancelled before it flew.

Block-B Engine Configurations

Flight No.	Engines	Vacuum Thrust[1] (kN)	Propellant (tons)	Burn time[2] (seconds)	Flow rate (tons/sec)	Total Impulse (mega N-s)
N1-3L	NK-15 (8)	12,349	505	110 + 24	3.96	1,506
N1-5L	NK-15V (8)	13,182	505	110 + 24	4.14	1,608
N1-6L	NK-15V (8)	13,182	505	110 + 24	4.14	1,608
N1-7L	NK-15V (8)	13,182	505	110 + 24	4.14	1,608
N1-8L[3]	NK-43 (8)	14,034	505	110 + 24	4.14	1,712

[1] All of the values in the table are for the stage as a whole.

[2] First value is the burn time for all 8 engines. The second value is for 4 engines to keep the acceleration positive (and the propellants at the bottoms of the tanks) while the Block-V engines are spooling up.

[3] Not flown.

NK-15

NK-15V

NK-43

Above, a comparison of the NK-15, 15V and 43. The true size of these engines becomes apparent once you see people next to them. To the left is Block-B of N1-3L with the NK-15 engine bells barely poking through the frame. The spacing obviously allows for much larger nozzles.

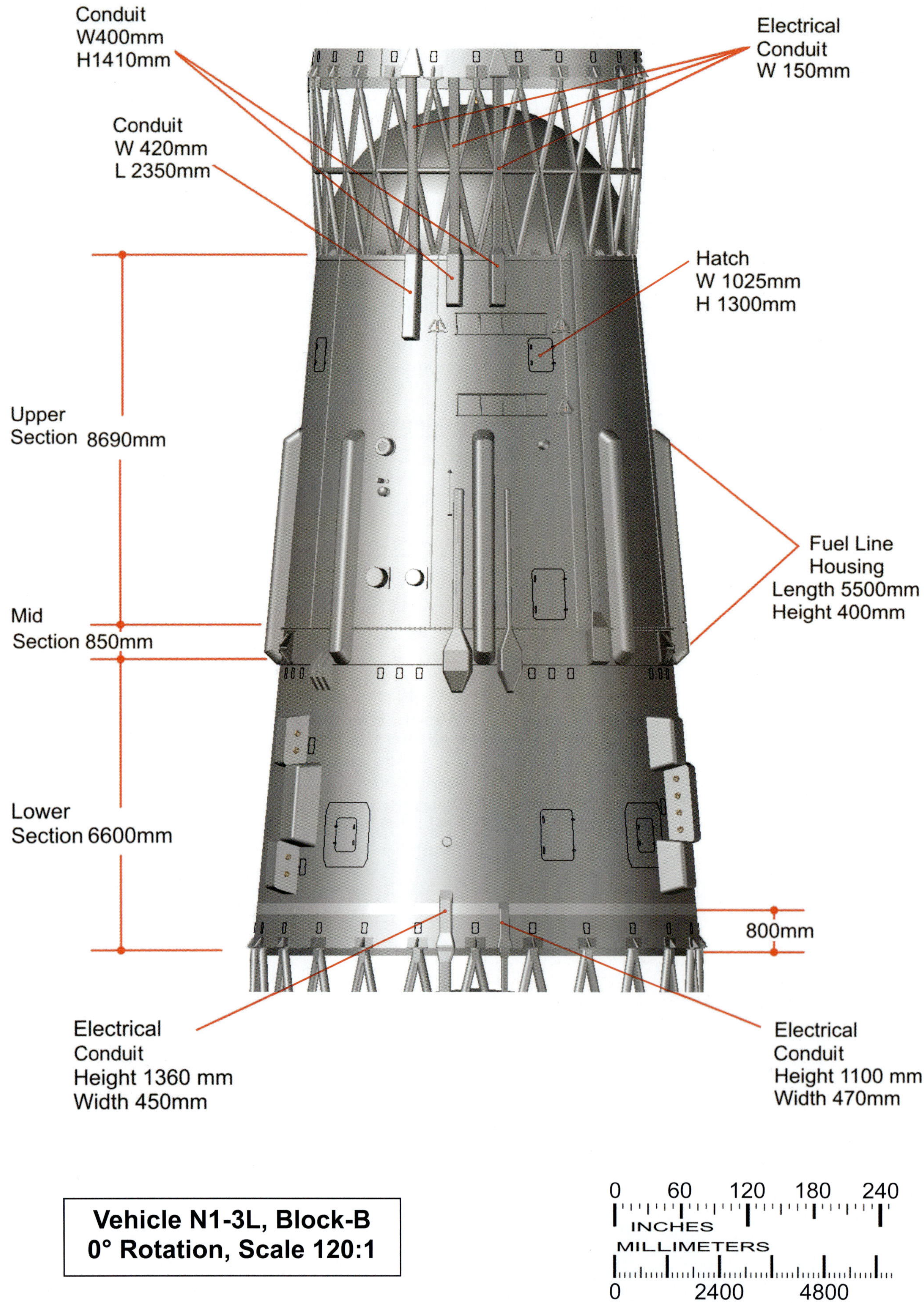

Vehicle N1-3L, Block-B
0° Rotation, Scale 120:1

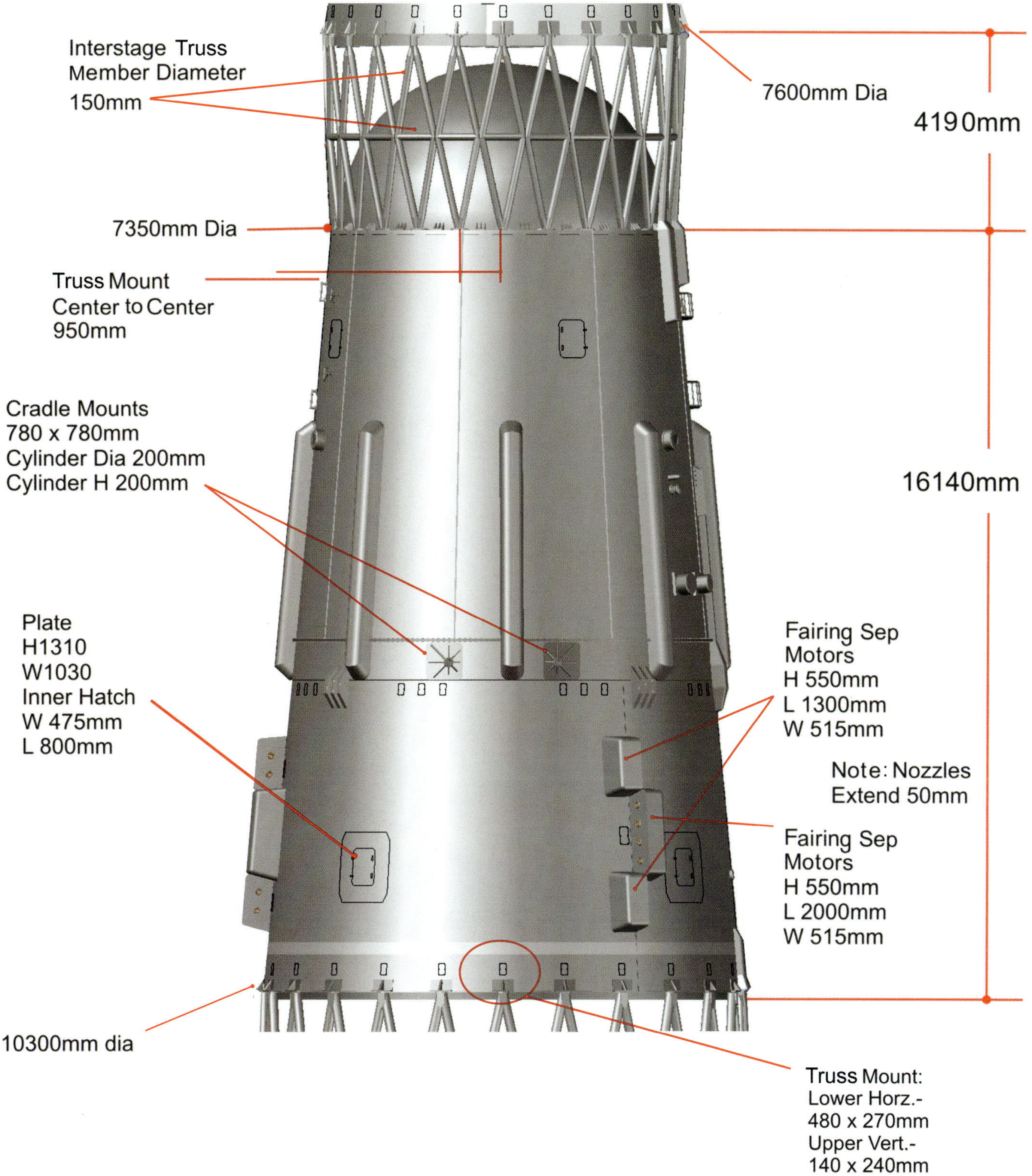

Vehicle N1-3L, Block-B
90° Rotation, Scale 120:1

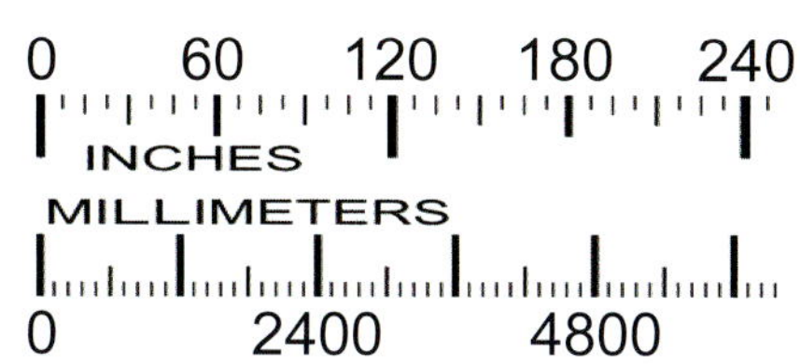

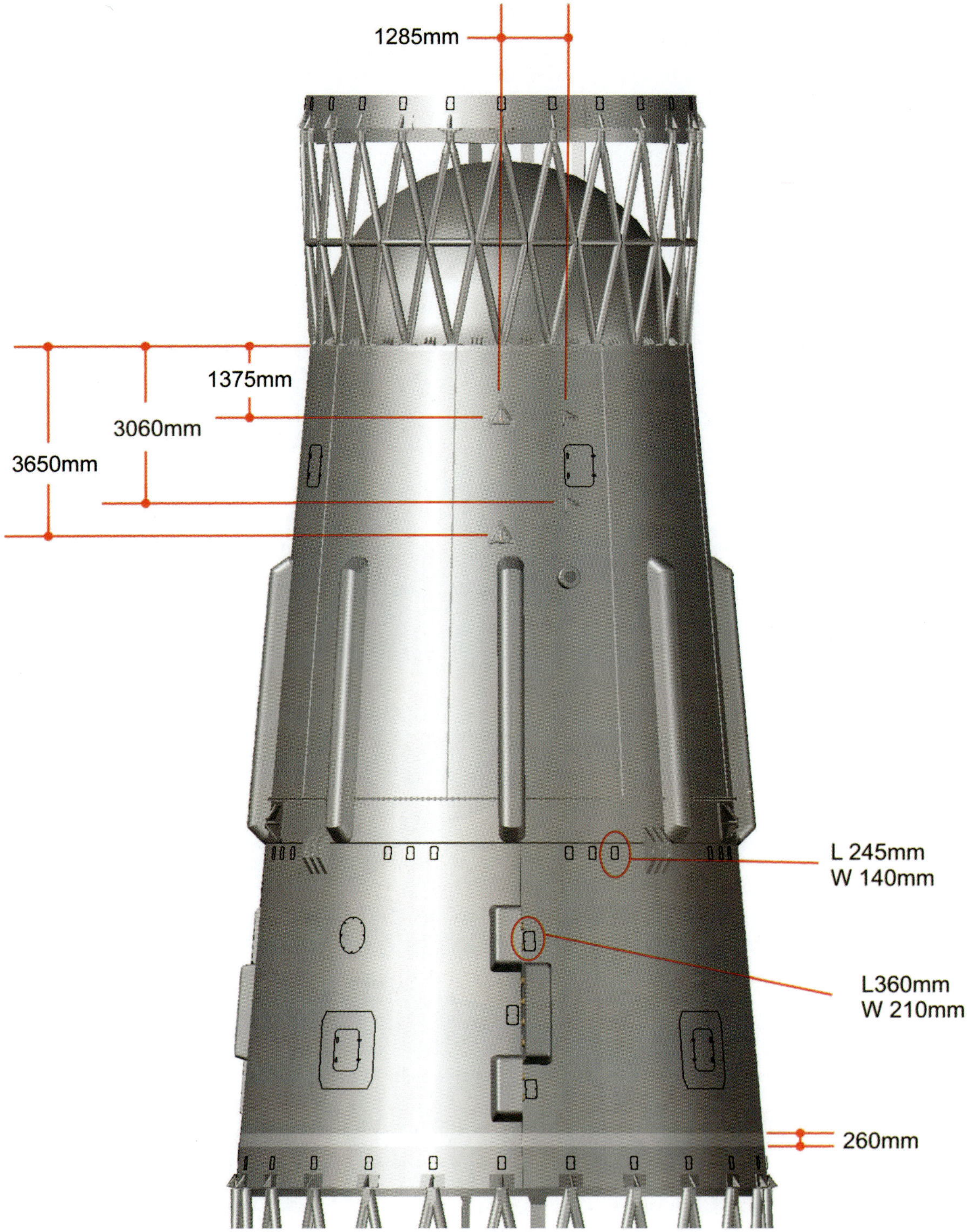

Vehicle N1-3L, Block-B
180° Rotation, Scale 120:1

0 60 120 180 240
INCHES
MILLIMETERS
0 2400 4800

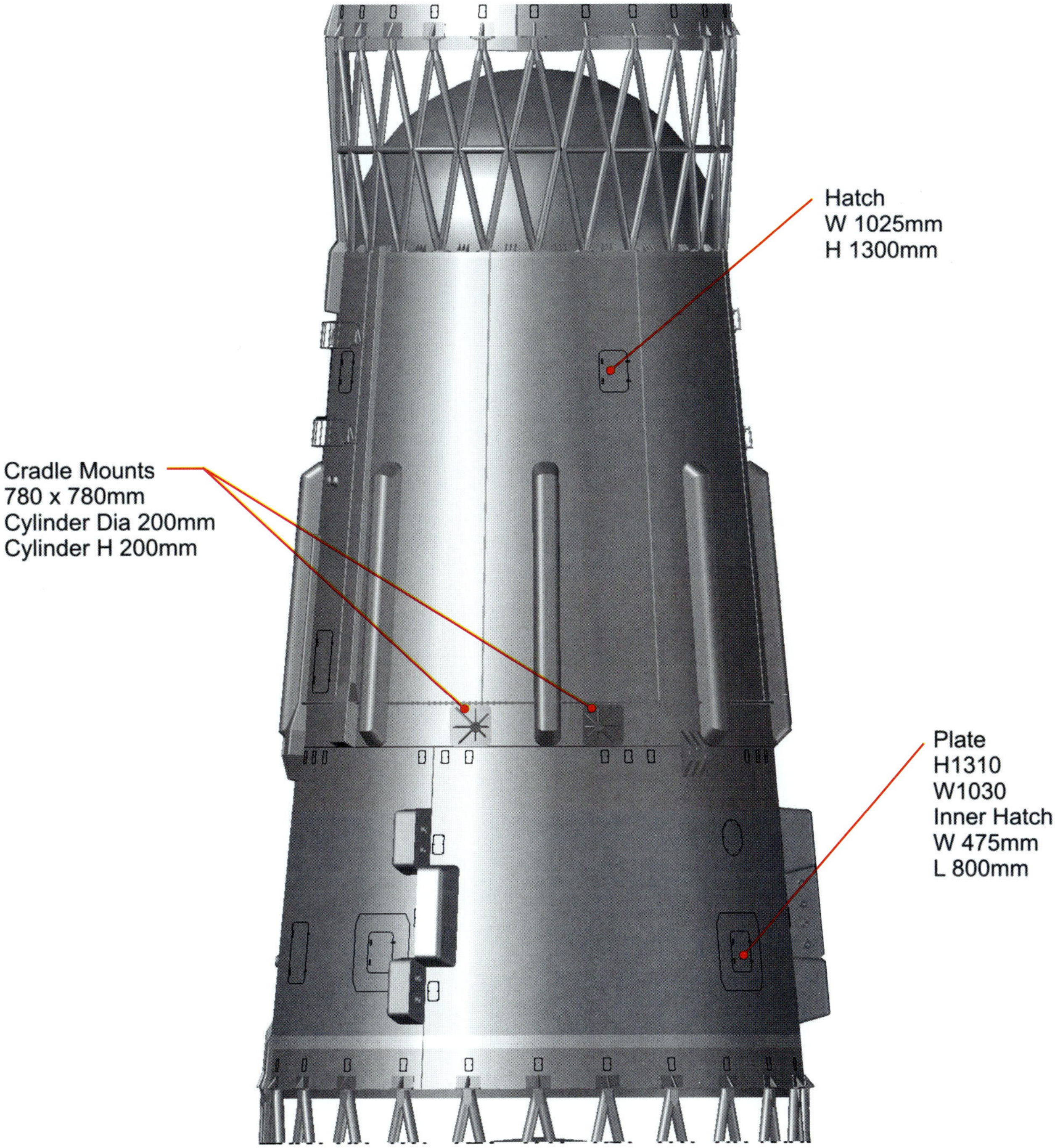

Vehicle N1-3L, Block-B
270° Rotation, Scale 120:1

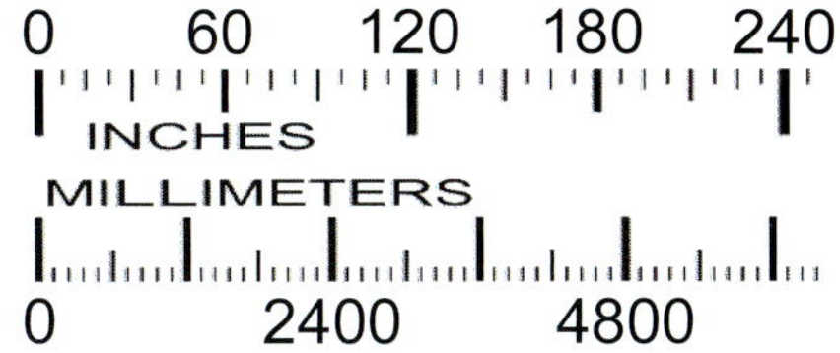

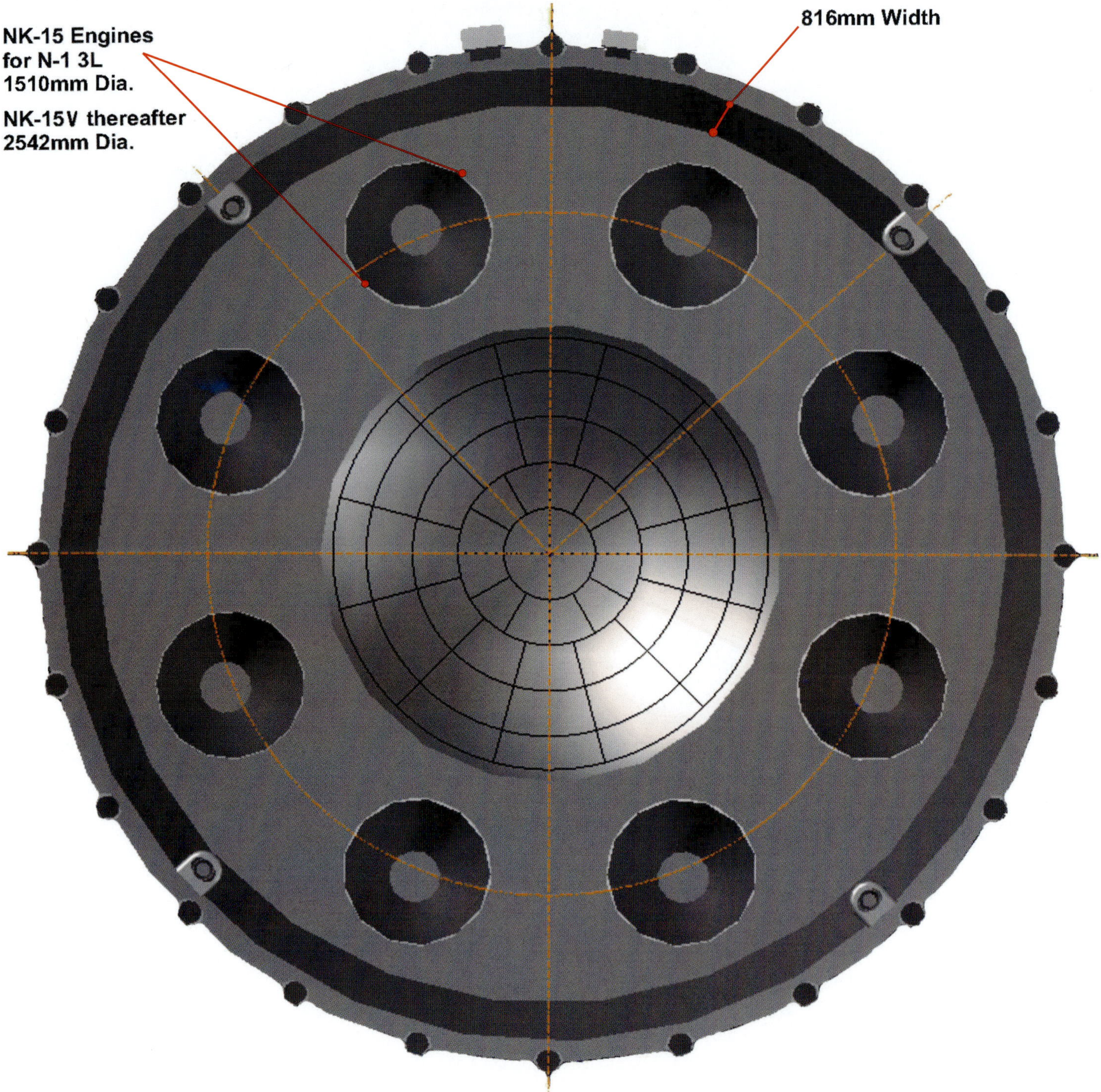

Vehicle N1-3L, Block-B
Bottom View, Scale 60:1

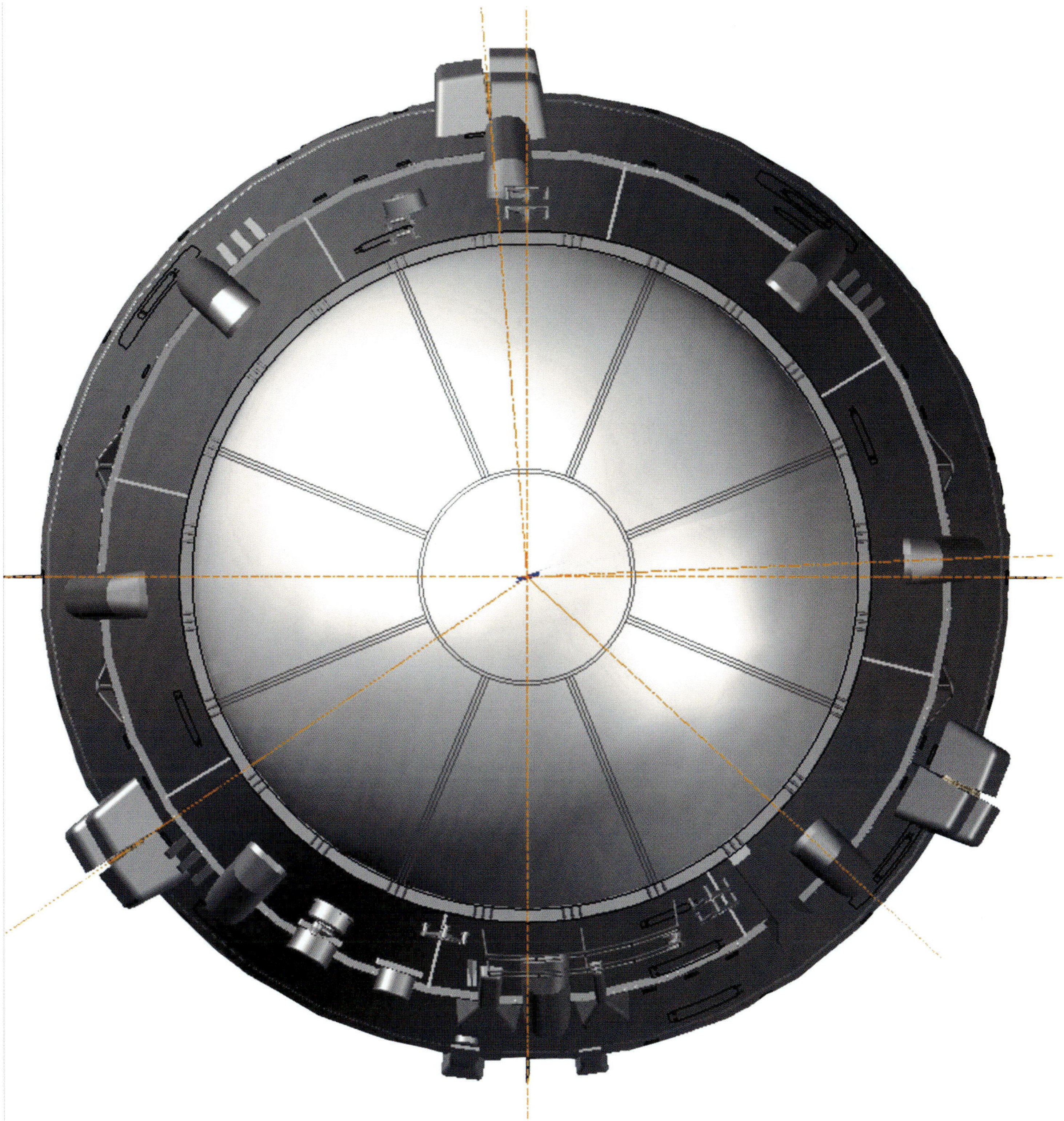

Vehicle N1-3L, Block-B
Top View, Scale 60:1

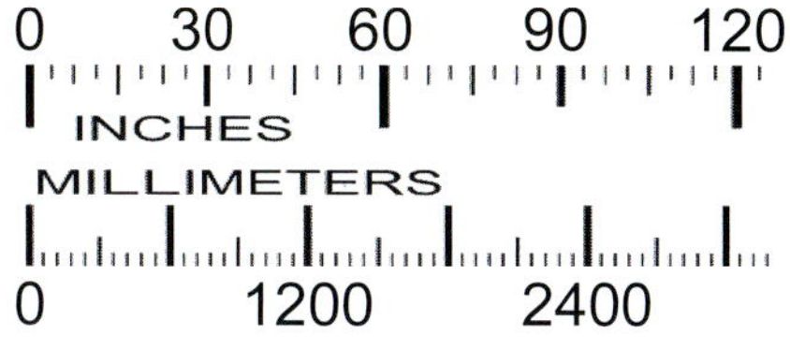

Block-B Engine Fairing Separation

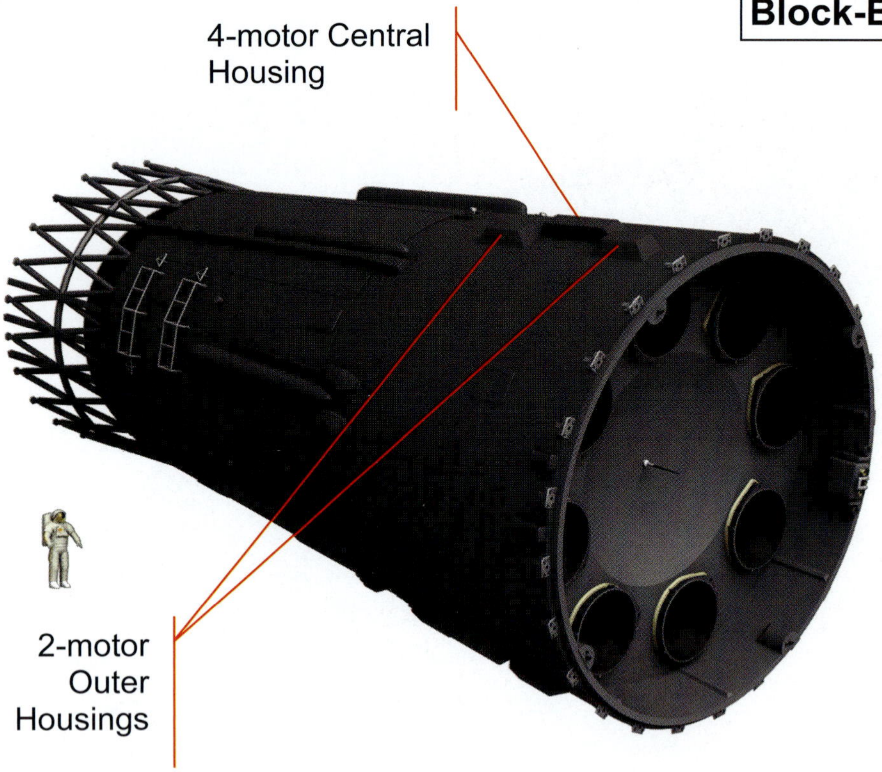

At staging, the aerodynamic fairing that covered the Block-B engines are jettisoned since the air is thin enough to not need them and the rocket engines need to be exposed for radiant cooling. Additionally, this removes about a ton of weight that the rocket no longer has to lift.

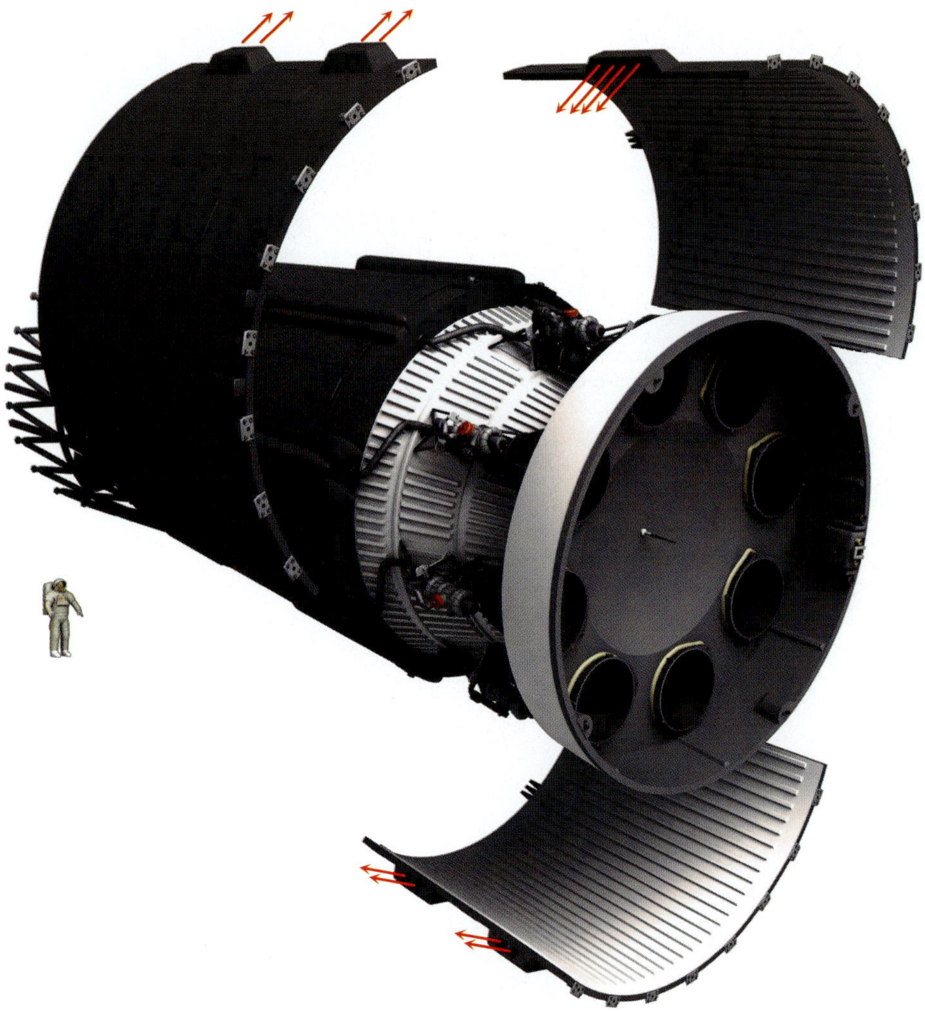

The fairing comes off in three identical segments.[4] Each segment has four separation motors on each edge meaning eight motors per segment and 24 total. To keep the thrust from all of these motors even, and not have their exhaust impinge on each other, the motors are divided up as shown with the four motors on one edge grouped together in a central housing, and the four motors on the other edge split into two housings on either side of its neighbor's central housing.

[4] The basic structure of each segment is the same but details, especially the engine access hatches, are different between the segments. This is driven by the fact that there are eight engines with the hatches spread over three fairing segments.

Vehicle N1-3L Block-B, After Fairing Separation, 270° Rotation, Scale 100:1

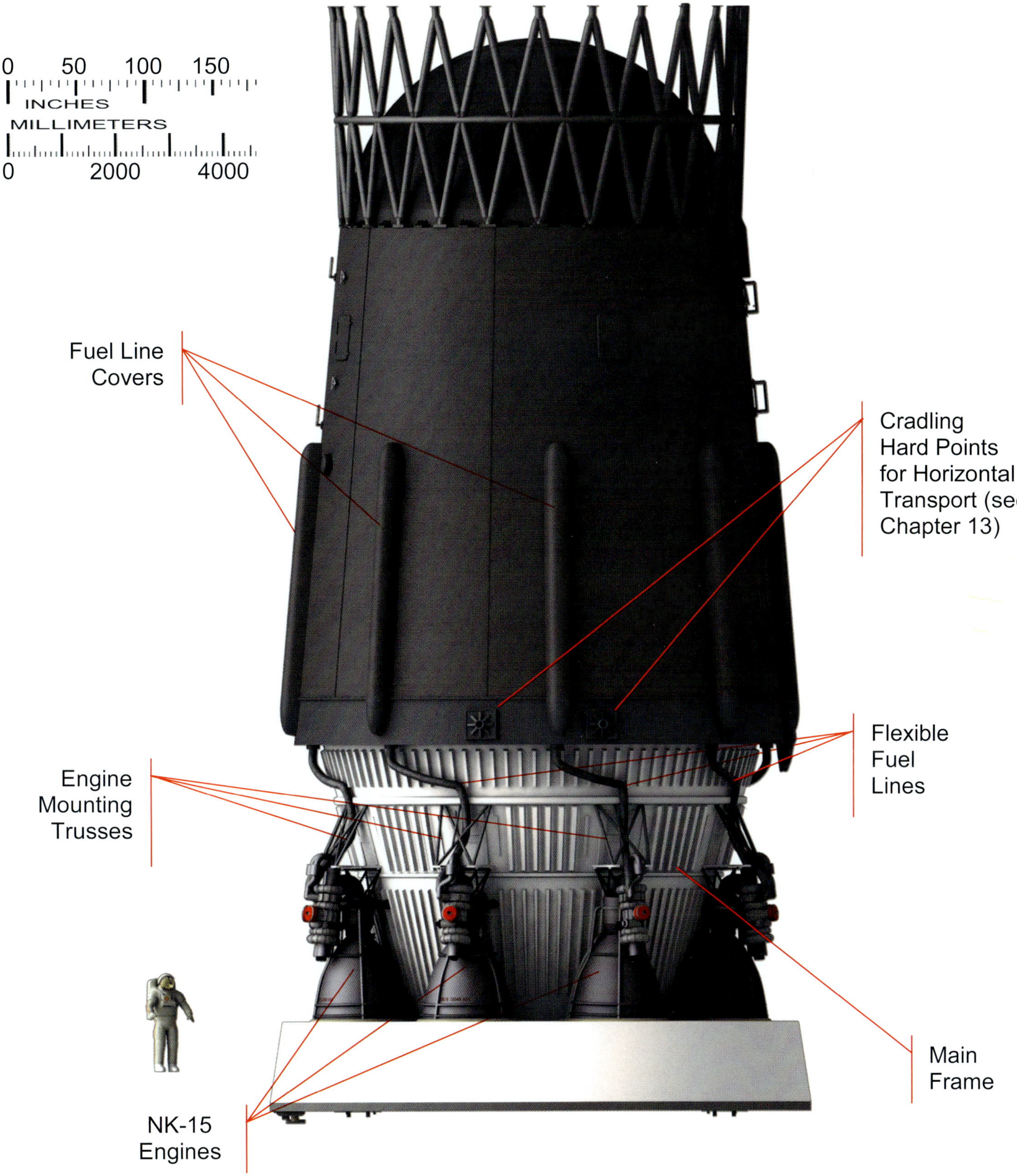

The squat conical ring surrounding the engines after the fairings are jettisoned looks somewhat out of place to those accustomed to the free-hanging engines on the second stage of the Saturn V, almost as if it had been accidentally left behind. It's used to fix the lower ends of the engine nozzles to keep them in alignment with the main frame. This very precise, strong arrangement is possible because the engines don't have to gimbal (swivel) to steer.

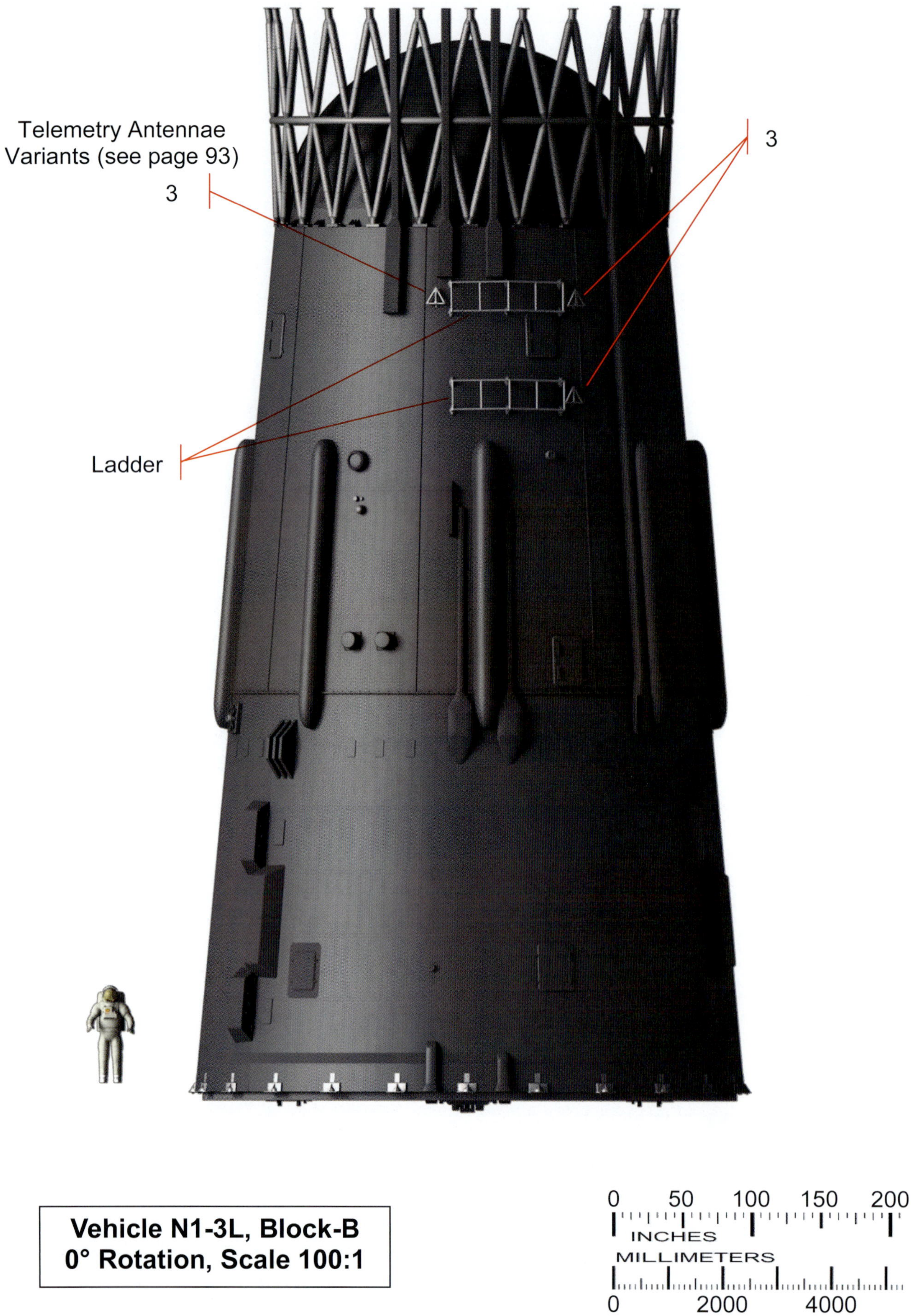

Vehicle N1-3L, Block-B
0° Rotation, Scale 100:1

Vehicle N1-3L, Block-B
90° Rotation, Scale 100:1

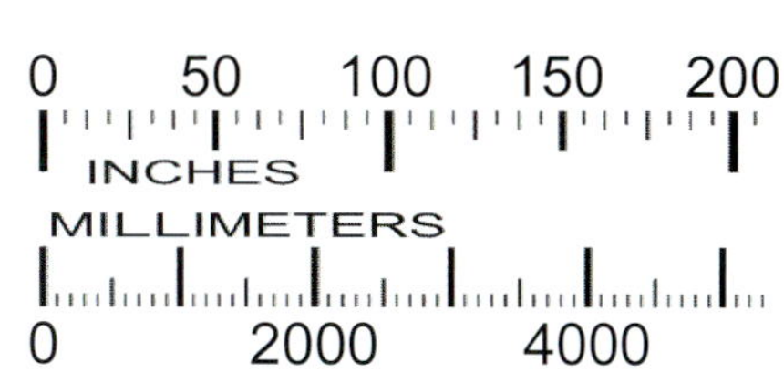

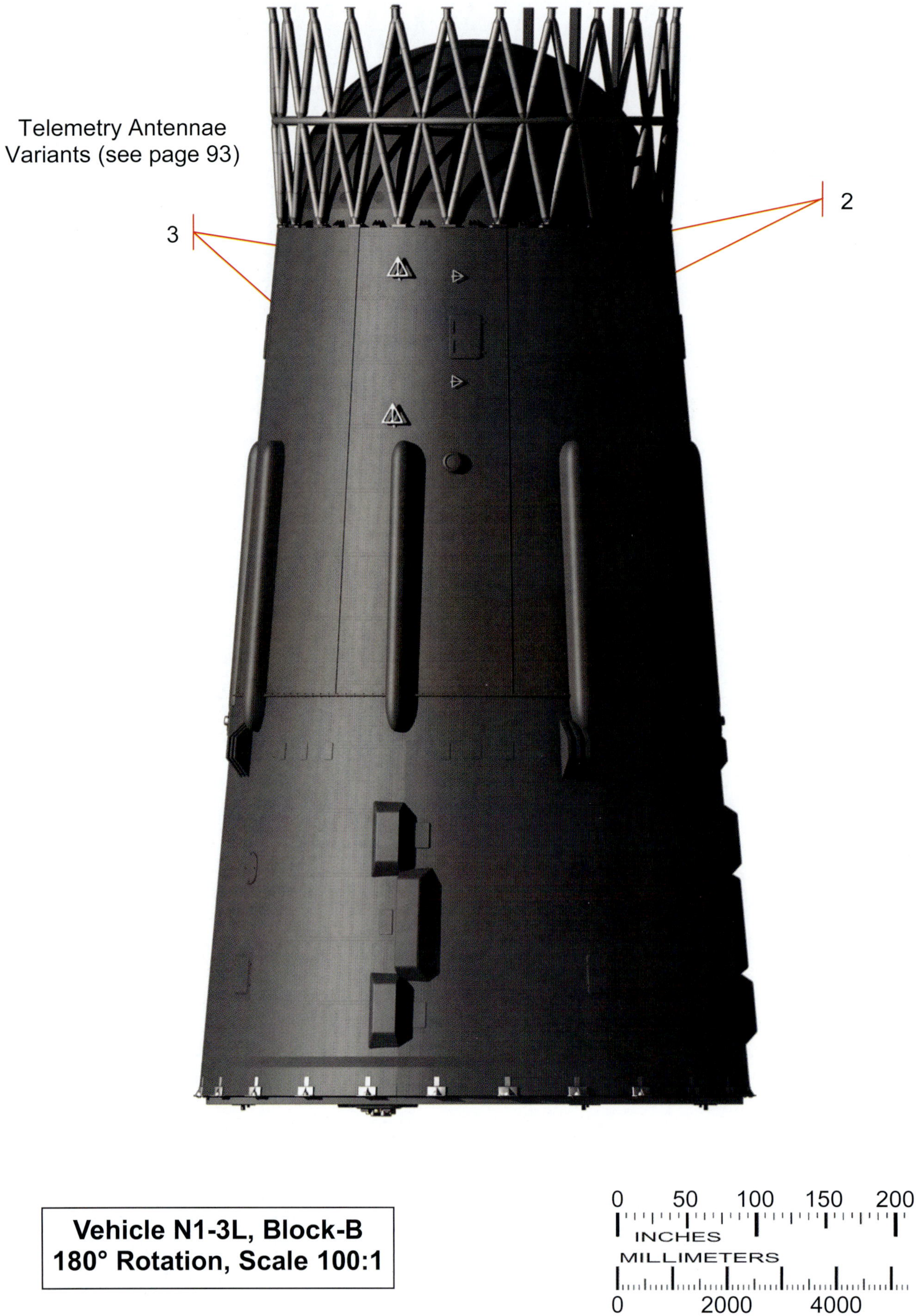

Vehicle N1-3L, Block-B
180° Rotation, Scale 100:1

Vehicle N1-3L, Block-B
270° Rotation, Scale 100:1

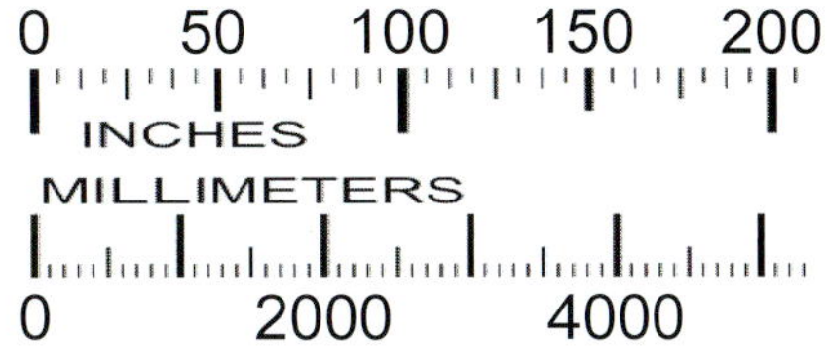

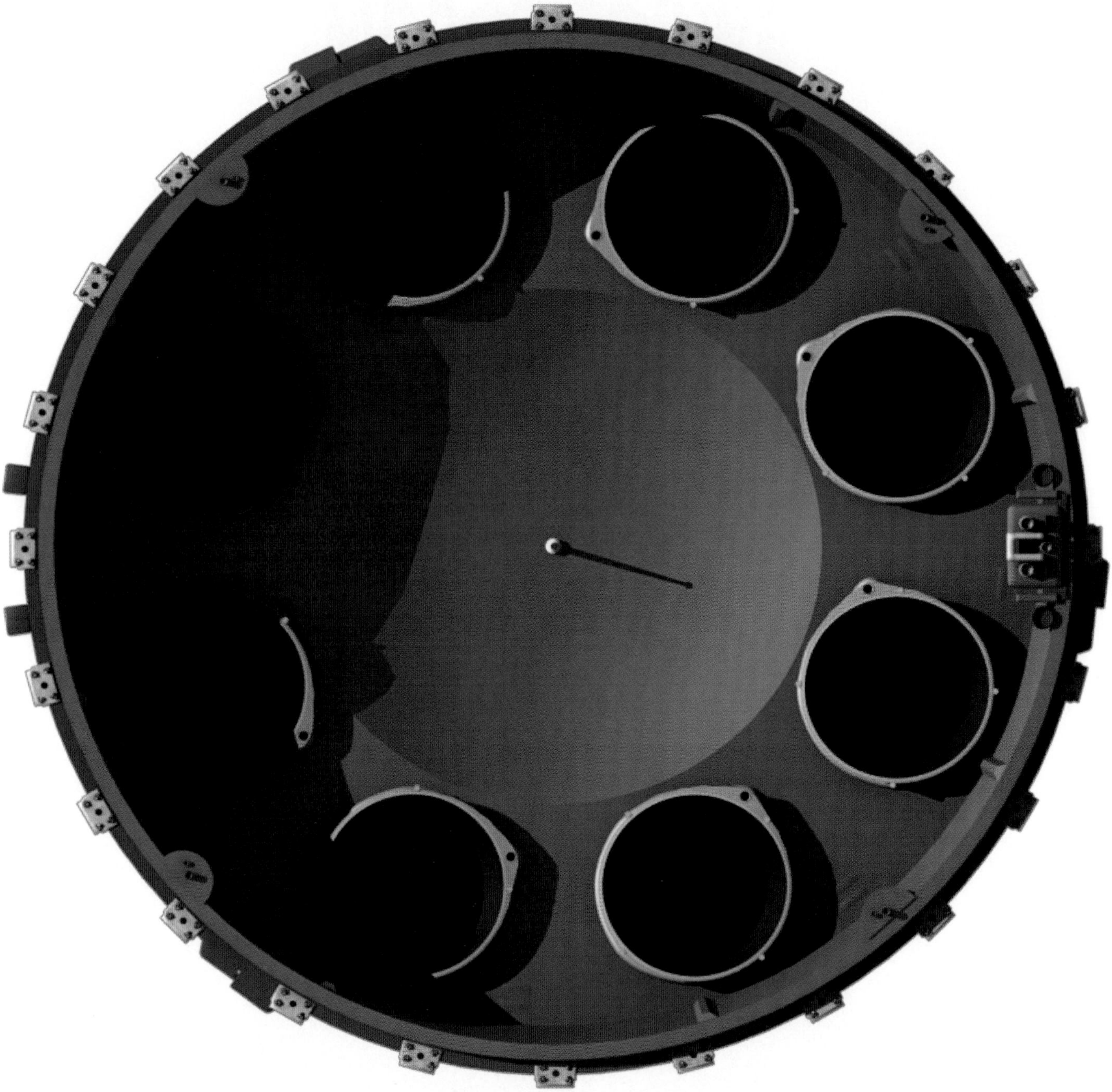

Vehicle N1-3L, Block-B
Bottom View, Scale 60:1

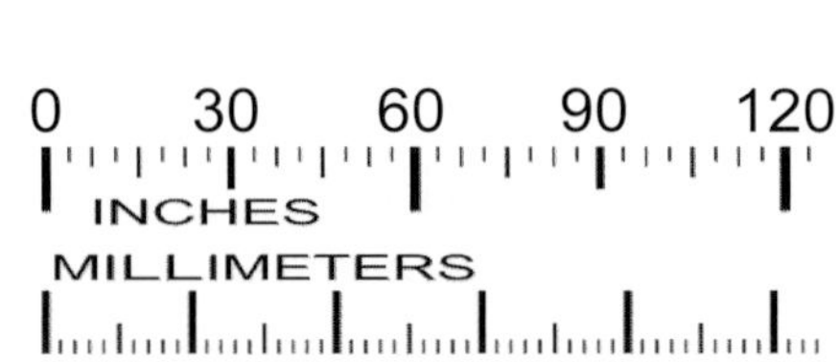

Vehicle N1-3L, Block-B
Top View, Scale 60:1

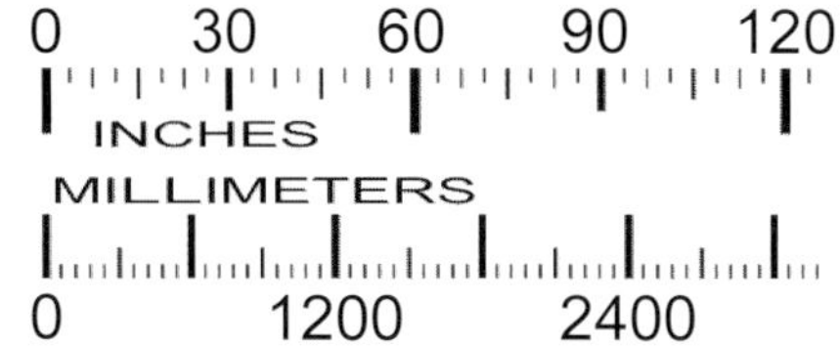

Vehicle N1-5L, Block-B Changes from N1-3L

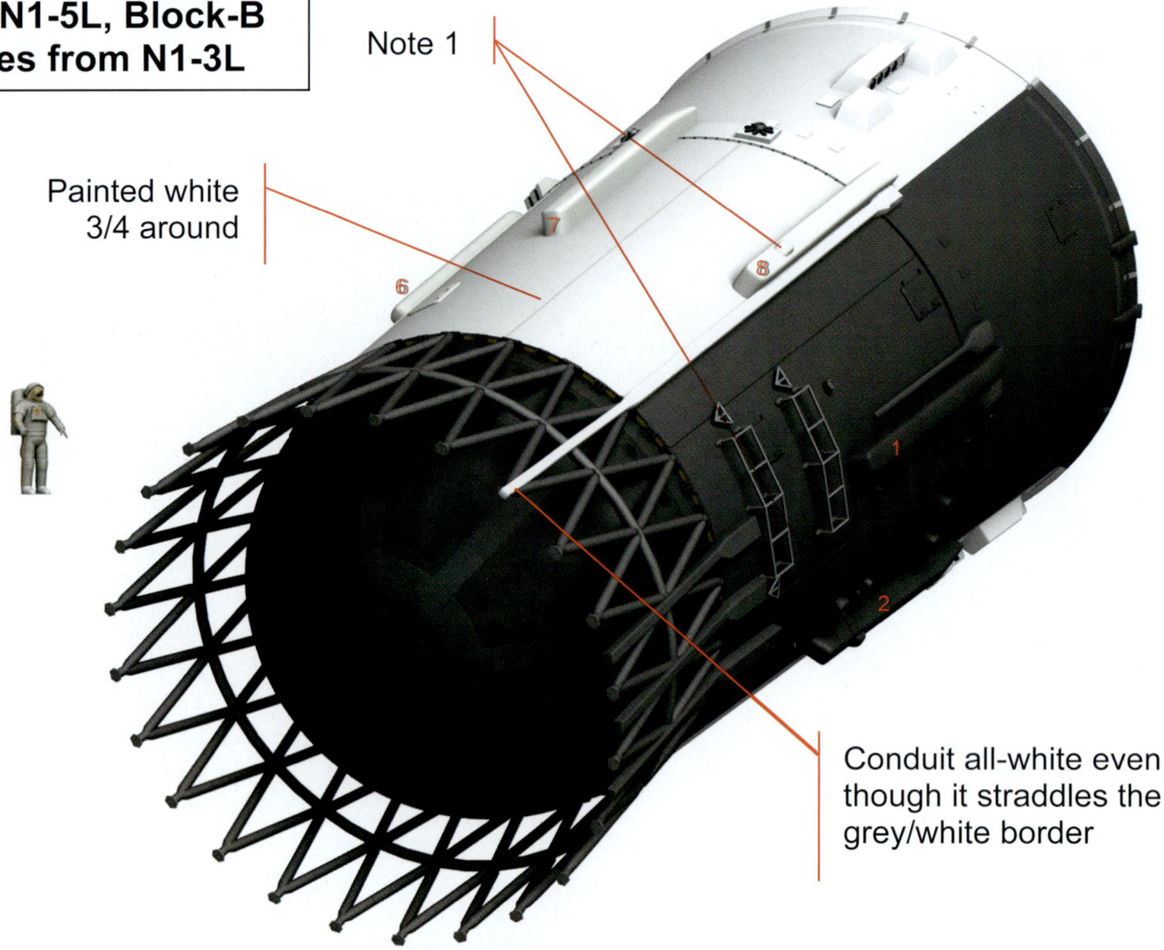

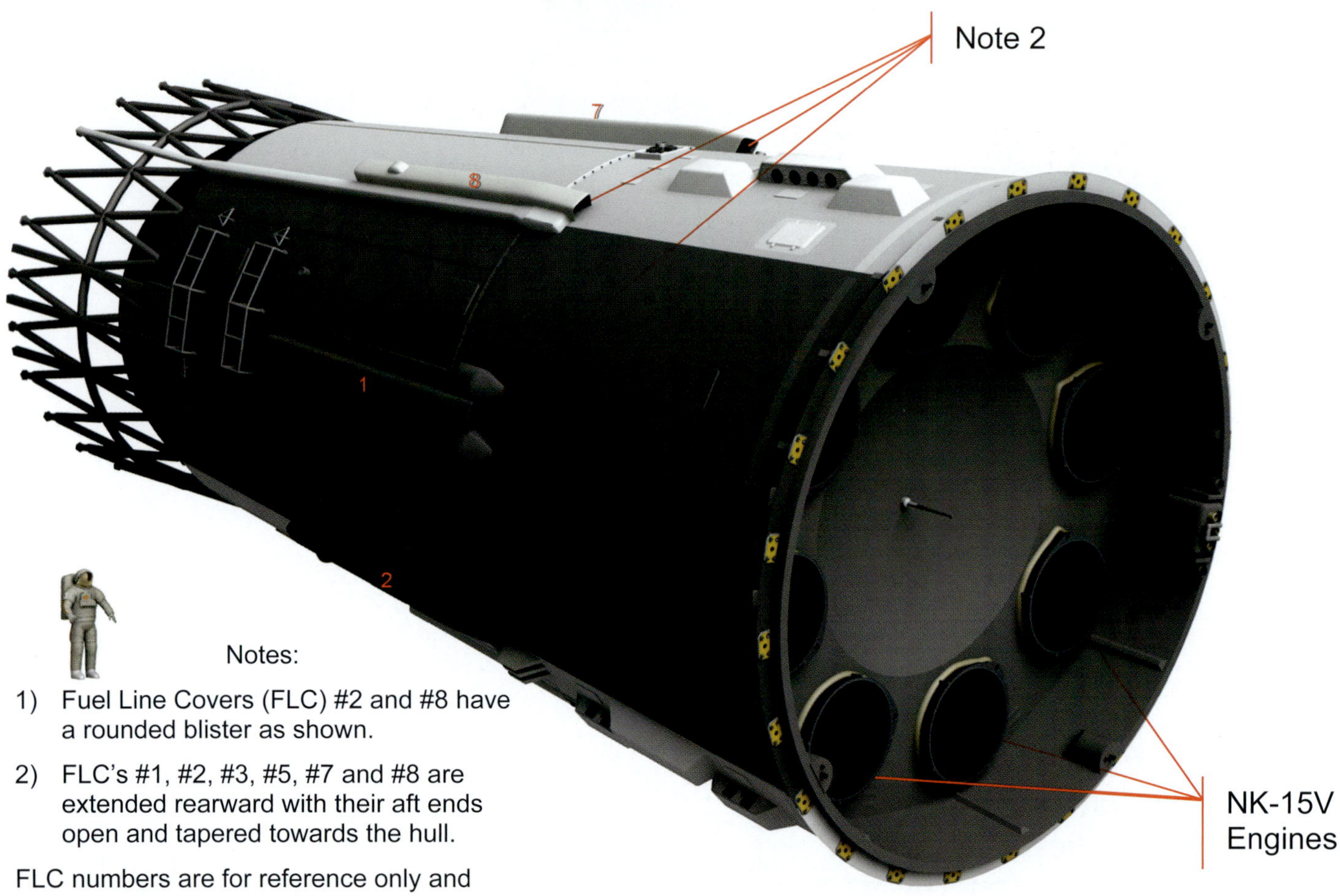

Notes:

1) Fuel Line Covers (FLC) #2 and #8 have a rounded blister as shown.
2) FLC's #1, #2, #3, #5, #7 and #8 are extended rearward with their aft ends open and tapered towards the hull.

FLC numbers are for reference only and are not official.

Vehicle N1-6L, Block-B Changes from N1-5L

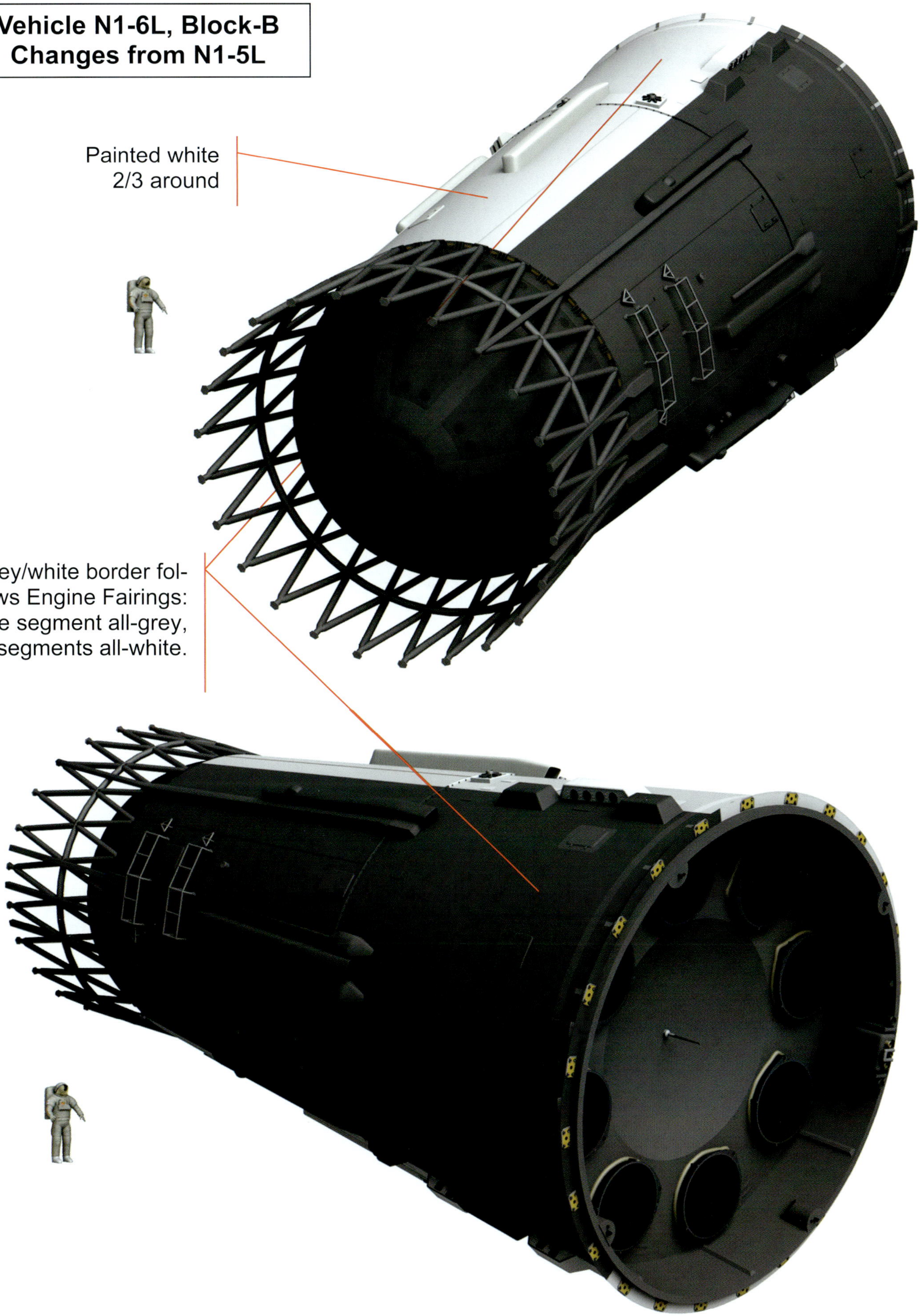

Vehicle N1-7L, Block-B Changes from N1-6L

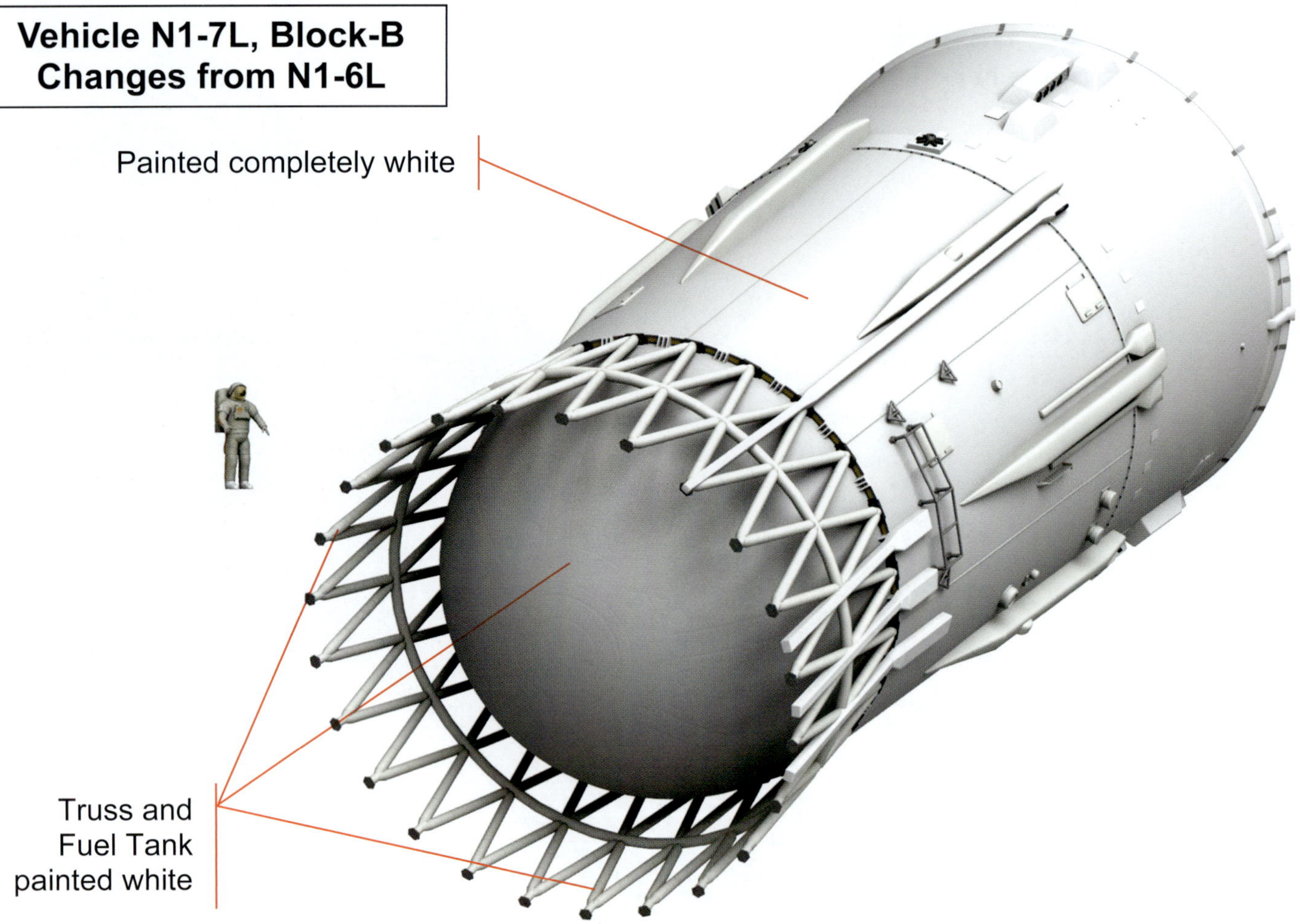

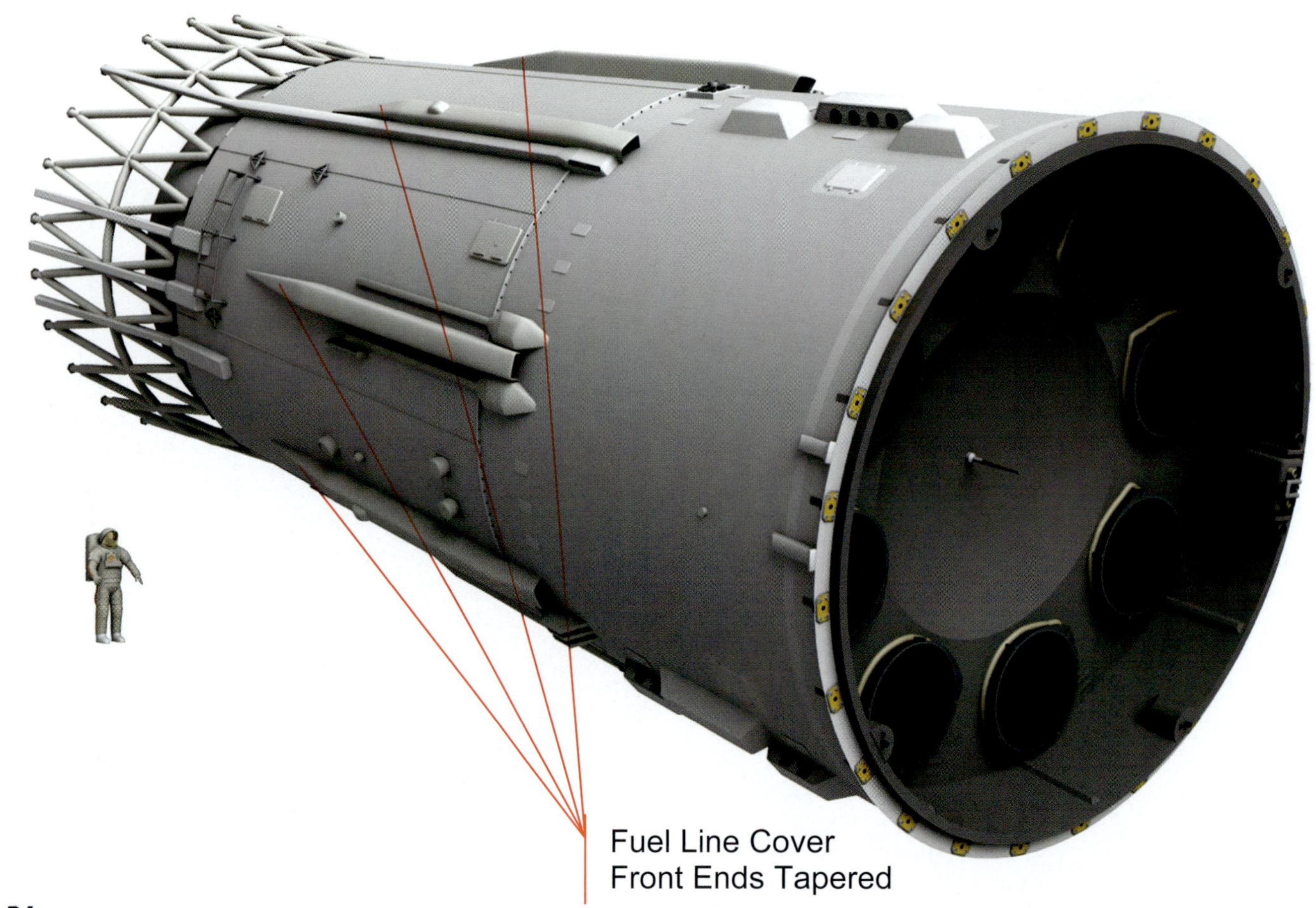

10: N-1 Block-V Reference Guide

10: Справочный раздел: блок В

CAD Drawings by ALEXANDER SHLIADINSKY
CGI Renderings by NICK STEVENS
Dimensions Applied by MATT JOHNSON

[Editor's note: This is the third stage of the N-1, so it is named with the third letter of the Russian alphabet, "ve," which translates phonetically to the English "V." Unfortunately, the Russian "ve" looks exactly like the English letter "B" which could lead to confusion. Rest assured that anytime you see "Block-B" in this book it's referring to the second stage and the only "ve" that you will see is in a Russian sentence, such as in the title above — JH]

Block-V was the top of the N-1 proper. Anything above it was technically part of the payload in much the same way that anything above the SIV-B stage of the Saturn V was payload. It used four of the vacuum rated NK-21 engines which were eventually to be replaced with the improved NK-39. N1-8L was built this way, but the program was cancelled before it flew.

Block-V Engine Configurations

Flight No.	Engines	Vacuum Thrust[1](kN)	Propellant (tons)	Burn time (seconds)	Flow rate (tons/sec)	Total Impulse (megaN-s)
N1-3L	NK-21 (4)	1,767	175	330	0.530	583
N1-5L	NK-21 (4)	1,767	175	330	0.530	583
N1-6L	NK-21 (4)	1,767	175	330	0.530	583
N1-7L	NK-21 (4)	1,767	175	330	0.530	583
N1-8L[2]	NK-39 (4)	1,609	175	375	0.466	603

[1] All of the values in the table are for the stage as a whole. [2] Not flown.

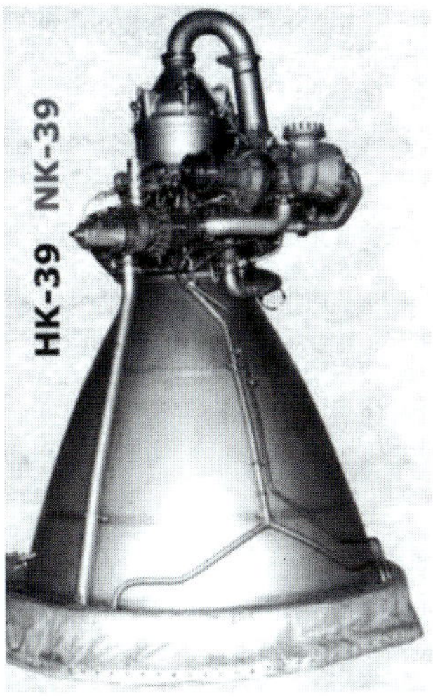

Above left, the NK-9V, progenitor of both the NK-21 used in Block-V, and the NK-19 used in Block-G (next chapter). Below left is the upgraded NK-39 which was scheduled to replace the NK-21 in this stage starting with N1-8L. Above, Block-V of N1-3L under construction.

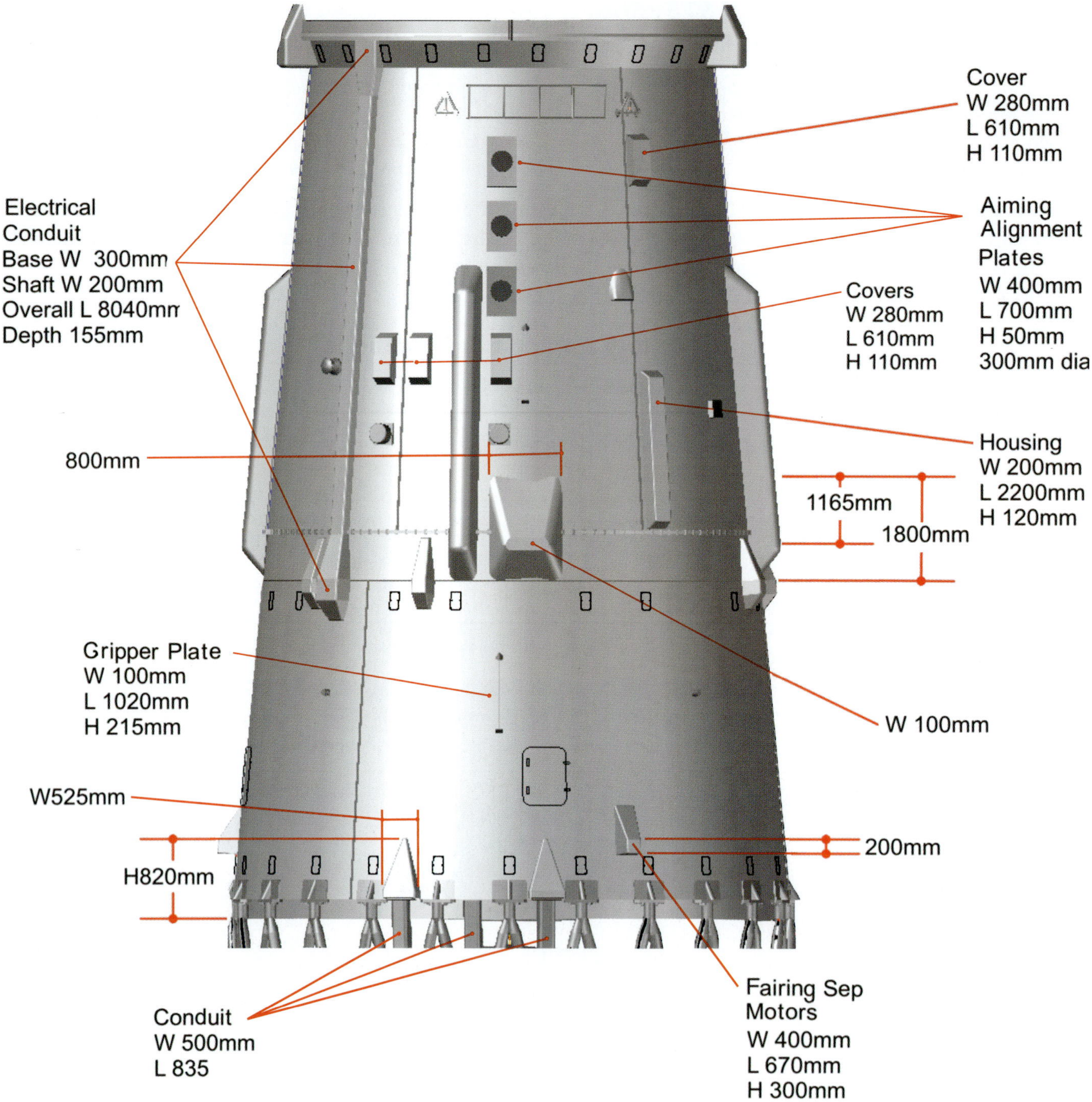

Vehicle N1-3L, Block-V
0° Rotation, Scale 80:1

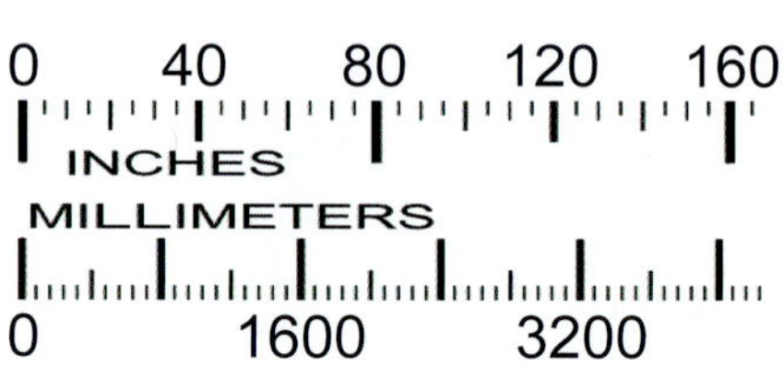

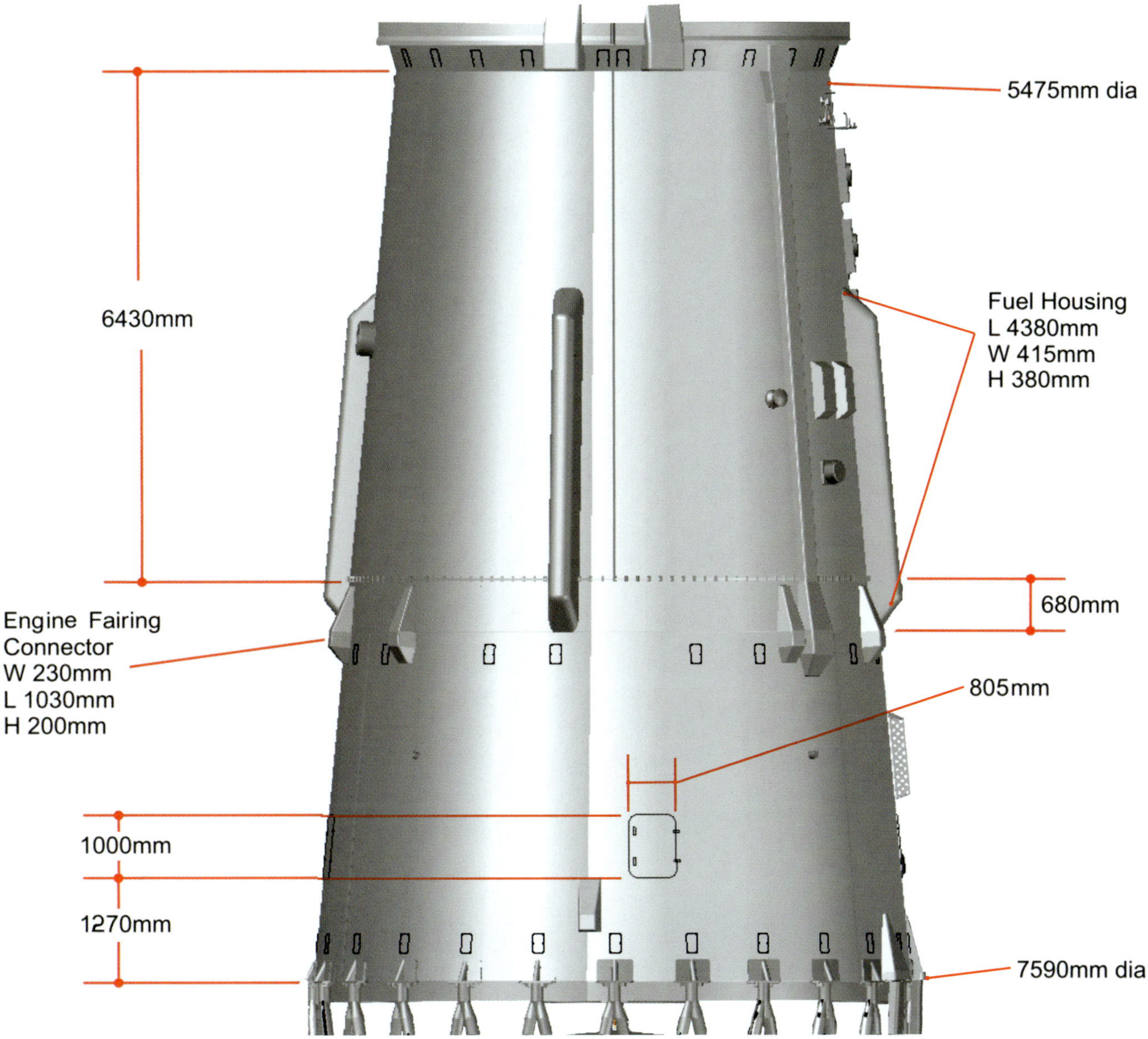

Vehicle N1-3L, Block-V
90° Rotation, Scale 80:1

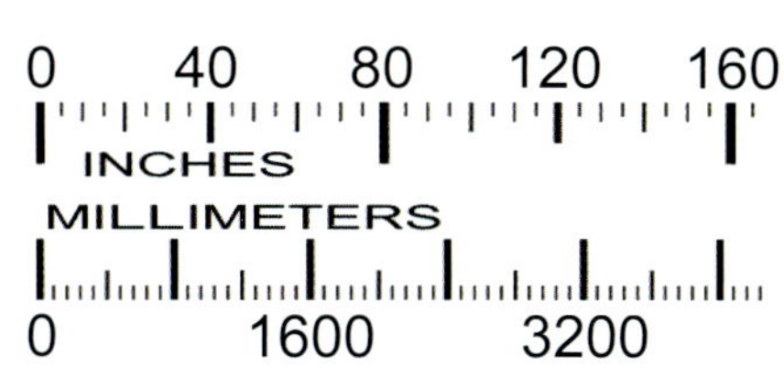

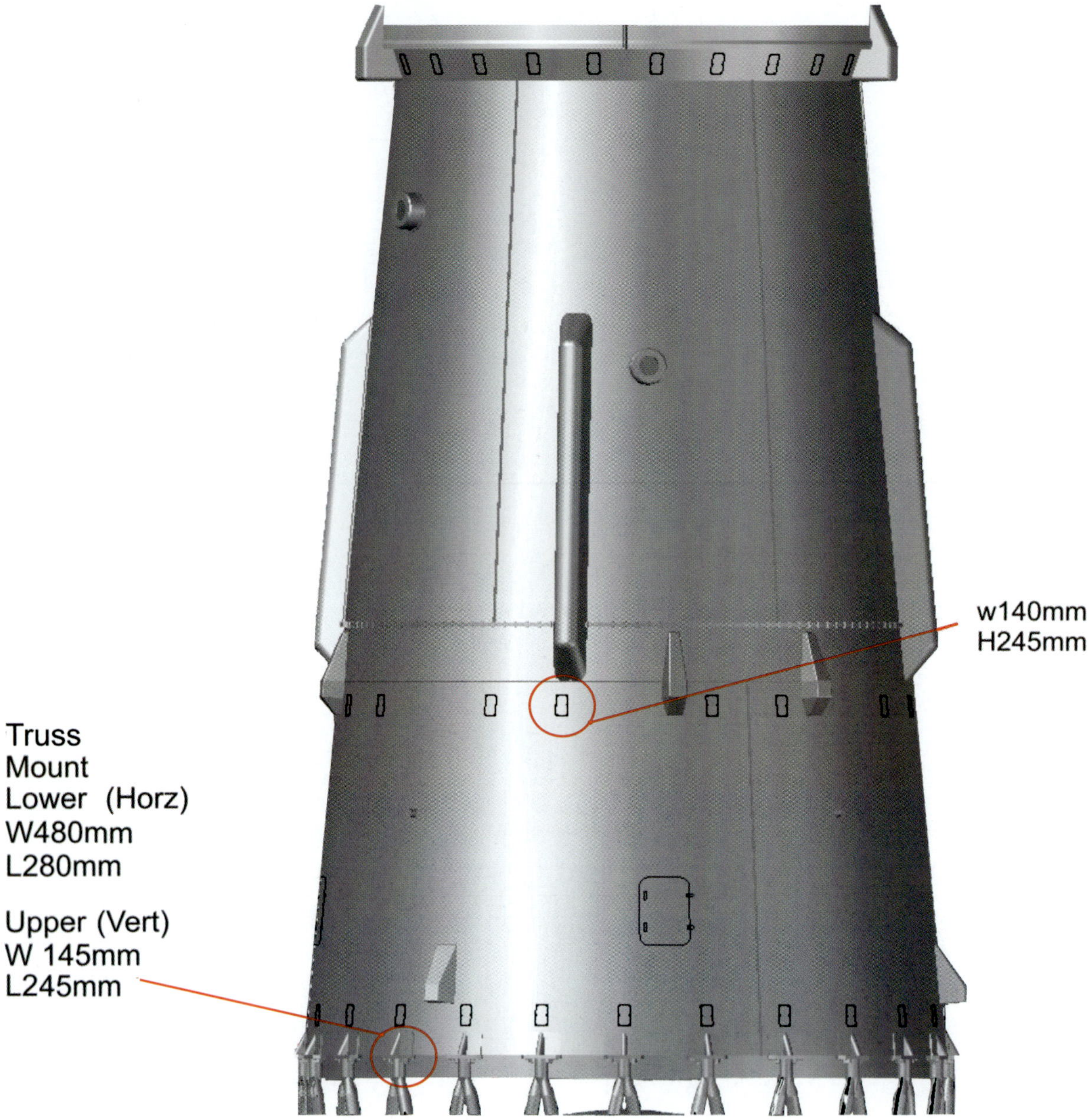

Vehicle N1-3L, Block-V
180° Rotation, Scale 80:1

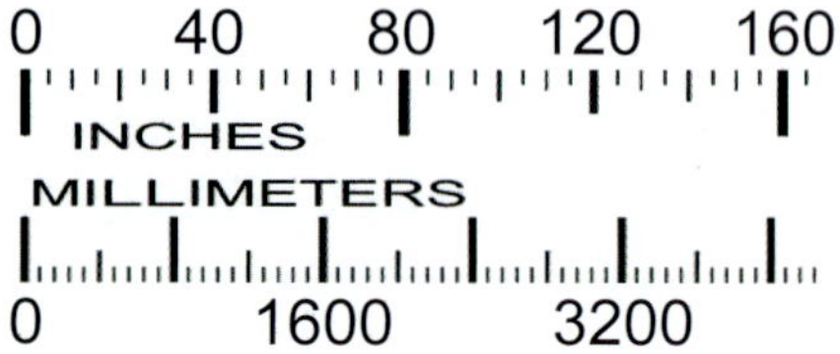

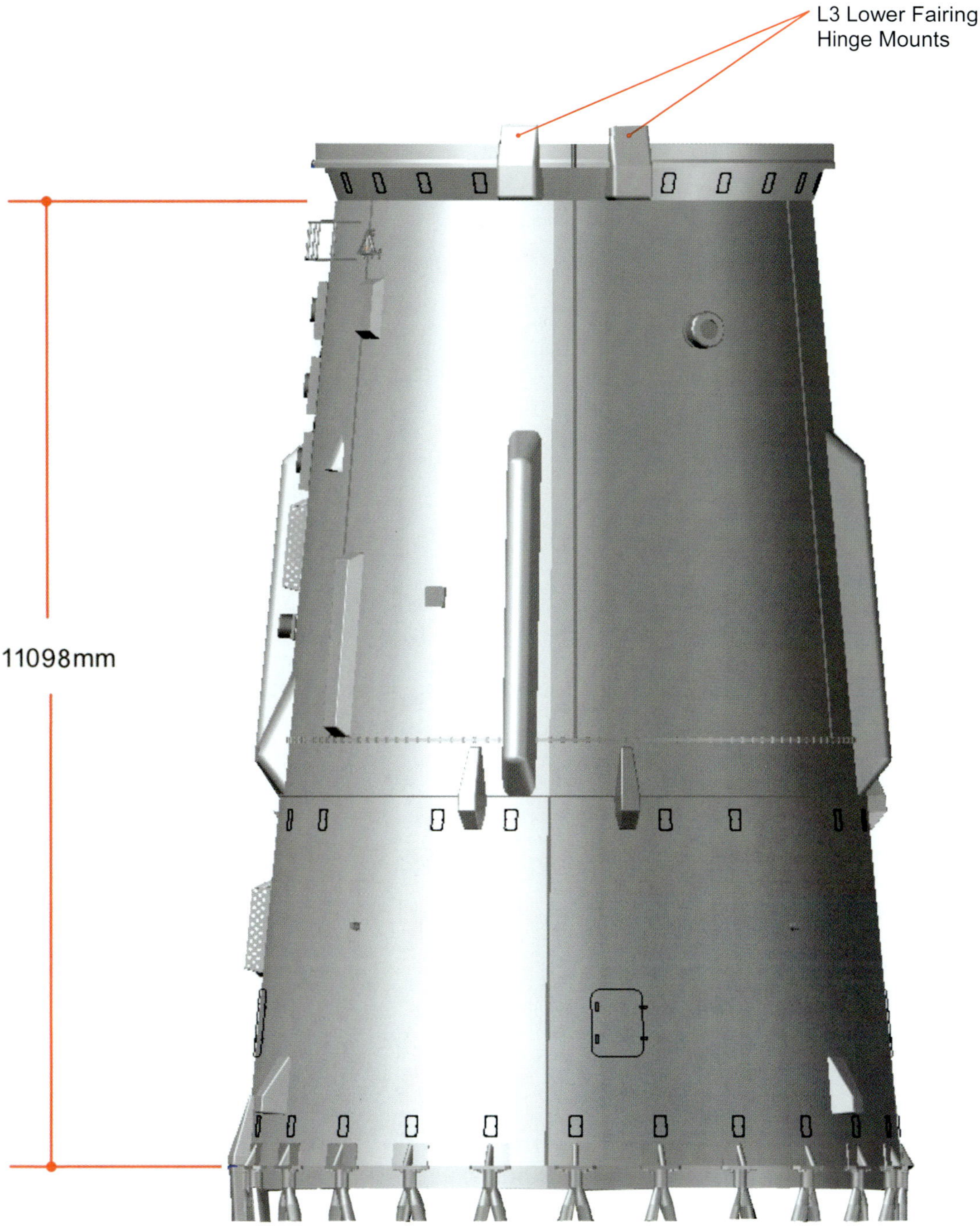

Vehicle N1-3L, Block-V
270° Rotation, Scale 80:1

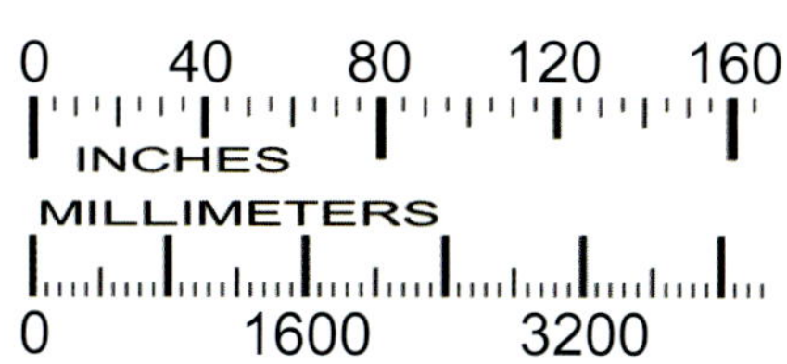

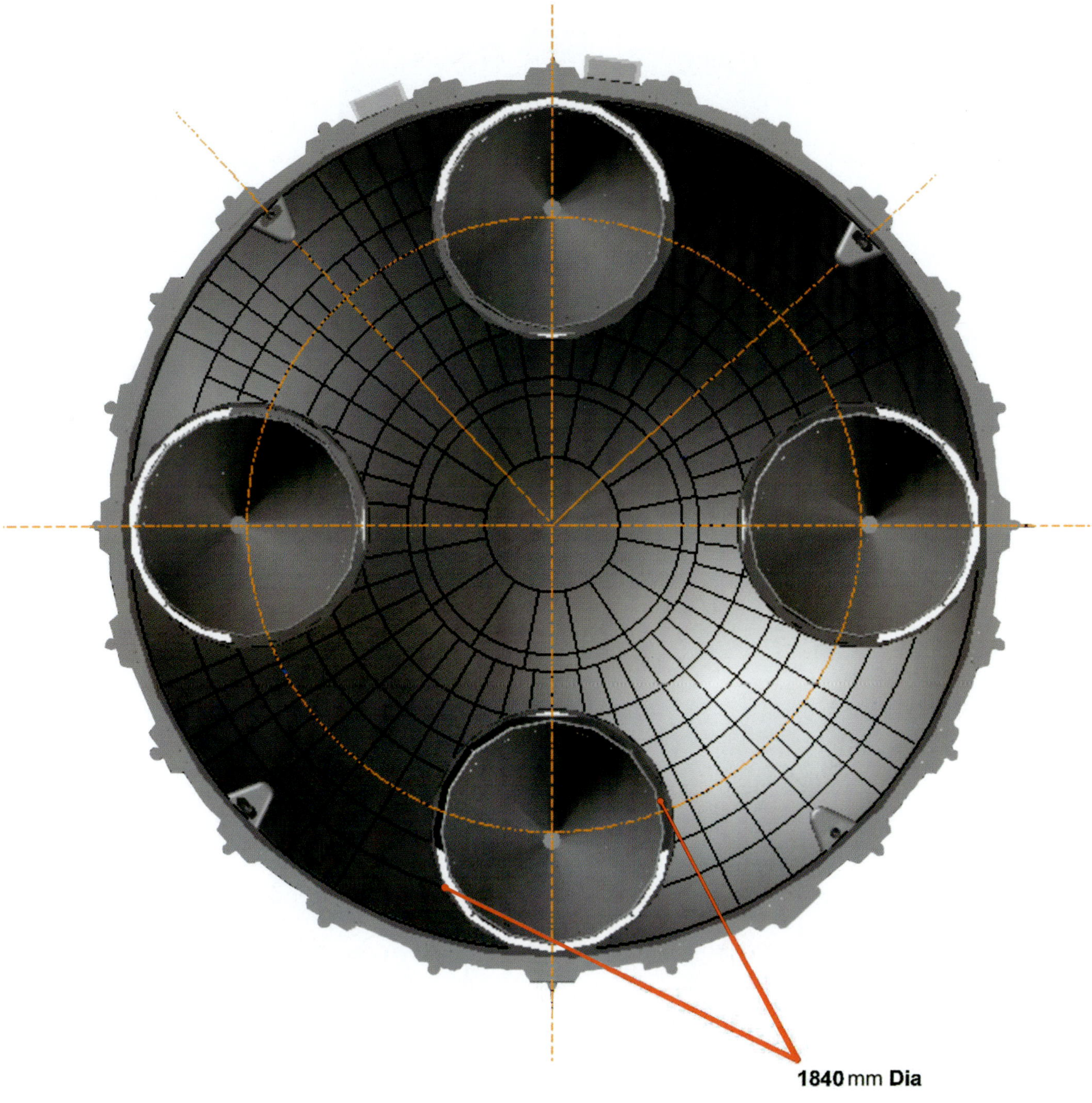

Vehicle N1-3L, Block-V
Bottom View, Scale 50:1

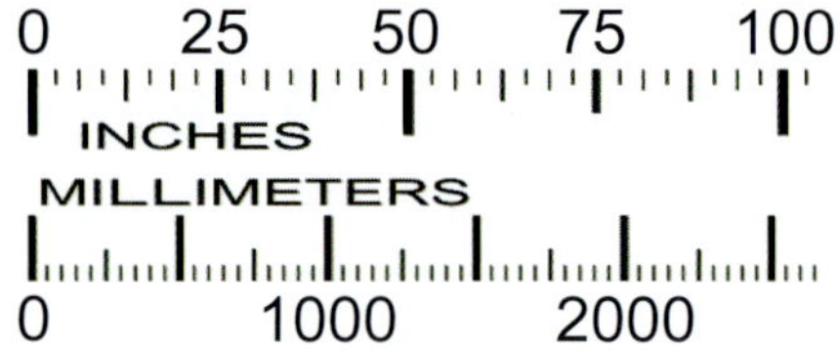

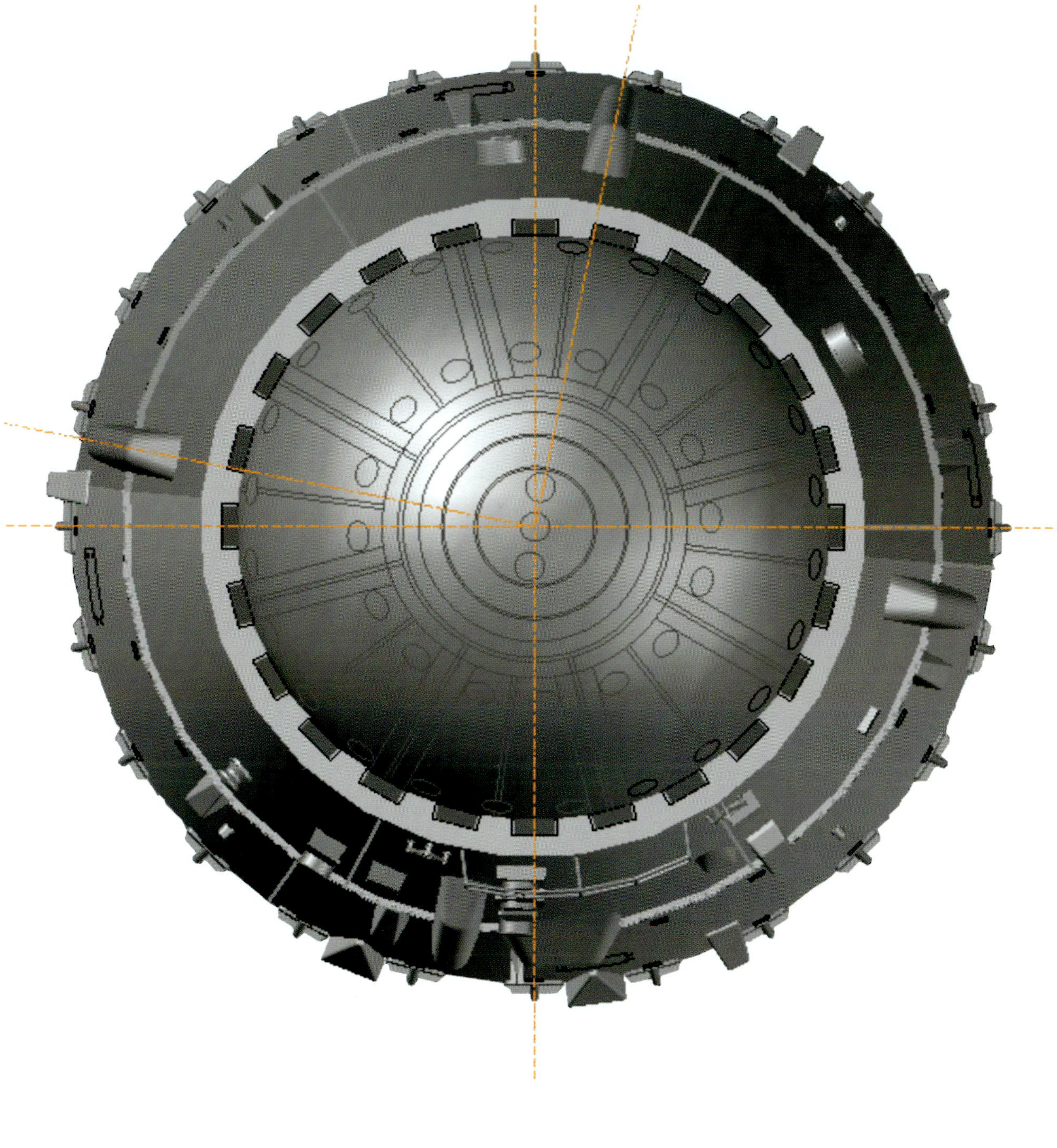

Vehicle N1-3L, Block-V
Top View, Scale 50:1

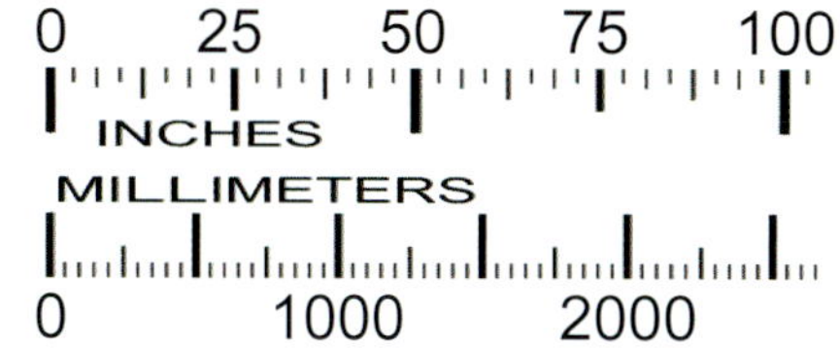

Block-V Engine Fairing Separation

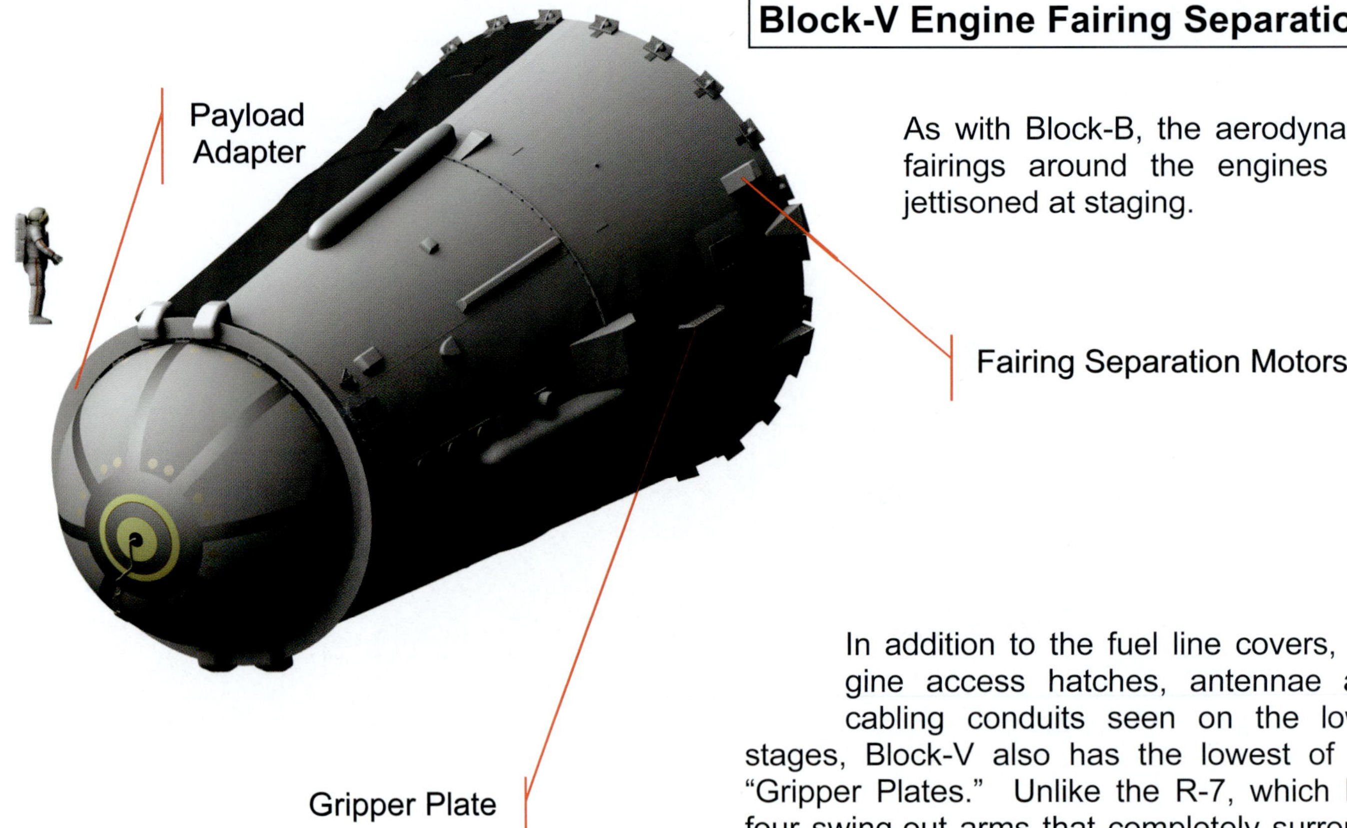

As with Block-B, the aerodynamic fairings around the engines are jettisoned at staging.

In addition to the fuel line covers, engine access hatches, antennae and cabling conduits seen on the lower stages, Block-V also has the lowest of the "Gripper Plates." Unlike the R-7, which has four swing-out arms that completely surround the rocket to stabilize it until the moment of launch, the N-1 used a single gantry tower, much like the Saturn V.

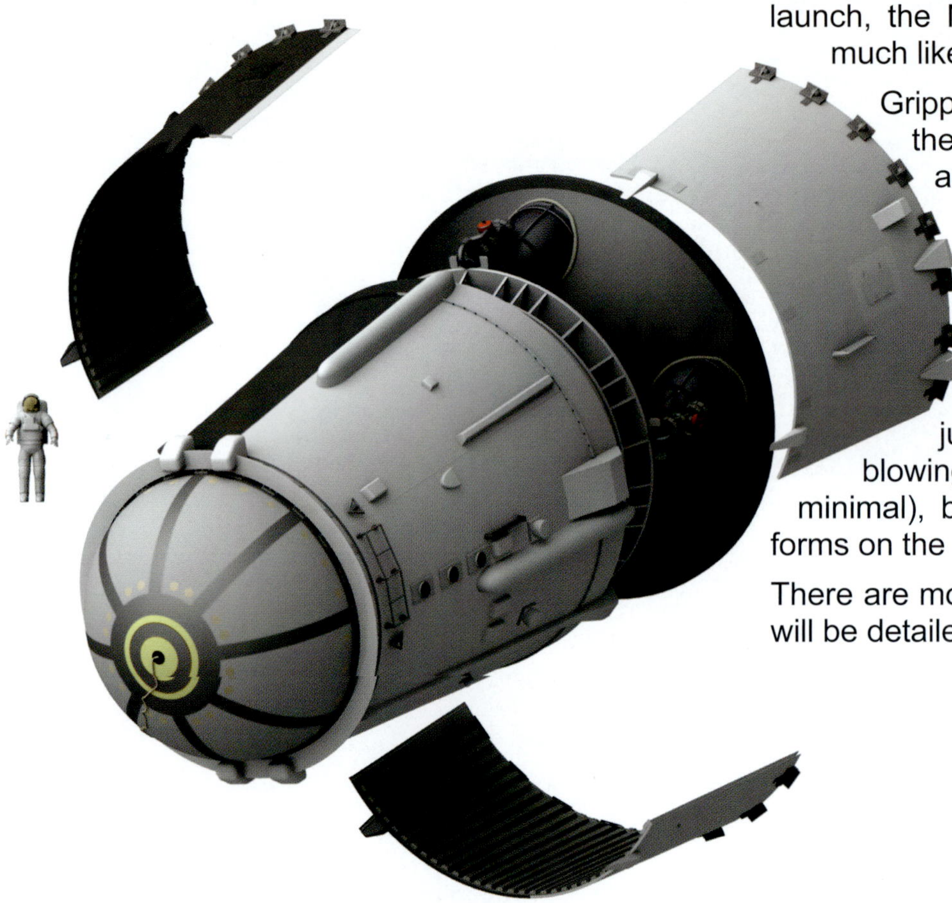

Gripper plates are special fixtures on the upper part of the vehicle where arms from the tower can hold on to the rocket to stabilize it until launch. In the high winds at Baikonur, the N-1 stack could sway back and forth up to 1.5 meters (5 feet) in the upper stages. Their purpose is not just reducing the risk of the rocket blowing over (which is actually quite minimal), but more to keep the work platforms on the tower lined up and stable.

There are more of these plates further up that will be detailed in later chapters.

Vehicle N1-3L Block-V,

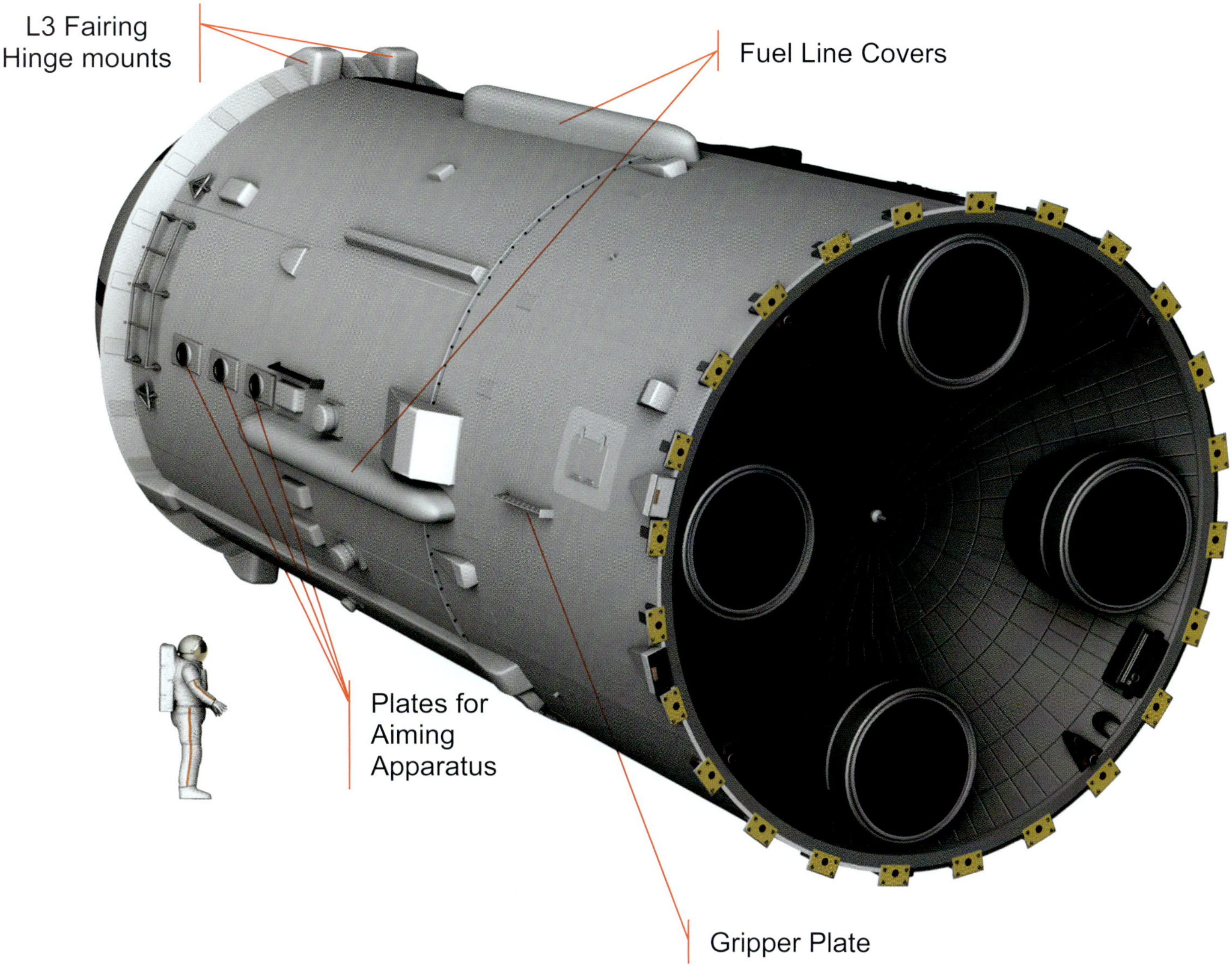

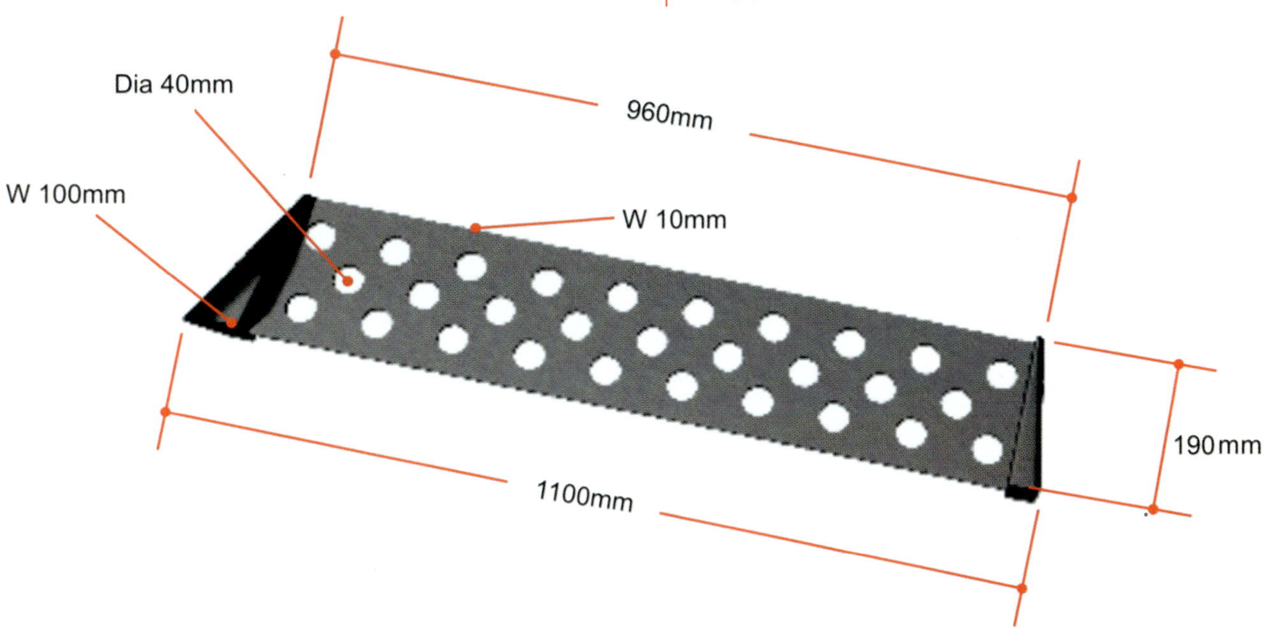

Gripper Plate Detail

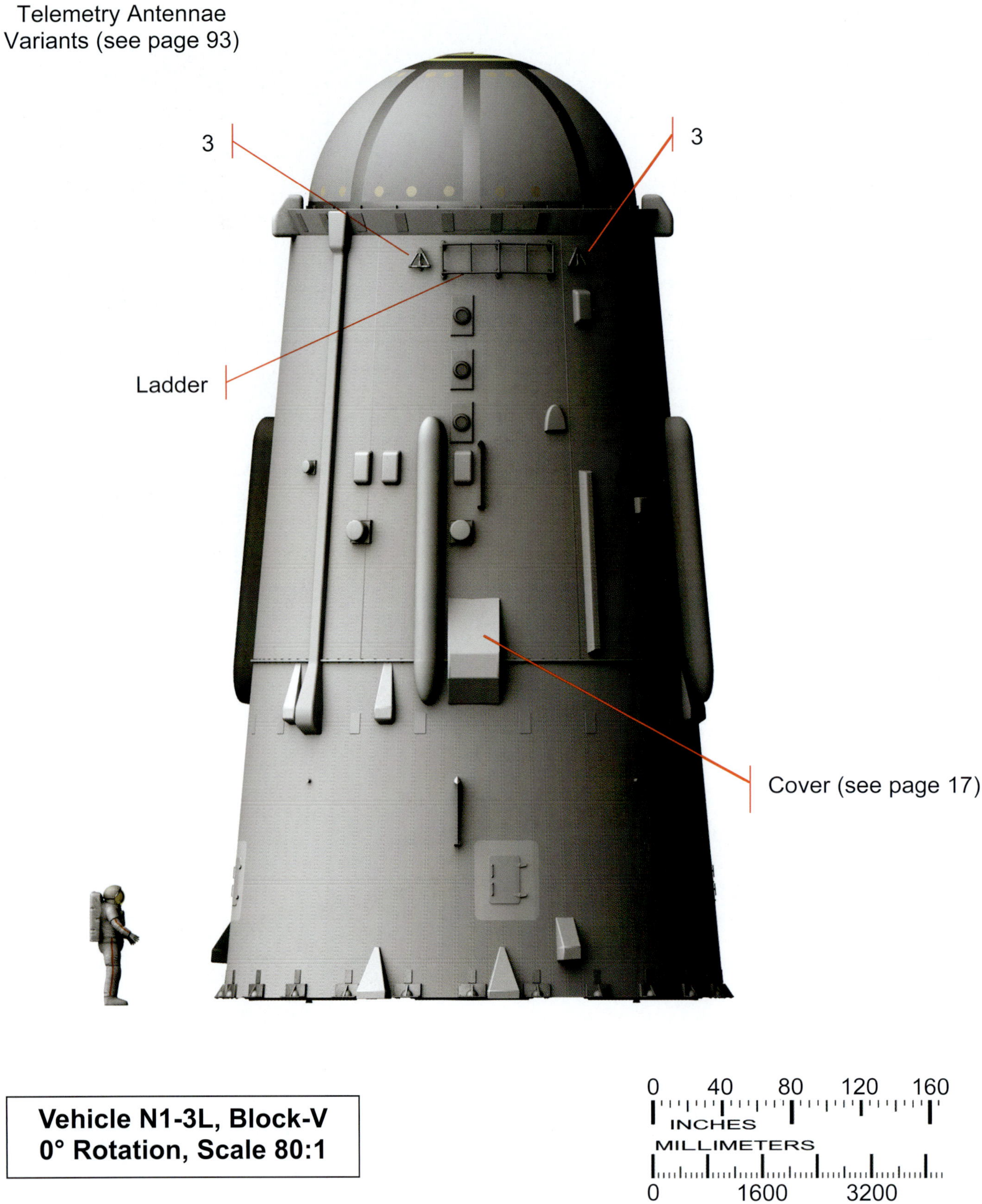

Vehicle N1-3L, Block-V
0° Rotation, Scale 80:1

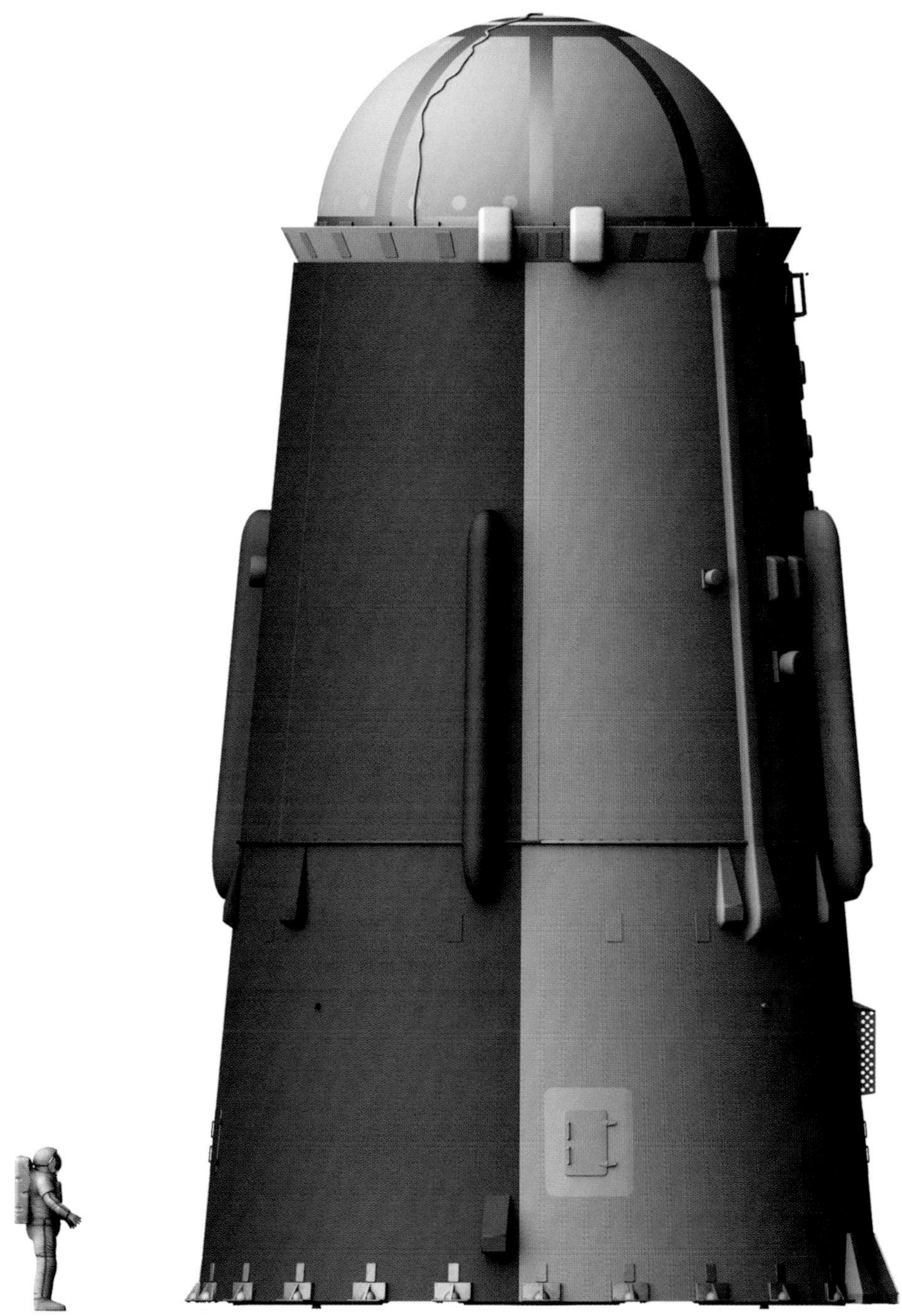

Vehicle N1-3L, Block-V
90° Rotation, Scale 80:1

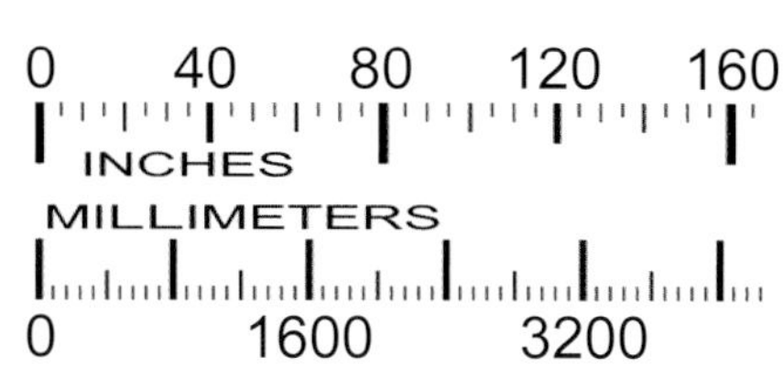

Vehicle N1-3L, Block-V
180° Rotation, Scale 80:1

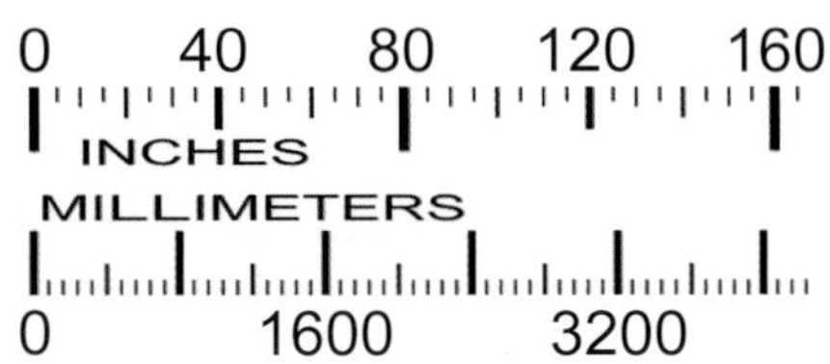

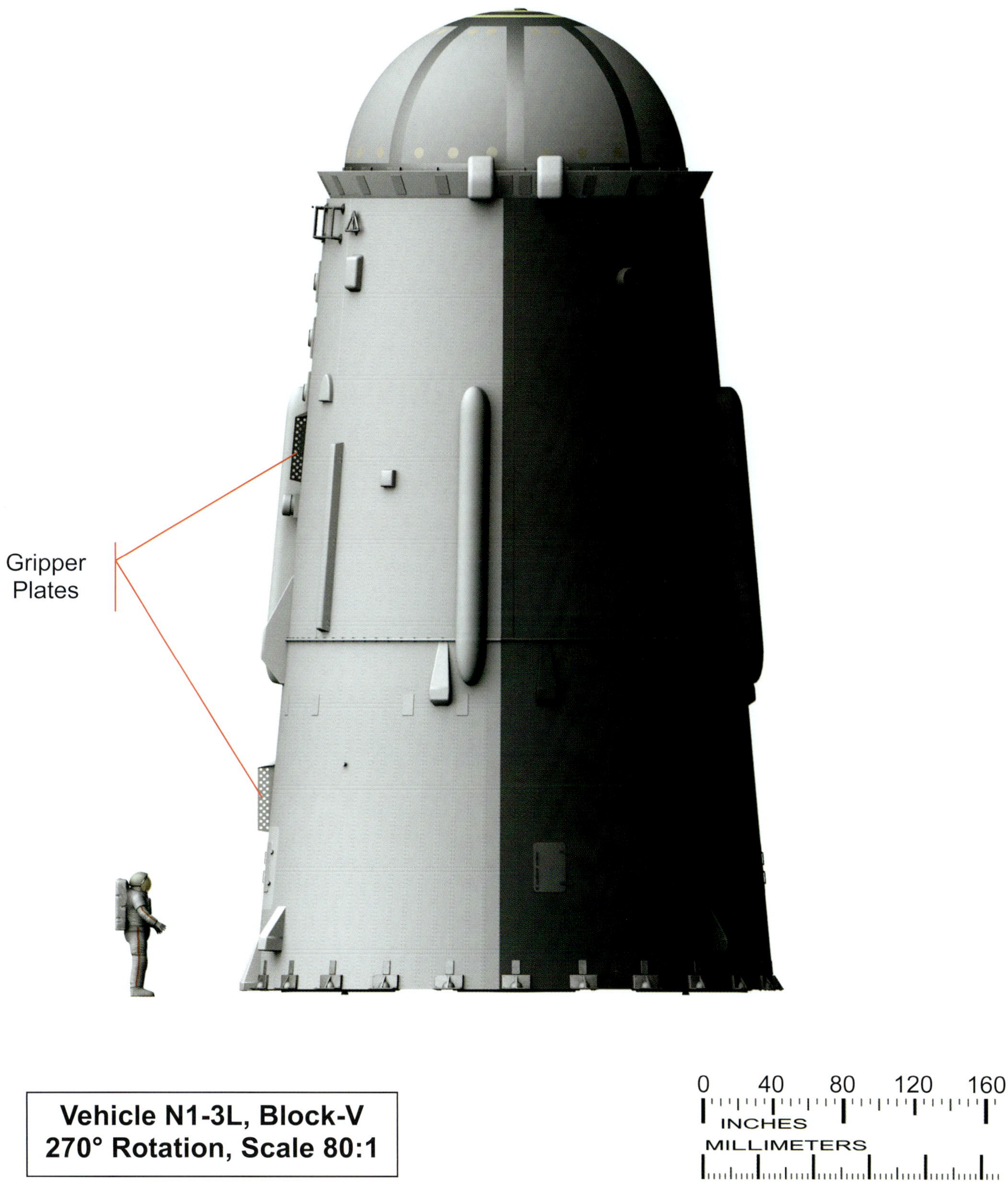

Vehicle N1-3L, Block-V
270° Rotation, Scale 80:1

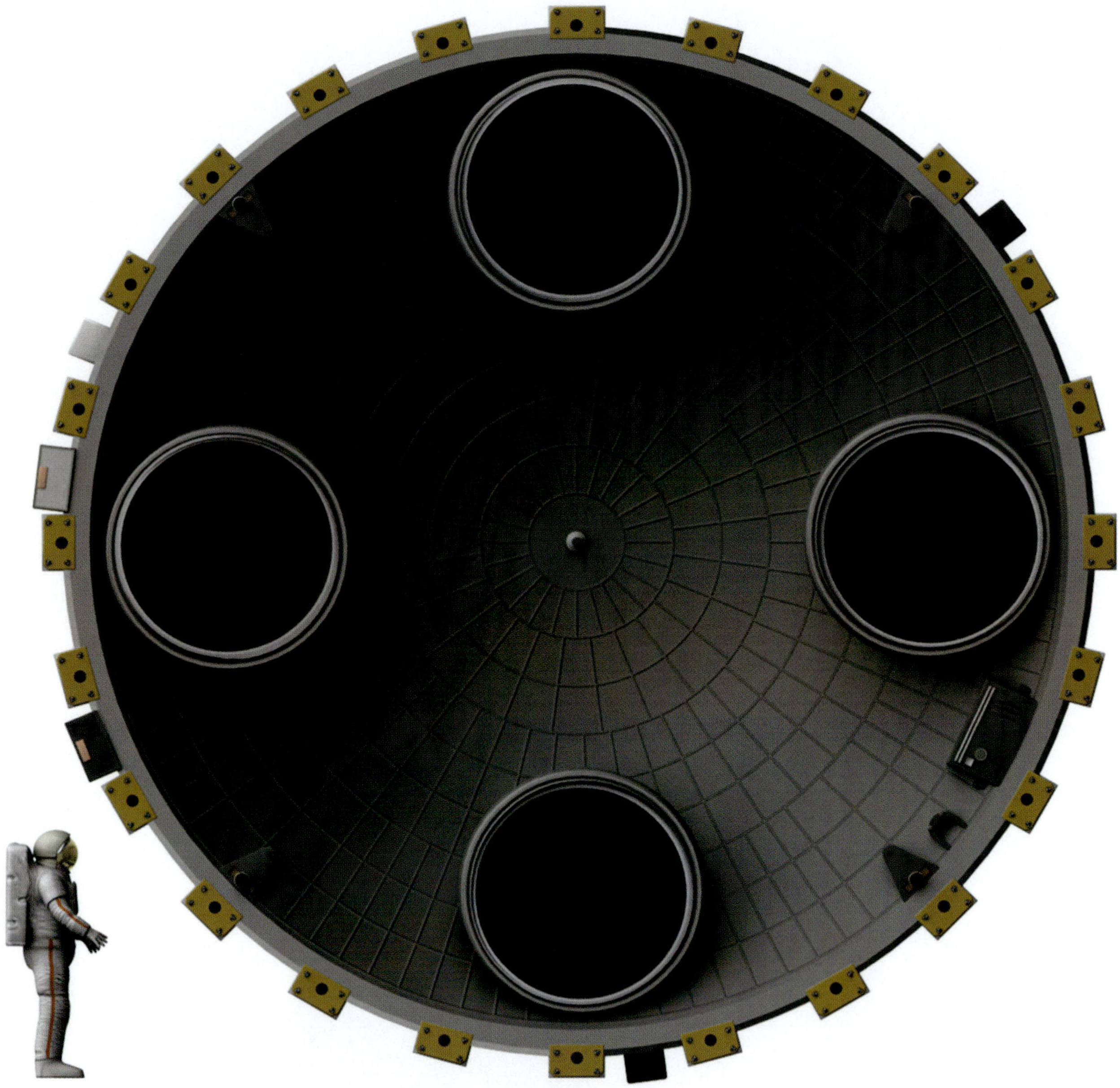

Vehicle N1-3L, Block-V
Bottom View, Scale 50:1

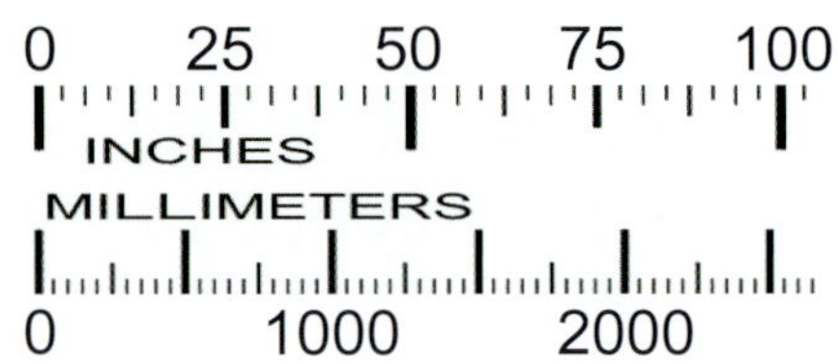

Vehicle N1-3L, Block-V
Top View, Scale 50:1

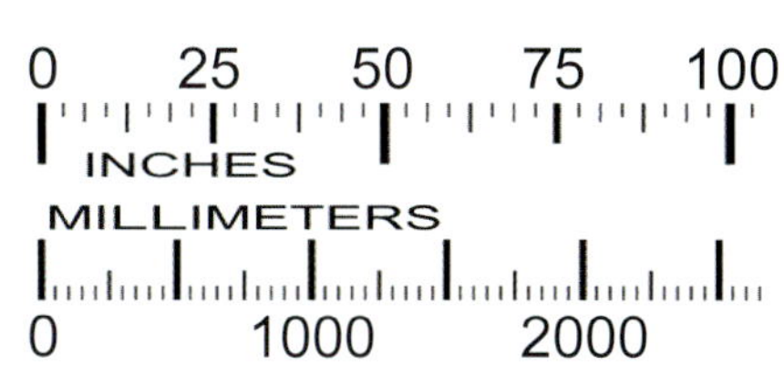

Vehicle N1-5L, Block-V Changes from N1-3L

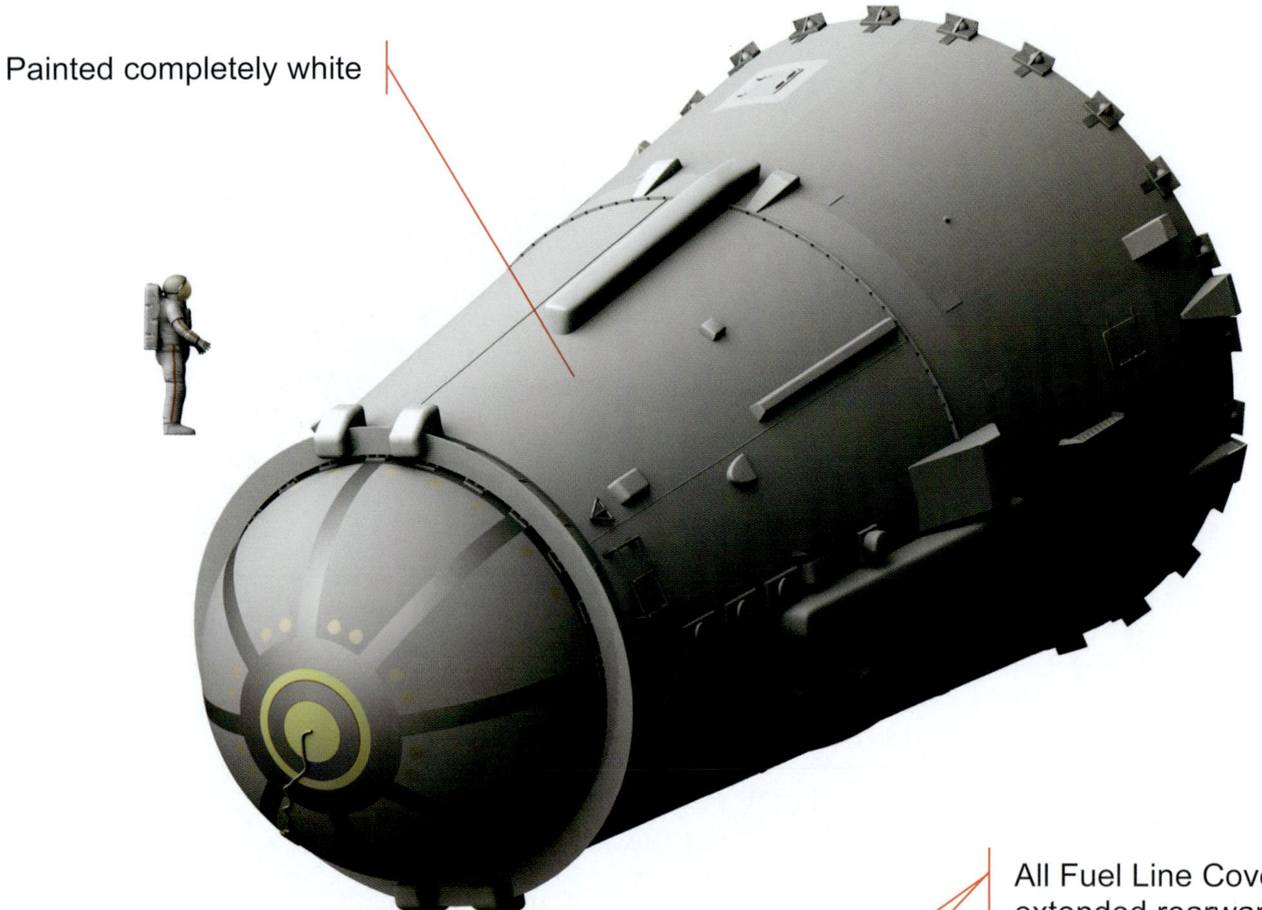

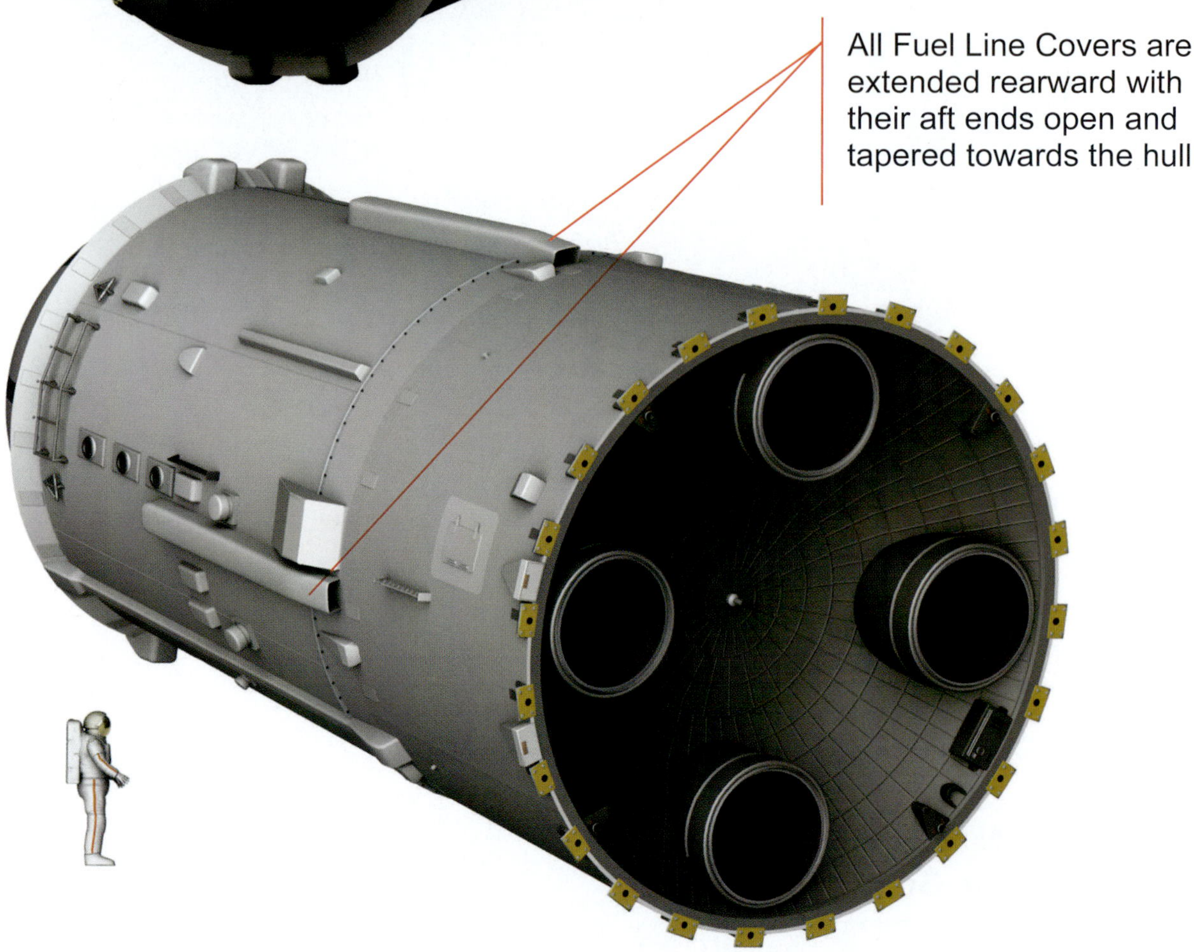

Vehicle N1-6L, Block-V Changes from N1-5L

No visible external differences between N1-5L and N1-6L

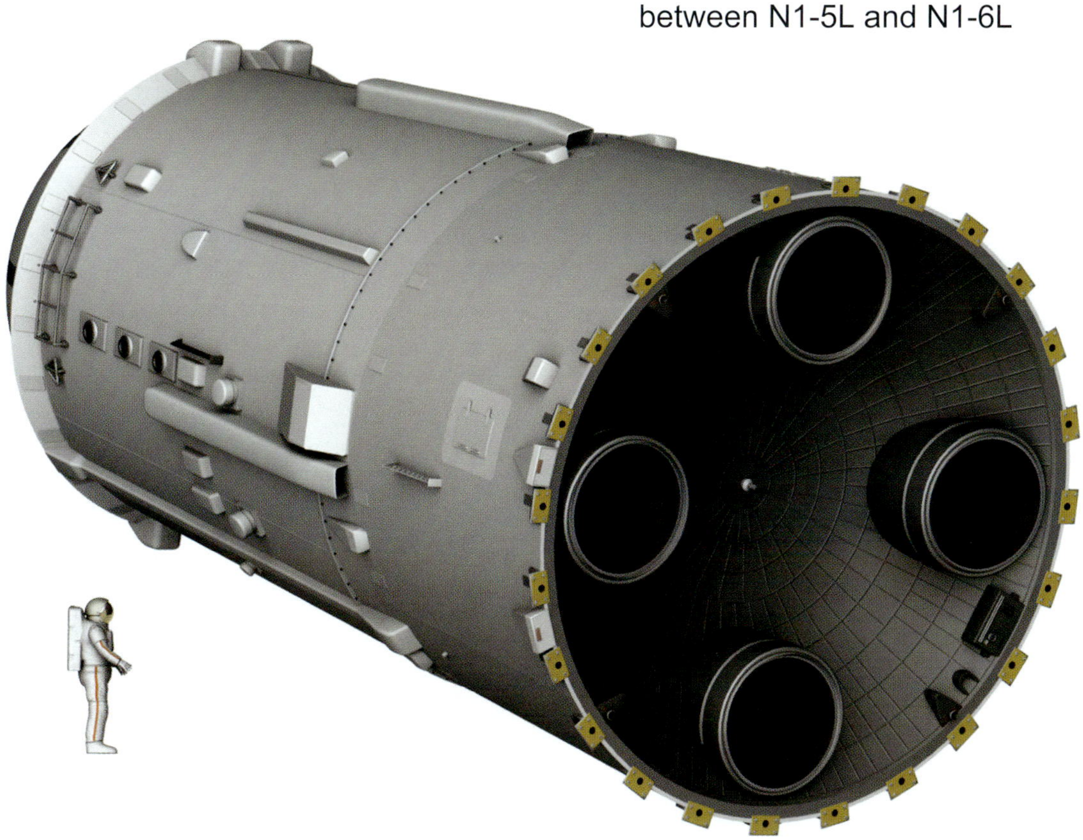

Vehicle N1-7L, Block-V Changes from N1-6L

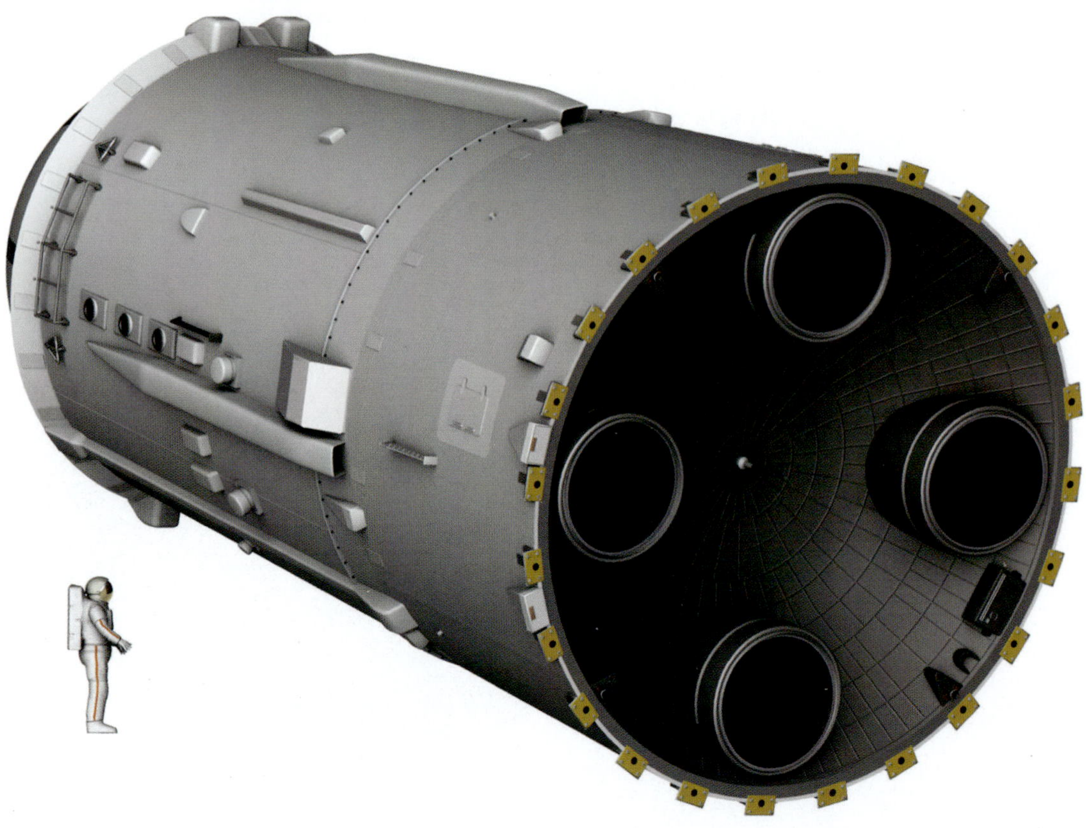

11: N1-L3 Complex Reference Guide

11: Справочный раздел: комплекс Н1-Л3

CGI Renderings and Line Drawings by NICK STEVENS
Dimensions and Notes by the Editor

[Editor's note: More confusing terminology. The "L3 Complex" is the moon ship. As was explained in Chapter 2, Soviet mission planners studied many different expeditions to the moon. The one chosen for the first manned landing mission was "Lunar Plan 3" or L3. "Complex," in the Russian space program, means the assemblage of vehicles to perform the mission (in the US this is called an "assembly" or a "stack"). Unfortunately for us historians, the first booster scheduled to fly was the third one built, 3L, leading to possible confusion between "N1-3L" and the "L3 Complex." Further mudding the waters, the Russian letter З ("ze") looks almost exactly like the number 3 used in both languages. You occasionally find references to the "LZ Complex" especially when the text was machine translated — JH]

The L3 Complex consists of four rocket stages, two of them manned, the connecting trusses and skirts and the large aerodynamic fairings that covers them all, topped by a large Launch Escape System (LES). This is too much to cover in a single chapter so we have arbitrarily split the descriptions into spacecraft components (this chapter) and the aero-fairings and LES (Chapter 12).

Of the four powered stages, the two manned craft (the LK lander and LOK orbiter) have already been covered in Chapter 3. Their only reference here will be where they fit into the L3 using their Complex designations Block-E (LK) and Block-I (LOK).

The other two stages, Block-G and Block-D, provide the muscle to get the Complex to the moon. Block-G does the Trans-Lunar Injection burn that gets the Complex out of Earth orbit while Block-D does any mid-course corrections, plus it performs both the braking burns to put the ship into lunar orbit and the initial descent burn that starts the LK on its way to the surface.

L3 Complex Engine Configurations

Stage and Flight No.	Engine	Vacuum Thrust (kN)	Propellant (tons)	Burn time (seconds)	Flow rate (tons/sec)	Total Impulse (megaN-s)
Block-G N1-3L, 5L,6L, 7L	NK-19	393	56	470	0.118	186
Block-G N1-8L[1]	NK-31	402	56	480	0.116	193
Block-D N1-6L,[2] 7L, 8L[1]	RD-58	82	15	600	0.024	49

[1] Not flown [2] N1-3L and N1-5L did not have a Block-D

Left to right: the NK-19 used in Block-G and the upgraded NK-31 which was scheduled to replace it starting with N1-8L; the RD-58 used in all variations of Block-D; the L3 Complex under construction.

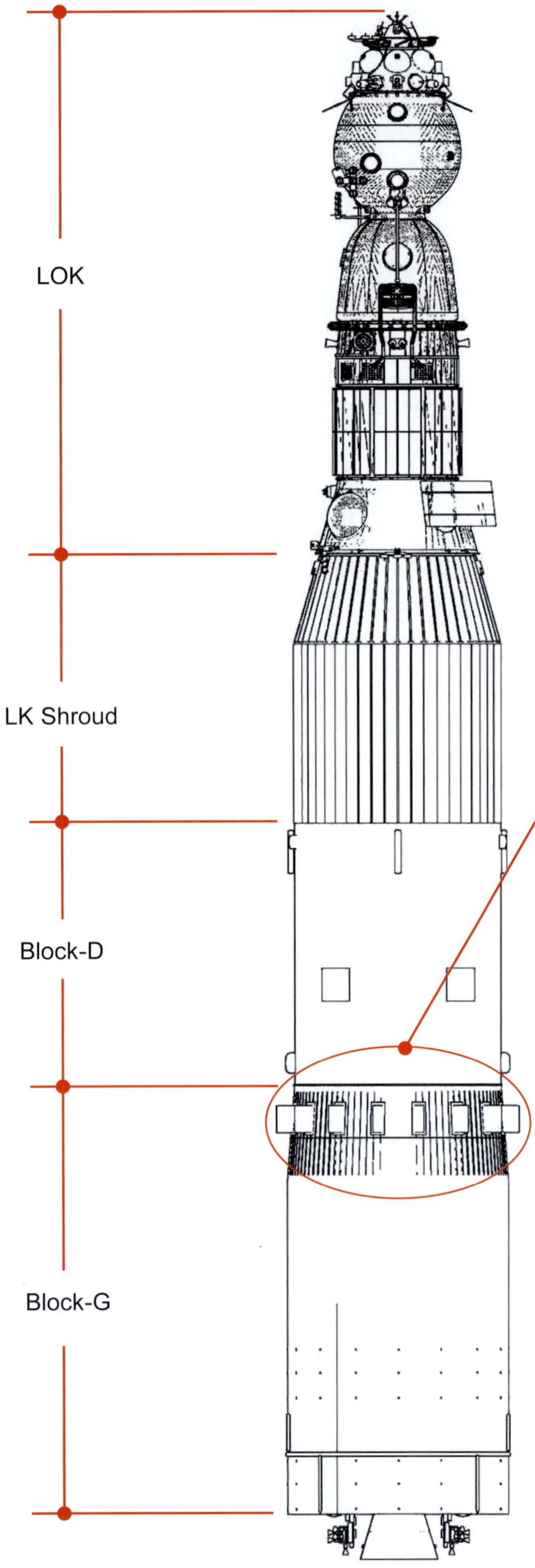

What Are Those Things?

The most prominent feature of Block-G is the ring of 16 large rectangular plates that surround the front end of the stage like a spiked collar. They are so distinctive that they must have an important purpose, but what might it be? Currently it's a mystery and as of this writing no one knows for sure. It has been the topic of spirited debate on websites, mostly in Russia and Eastern Europe, but no explanation has been found in any sort of definitive documentation. However, that hasn't stopped some rampant speculation, which we'll get to in a moment.

Physically, they are rectangular when seen from the side, some 500mm (20") long and protruding some 200mm (8") beyond the maximum diameter of Block-G. Even though they emerge from the tapering transition section between Block-G and Block-D, their outer edge is parallel to the main axis of the rocket. Seen from the front, they have a triangular cross-section, being about 200mm wide where they attach to the transition section, tapering to about 50mm (2") near the tip where they are capped off with a radius rather than going all the way to a point.

The first speculation as to their purpose is usually antennae of some sort. After all, the three stages of the N-1 proper were festooned with telemetry antennae. The problem with this is that they are a very odd shape and arrangement for transmitting information.

L3 Complex
0° Rotation, Scale 125:1

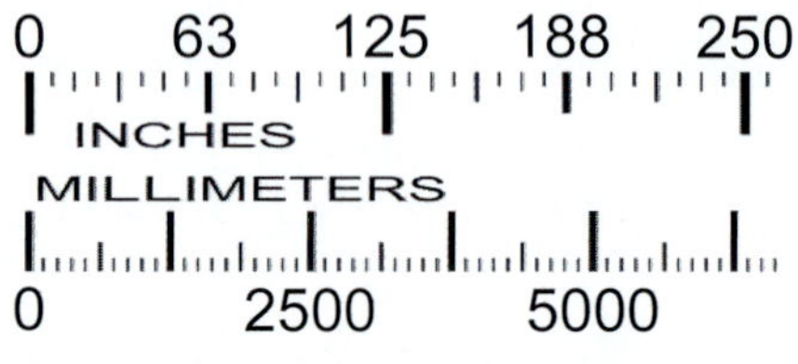

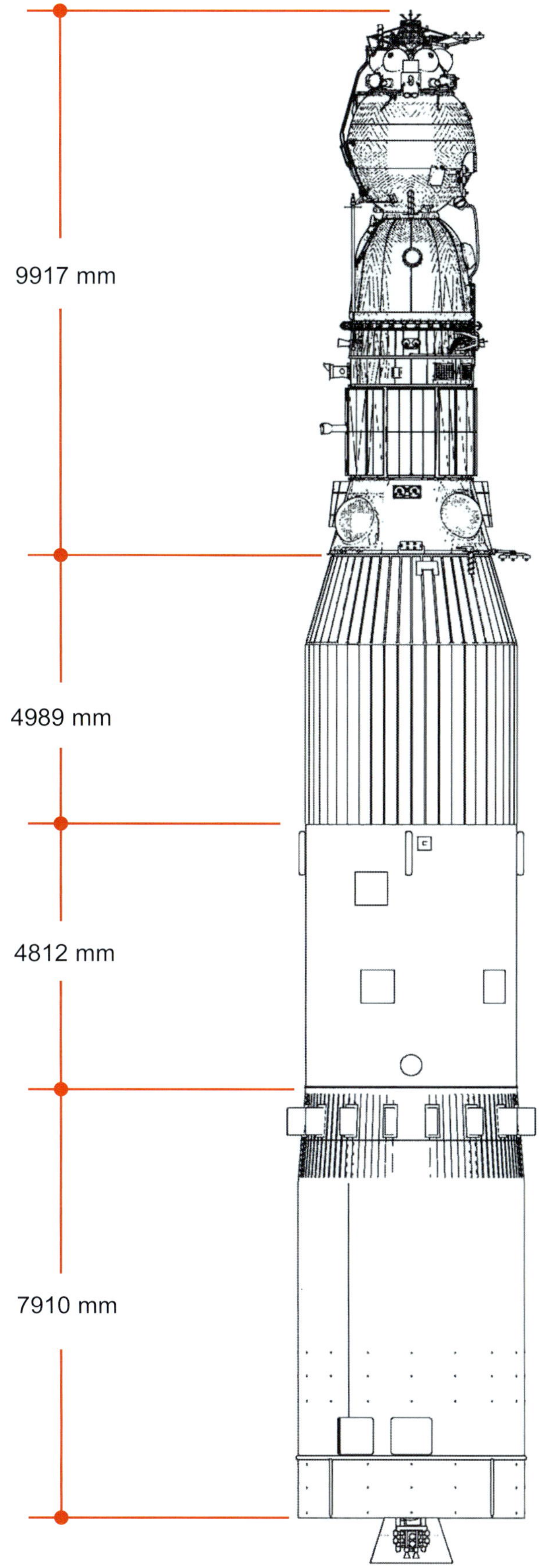

Another guess is that they are radiators to remove heat from components in the stage. It's hard to imagine, though, just what components need cooling this way. Directly in from the plates is actually the nozzle of the Block-D engine. Down a little further is the dome of the liquid oxygen tank. While that would obviously benefit from extra cooling, this isn't the stage that needs it. The Block-G stage is used and jettisoned within a few hours of liftoff. It's Block-D that has to keep its LOX cold for nearly five days. Also, if they were radiators, they would be painted black to maximize their "black-body coefficient."

The most mundane explanation is that they are spacers used to locate the outer fairing during liftoff. Again, the dimensions don't support this as the distance from the outside surface of Block-G to the inner surface of the lower fairing is about 500 mm (20"), more than twice as far as the plates protrude. Besides, if they really were for locating the fairing, you'd think they would be attached to the fairing itself so that their weight would be jettisoned along with the fairing when it is no longer needed.

Hopefully, someone reading this will have the documented answer as to what these plates are for and can resolve the mystery! - Ed.

L3 Complex
90° Rotation, Scale 125:1

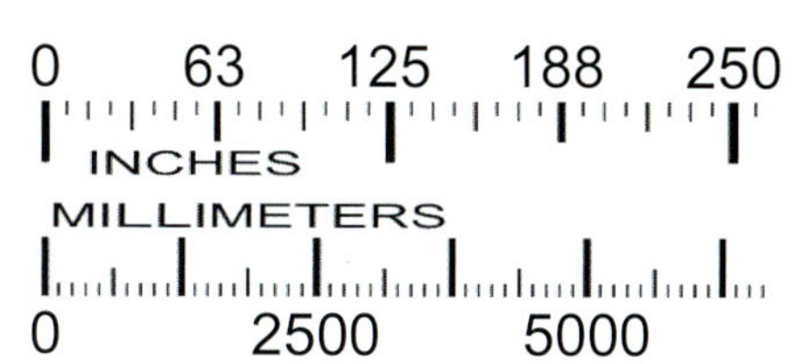

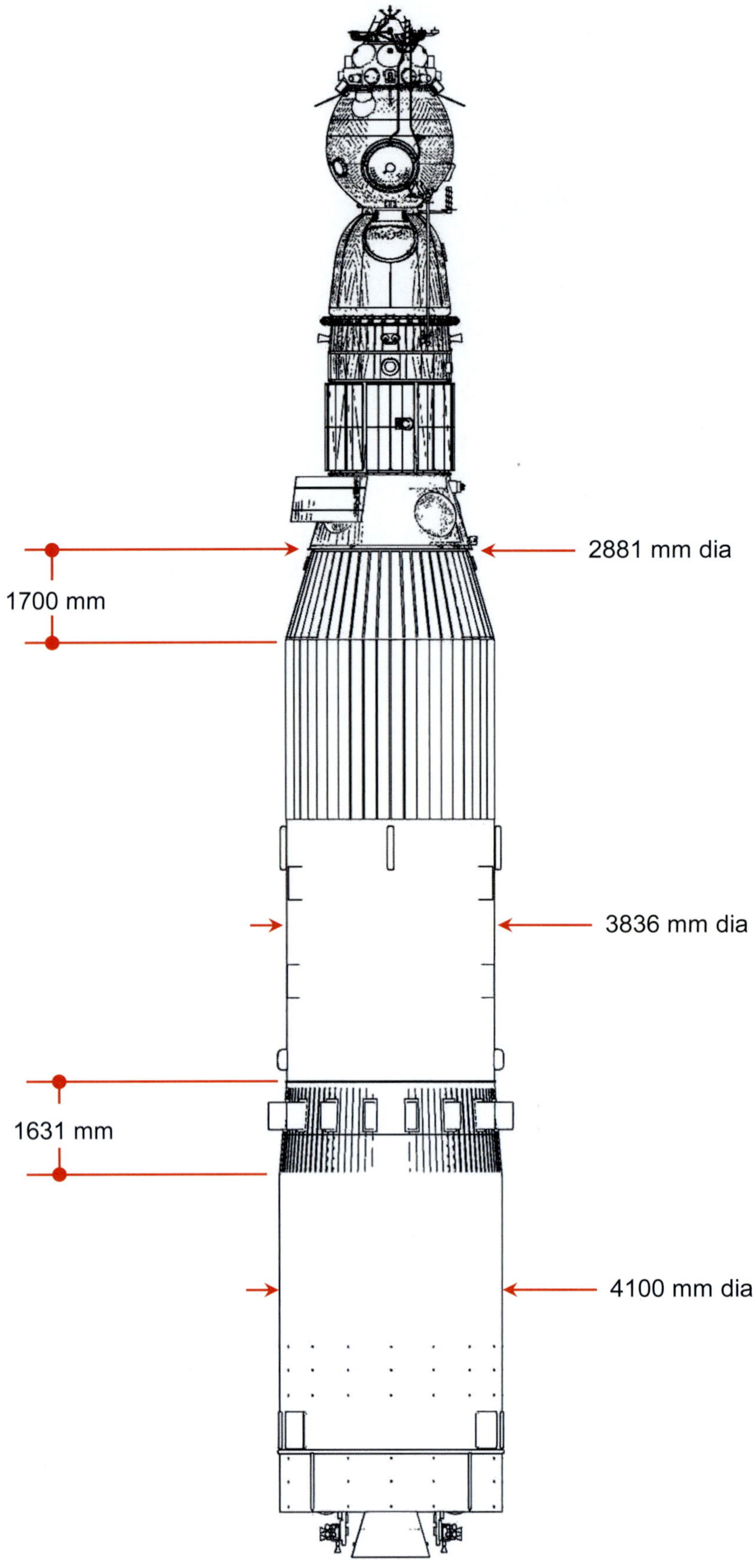

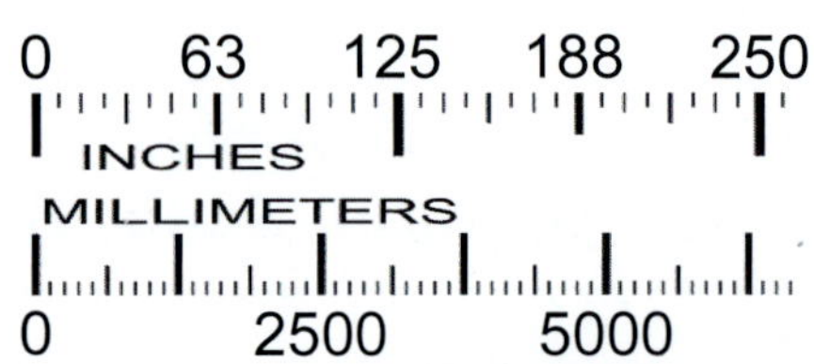

L3 Complex
180° Rotation, Scale 125:1

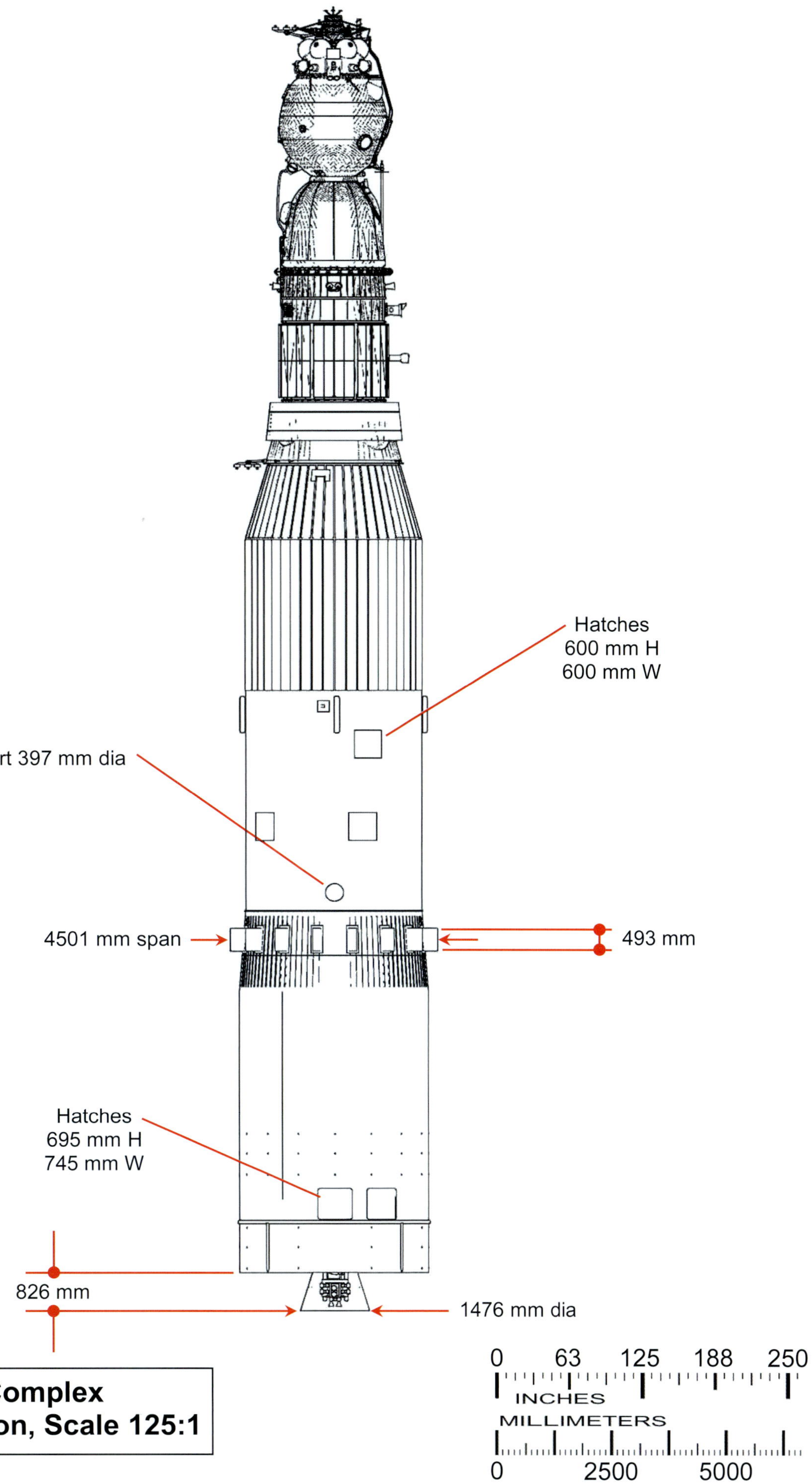

L3 Complex
270° Rotation, Scale 125:1

Block-D Design

Block-D is arguably the most significant and successful development of the Soviet moon program. Even more than the NK-33 engine and its derivatives, which didn't start flying until the 21st Century, the Block-D assembly has flown almost continually since the mid-60's on a wide range of missions. Compact, relatively simple and very reliable, it has been used as a "space tug" on nearly 200 missions and is still used today as the top stage of the Proton booster.

As used on the L3, Block-D starts at the bottom with an adapter that connects it to Block-G and supports everything above it. This support continues up with the outer skin (also load-bearing, although most versions of Block-D use an outer truss instead of a skin) and winds up at a truss on the top that supports the LK lander.

The internal components, i.e. the propellant tanks and engine, float within the skin. Attached to the top under the LK truss is a support frame for the spherical liquid oxygen tank (it's the part that looks like a washtub in the diagram). This frame both supports the tank and insulates it from the surroundings in exactly the same way the "thermal bridges" do in the main booster stages (see page 90).

Below this frame is another truss connecting it to the fuel tank and RD-58 engine. The kerosene fuel tank is toroidal (donut-shaped) which saves space since the top of the RD-58 fits in the central hole. This lets the pipes from both the fuel and oxidizer tanks connect to it directly without the external fuel lines needed on the N-1 stages. You'll notice that the fuel tank is not sitting level. Unlike the spherical LOX tank that has a single low spot at the bottom for the outlet, a torus has no single low point so the tank is tilted slightly to make sure all the fuel makes it to the outlet.

Note that when the lower stages are firing, these internal components hang from the top of the skin with the Top Frame and Lower Truss in tension. When the RD-58 is firing, these are now in compression, transferring the thrust of the engine up to the LK truss and through the LK shroud (not shown in this diagram, see foldout opposite) to the LOK at the very top of the stack.

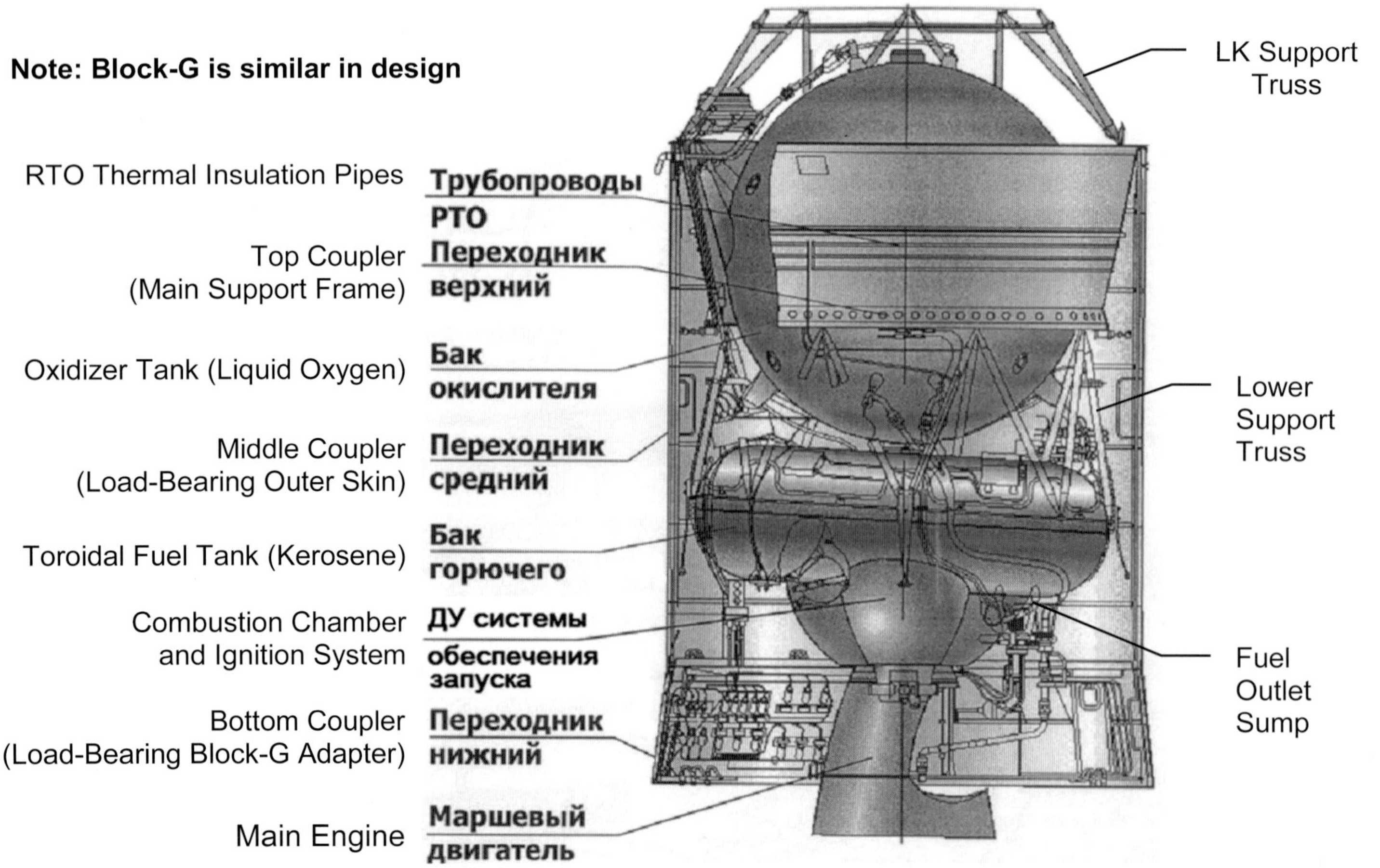

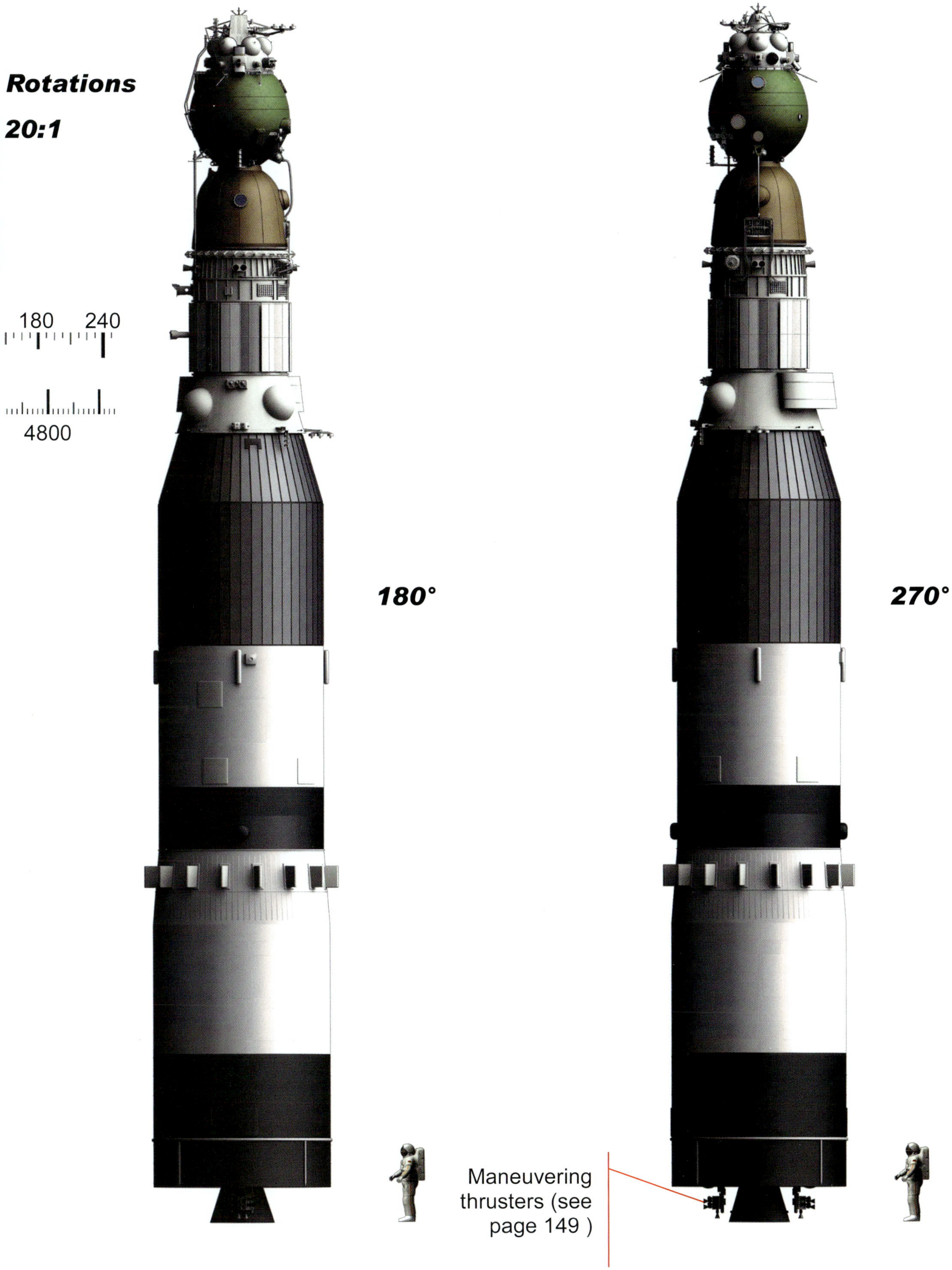
Rotations
20:1
180 240
4800
180°
270°
Maneuvering thrusters (see page 149)

L3 Complex (Exploded View)
Scale 80:1

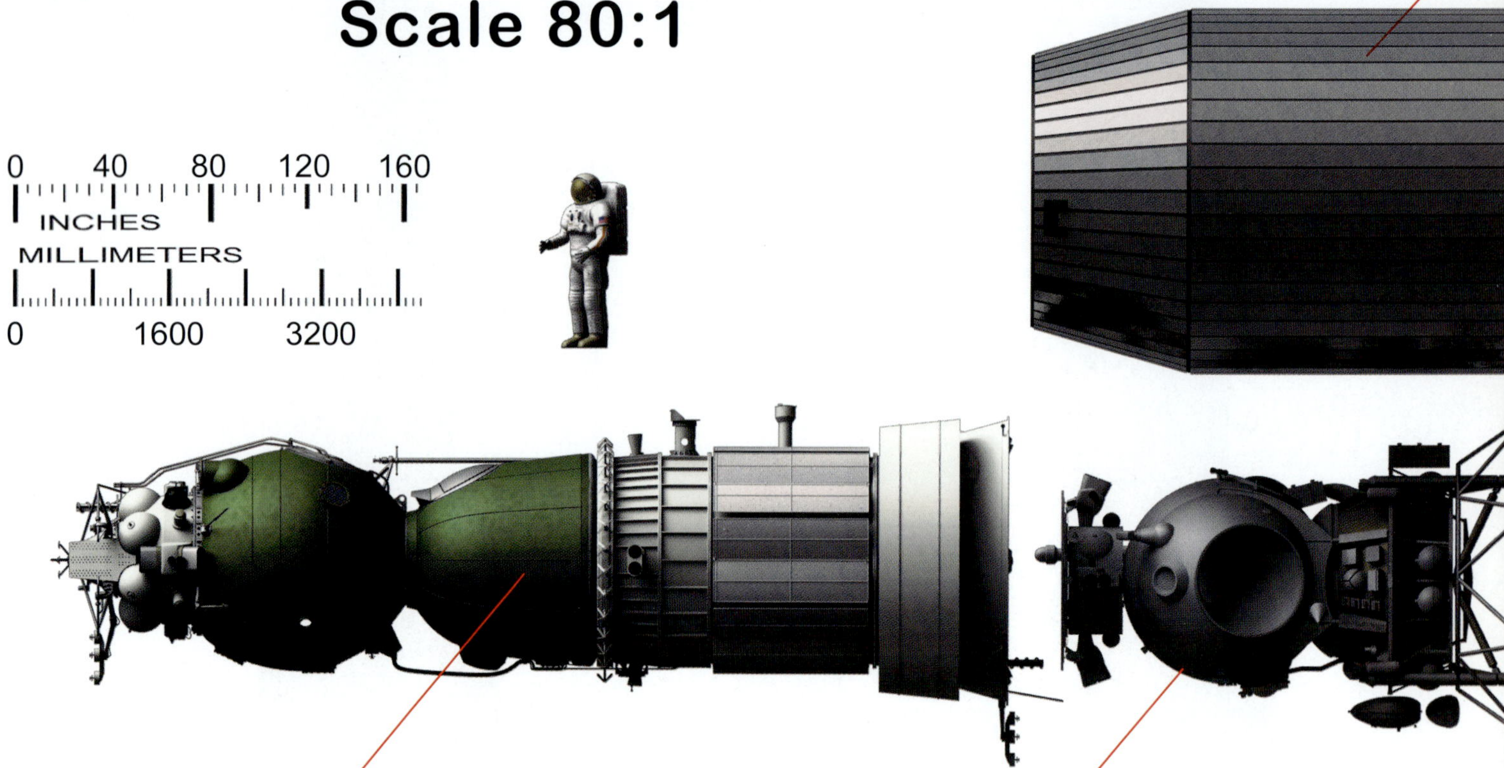

Block-I
(LOK Orbiter)

Block-E
(LK Lander)

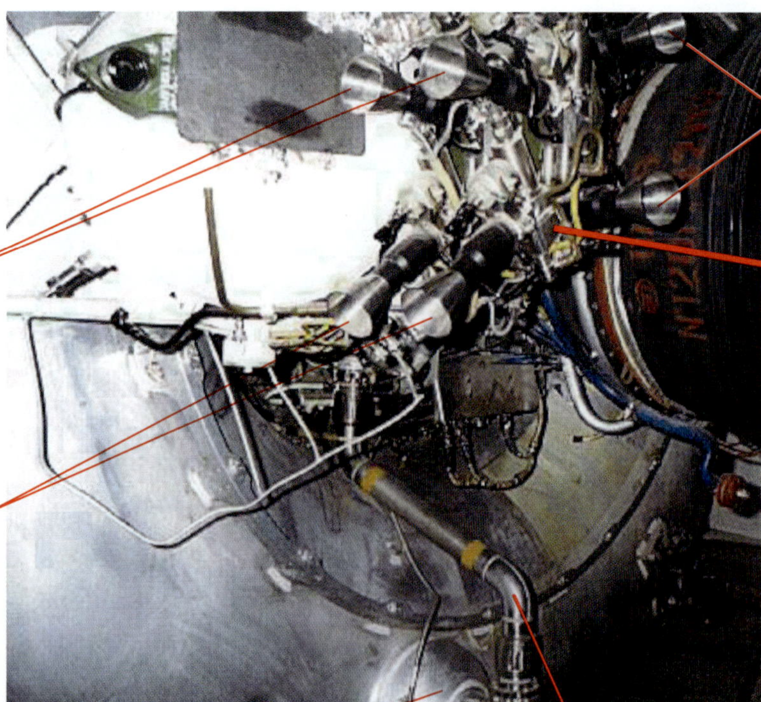

Ullage Thrusters*

Yaw Thrusters

Roll/Pitch Thrusters (a second pair of these with the nozzles pointed up, is just off the top of the frame)

Fuel Sump (bottom of toroidal fuel tank)

Main Fuel Line

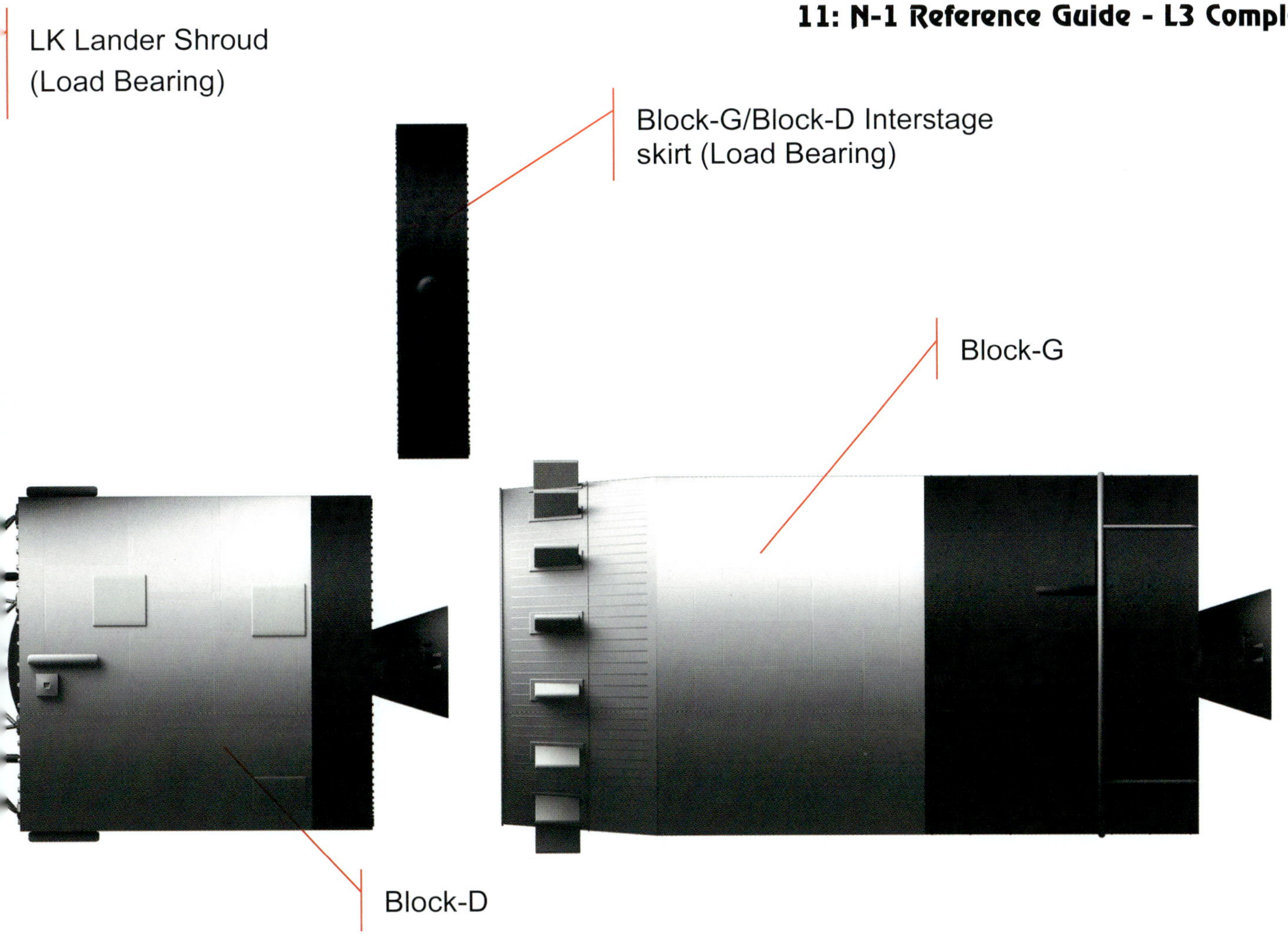

© Mark Wade

Close-up of the RD-58 in Block-D. The detail (far left) shows the orientation/ullage* thrusters that control the stage during coast, restart, and maneuvering.

Some of the load-bearing structure of the lower adapter is visible where the insulation has been removed.

* Ullage is a very old term from the brewing and wine industry that refers to the air space above the liquid in a cask or barrel. Same in rocketry, it's the amount of space above the propellant in a tank after a partial burn.

If an engine has to be re-started, and in the case of the Block-D that's five or more times, there's a problem because in zero-g the propellants are just floating around in the tanks and have no impetus to enter the pump inlets.

The solution chosen by the L3's designers (there are others) is to have four small thrusters that will accelerate the Complex slowly and force the propellants to settle at the bottom of the tanks, at which point the pumps can be started and the main engine ignited.

L3 Complex:

Scale 1

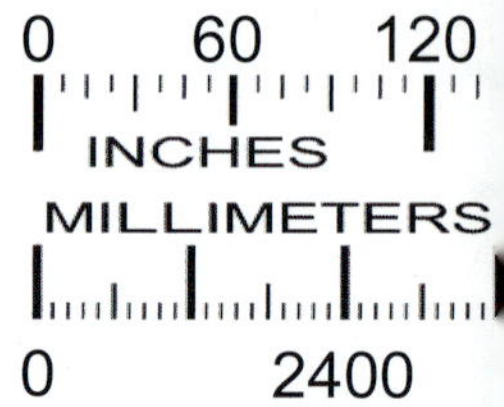

0°

90°

Rear Isometric
3/4 View

Front Isometric
3/4 View

Front

Rear

0°

90°

180°

270°

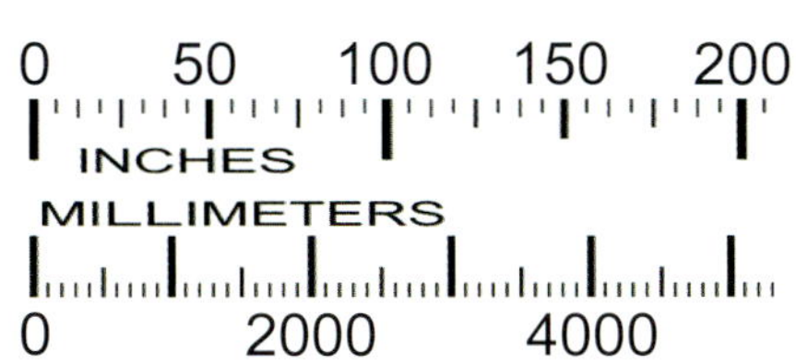

Block-G Views, Scale 100:1

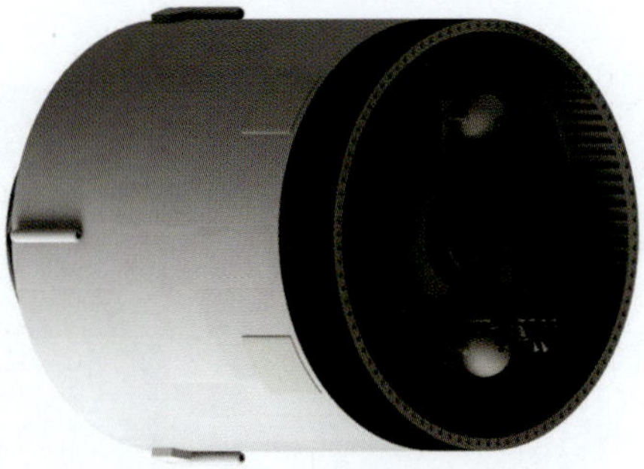

Rear Isometric
3/4 View

Front Isometric
3/4 View

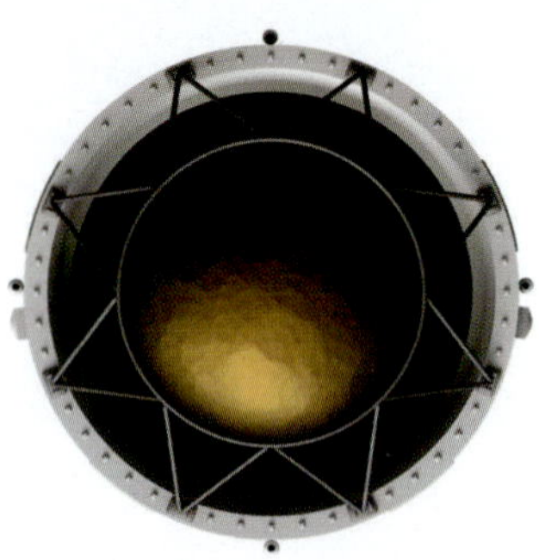

Front

Rear

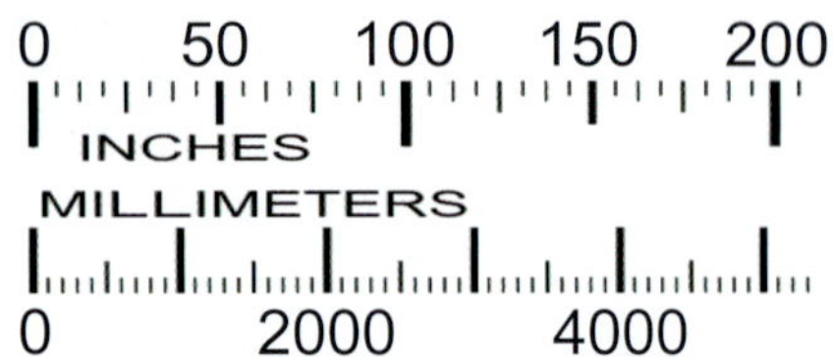

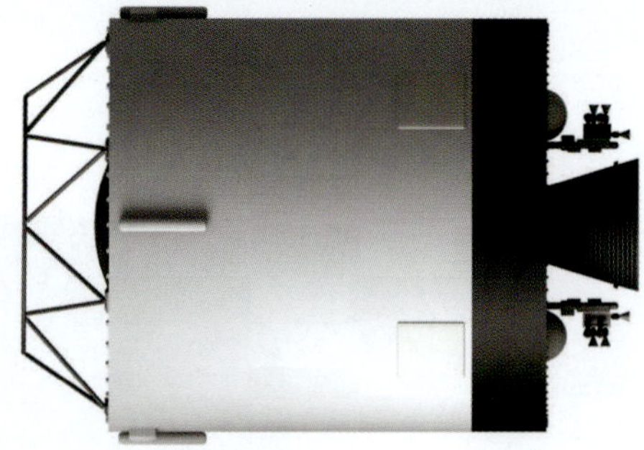

0°

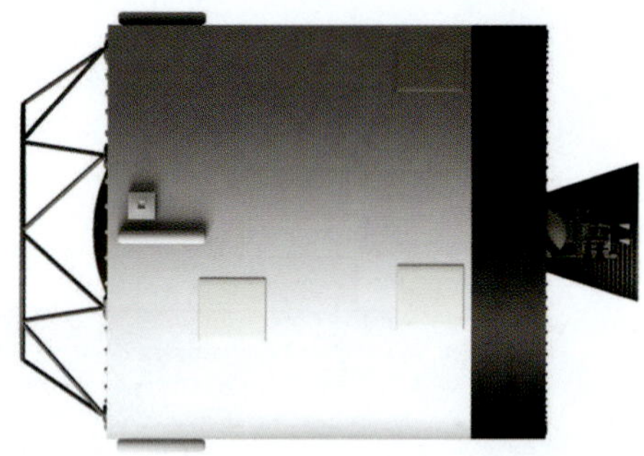

90°

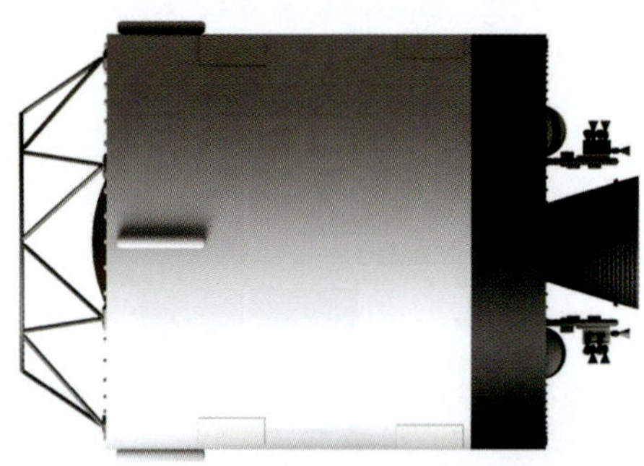

180°

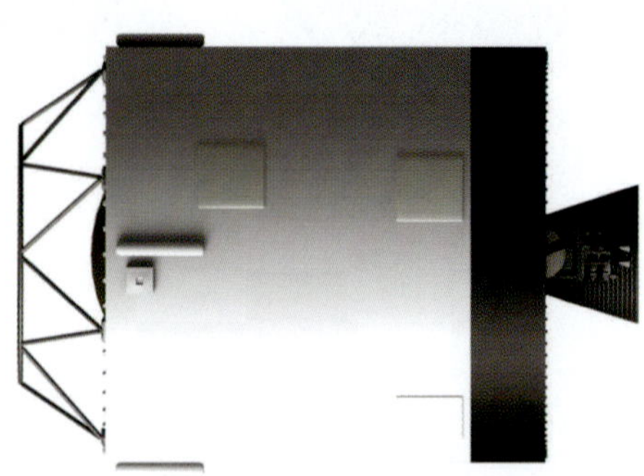

270°

Block-D Views, Scale 100:1

12: N1-L3 Fairing/LES Reference Guide

12: Справочный раздел: обтекатель Н1-Л3/САС

CAD Drawings by ALEXANDER SHLIADINSKY
CGI Renderings by NICK STEVENS
Dimensions Applied by MATT JOHNSON

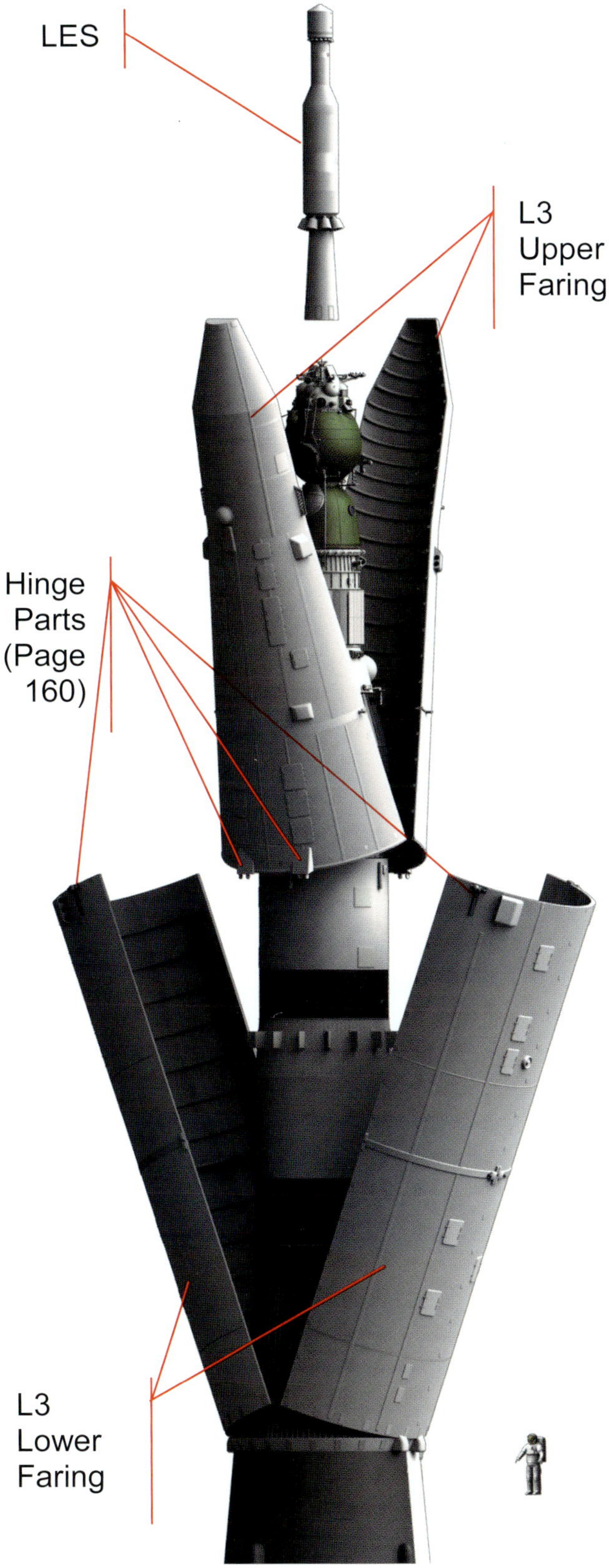

The L3 Fairing Assembly consisted of three major sections: the Lower Fairing, the Upper Fairing and the Launch Escape System (LES).

The Upper and Lower sections each came off in two pieces with their split lines offset 90° from each other for reasons that will be discussed on page 160. The Lower Faring covered the Block-G and Block-D stages, while the Upper Fairing covered the LOK and LK modules. Above the Upper Fairing was the LES tower that would pull the LOK to safety in the case of an emergency.

There is some question as to why such a complicated arrangement was necessary in the first place. Obviously the LOK, with its asymmetrical shapes and large amount of drag-inducing hardware attached to its outer surface, required some sort of cover during the initial boost. However, the smooth-sided Block-G and Block-D stages along with the conical Block-V/G adapter (see illustration next page) and the similar LK Shroud make perfectly good aerodynamic shapes. They were much like their counterparts on the Saturn V (the conical Stage 2/3 transition and the shroud around the LM) which needed no further covers. With the Soviet designers trying to shave every gram from the payload, it seems that all this extra structure is a wasteful extravagance.

The reason for all these covers is in the way the LES worked. In the Apollo system, the Command Module in which the astronauts rode was at the very top of the stack with only the escape system above it. In case of an emergency, the LES needed to pull only this top module free. In the L3, not only was the LOK covered by a large faring, but the cosmonauts weren't even in the top Utility Module of the LOK. The acceleration couches were actually in the lower Reentry Module. In case of an emergency, the tower had to pull not only the capsule with the crew in it to safety, but the entire LOK assembly. The fairings, therefore, were not just simple aero covers, but had to be strong enough to take all of the significant thrust from the LES (remember it was attached to the Upper Fairing, not the LOK inside) and transfer it to the spaceship as it is carried to safety.

This is also the reason the fairing was divided into upper and lower sections. A single large clamshell fairing would certainly have been possible, but in the case of an emergency abort, the lower portion covering the Block-G/D represent a lot of mass that didn't have to be pulled to safety.

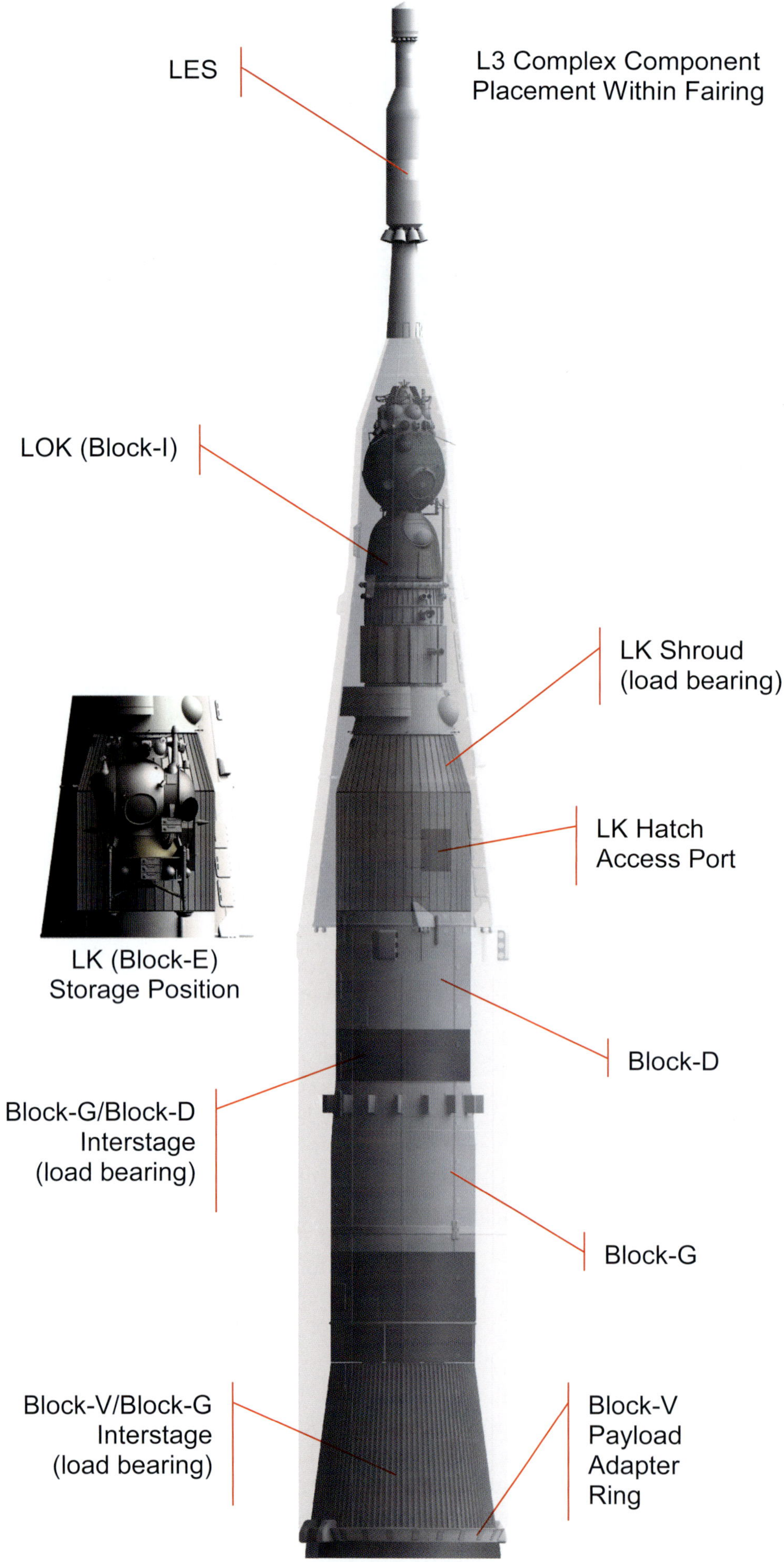

L3 Complex Component Placement Within Fairing

Frames from a video of an L3 Upper Fairing separation test. The motors firing in the upper left frame are similar to the ones shown in the debris field at the bottom of page 65.

Note the high-fidelity mockup of the LOK and LK shroud that is exposed as the fairing halves fall away.

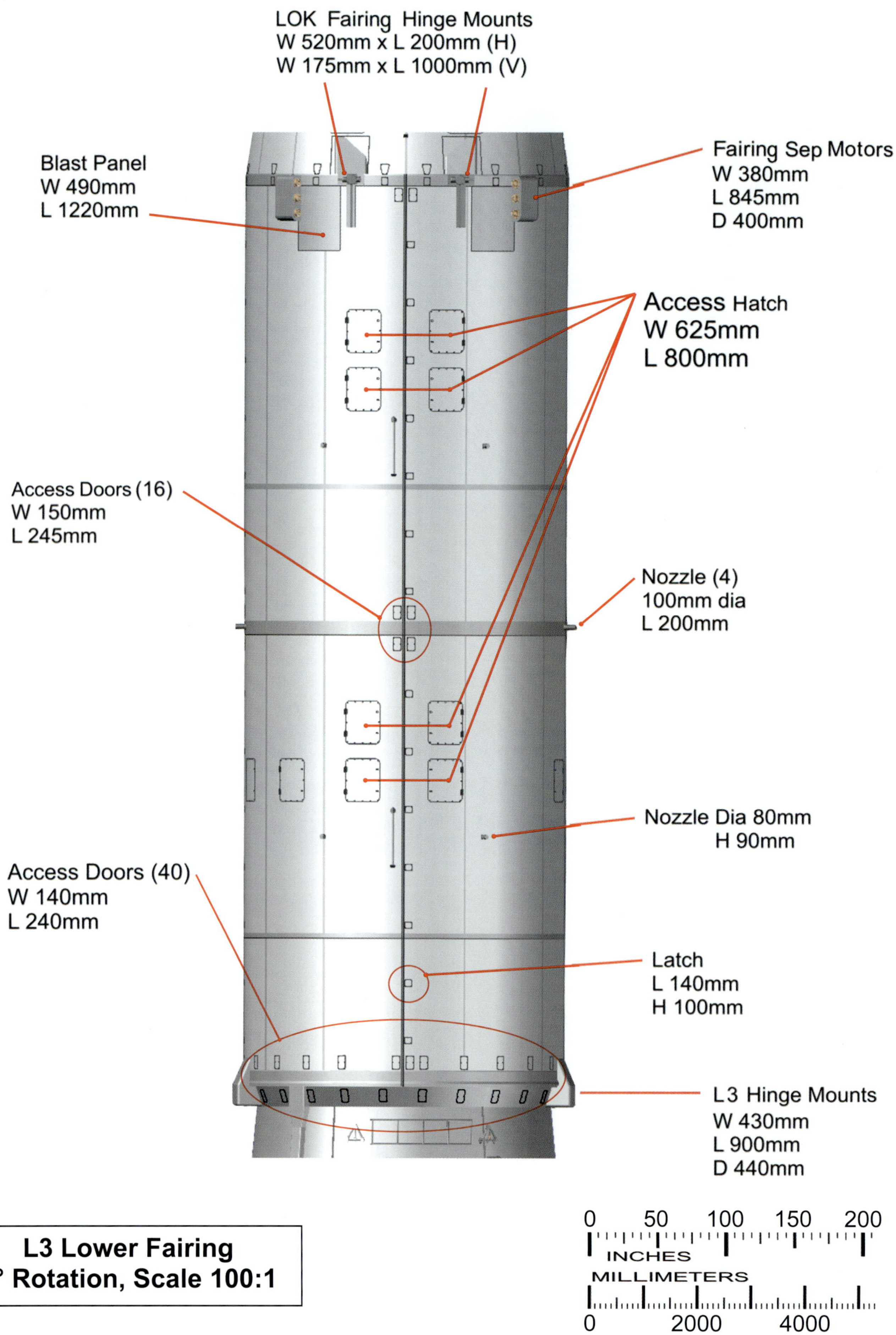

L3 Lower Fairing
0° Rotation, Scale 100:1

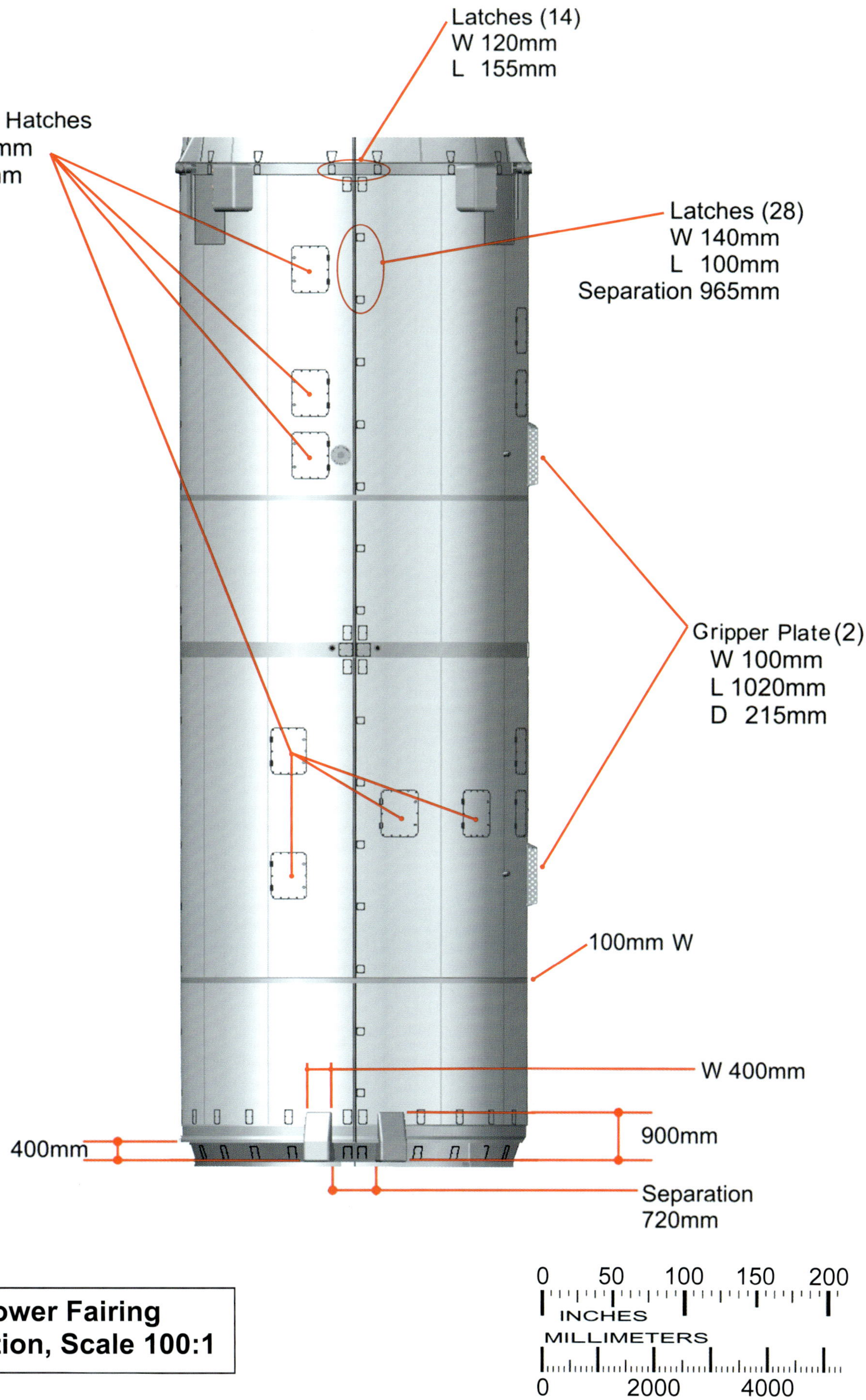

L3 Lower Fairing
90° Rotation, Scale 100:1

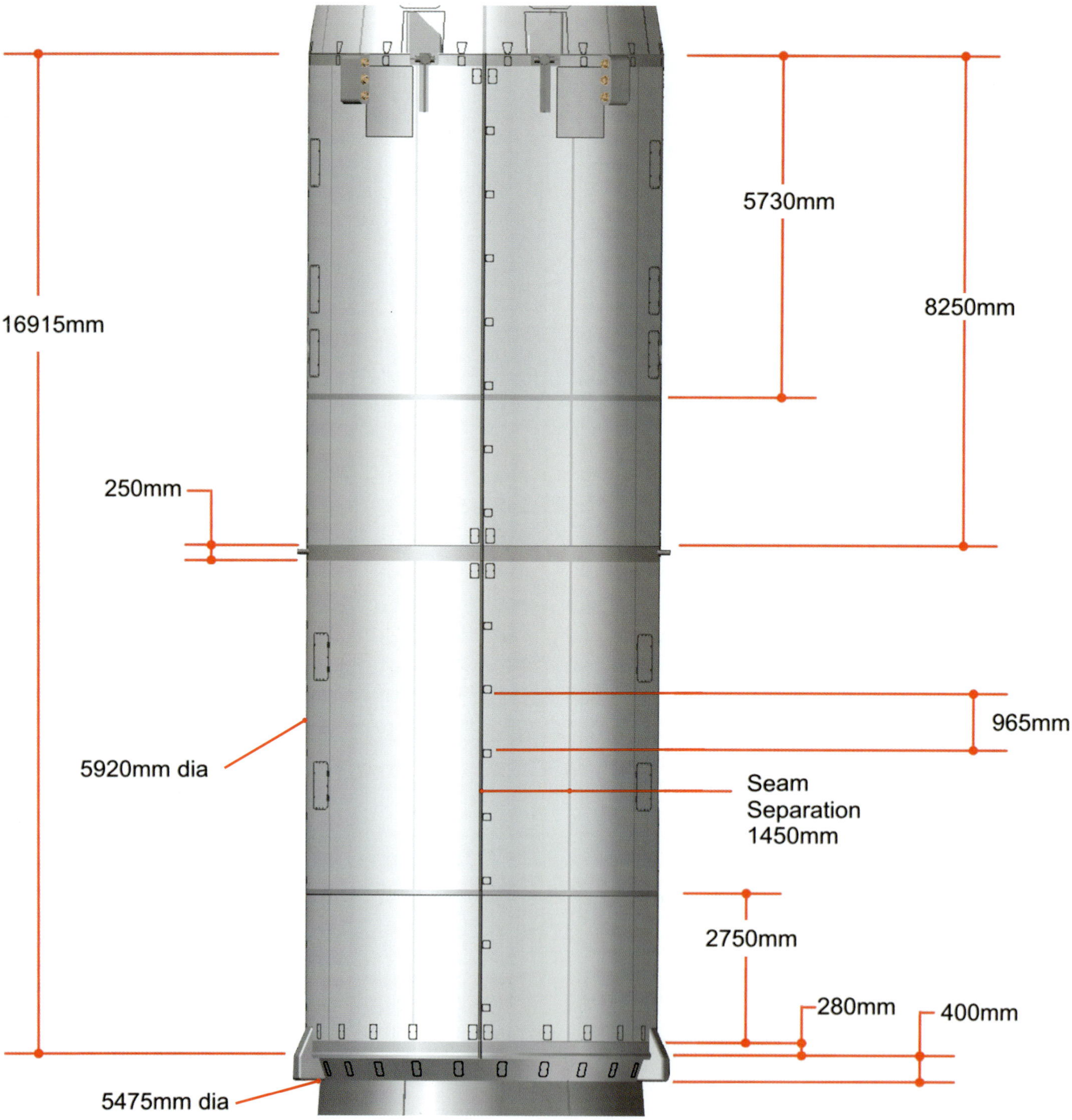

L3 Lower Fairing
180° Rotation, Scale 100:1

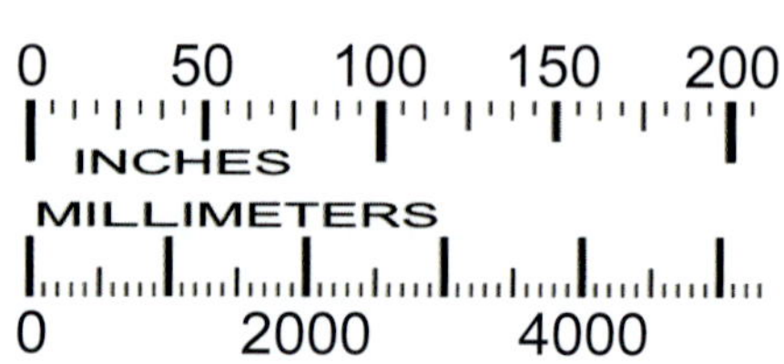

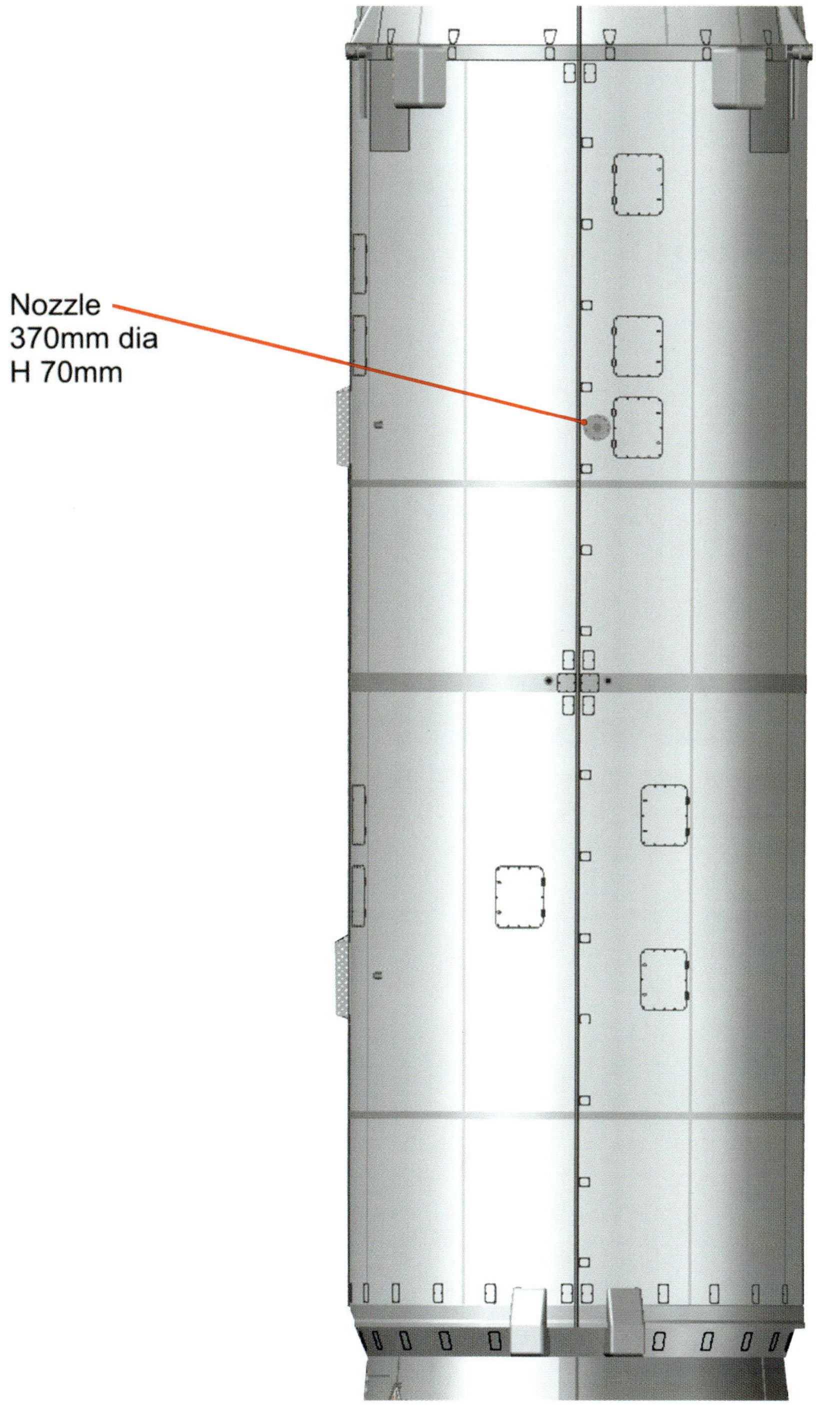

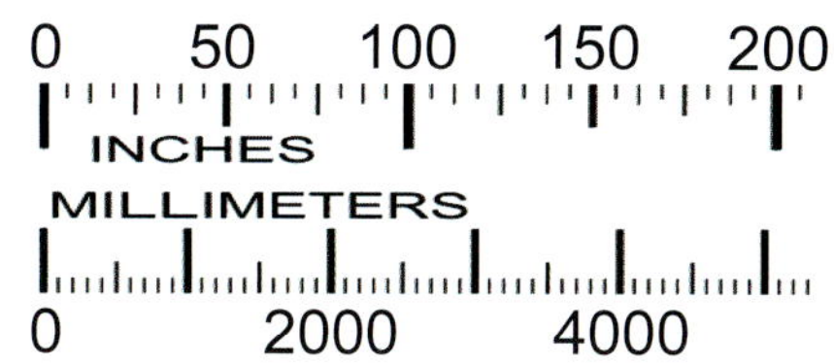

L3 Lower Fairing
270° Rotation, Scale 100:1

Fairing Hinges

The hinge that joined the Upper and Lower Fairing sections together did more than just provide a way for them to pivot apart. They were actually structural members that helped hold the entire fairing assembly together until it was time to jettison them.

There were two hinge assemblies, one at the bottom of each Upper Fairing. The illustration on page 153 shows the location of the hinge parts on one side (the assembly was duplicated on the other side). Each assembly had two mirror image subassemblies. The actual pins and pivots were the same, but the mounts were mirrored to let them attach to the skins of the fairings in a way that kept the hinge pins along the same axis (necessary for the hinge to work!).

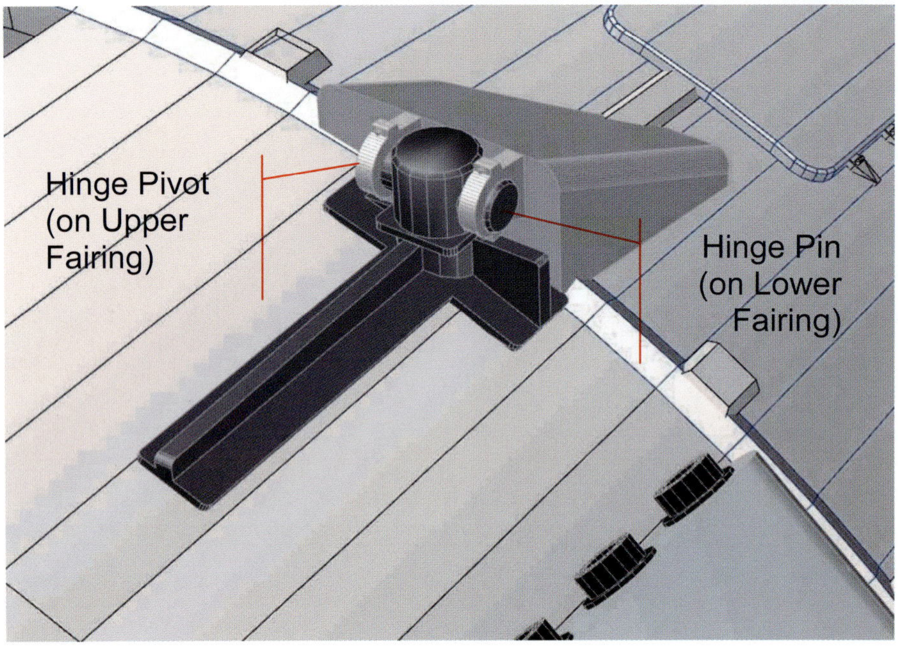

This is one of the reasons why the split lines of the two fairings are offset 90°. The two hinge pivots are on the same half of the Upper Faring while the two hinge pins are on either side of the split line of the Lower Fairing. In this way, the structure of the Upper Fairing helps hold the Lower Fairing together during the boost phase. Likewise, the LES tower monocoque (see page 167) helps hold the halves of the Upper Fairing together until they are jettisoned in rapid sequence: LES/Upper/Lower.

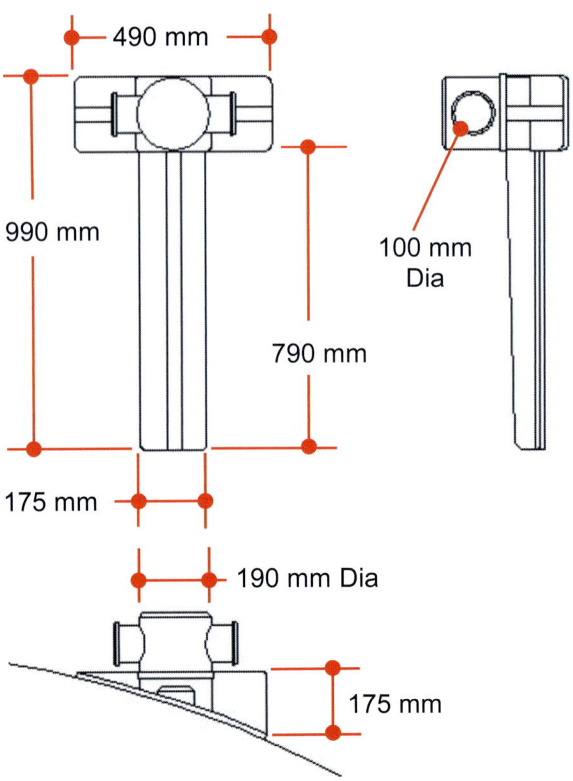

Hinge Pin, Scale 20:1

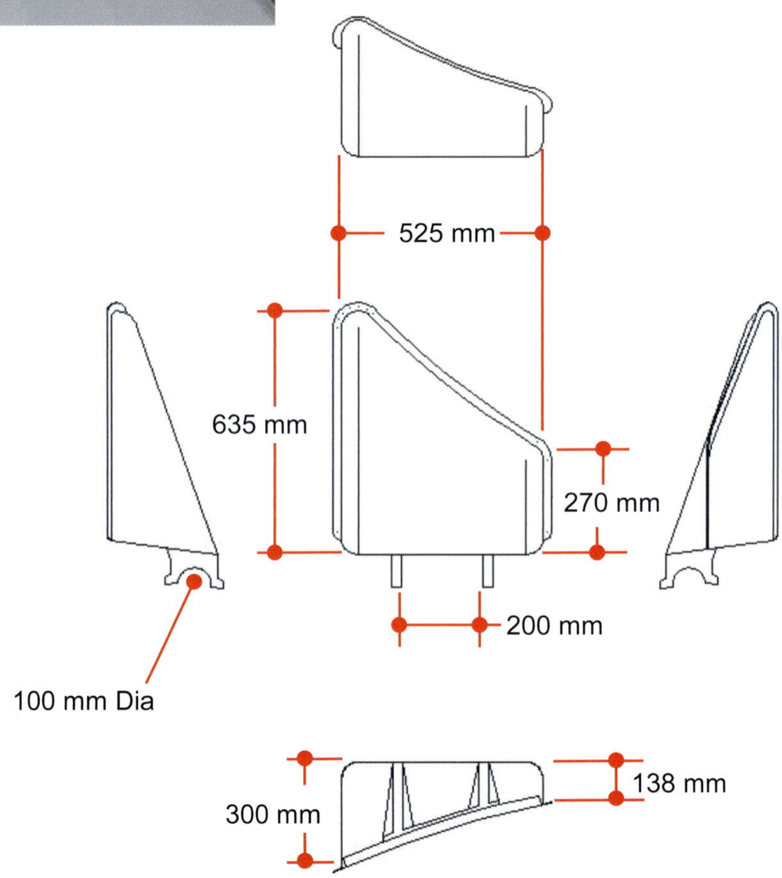

Hinge Pivot, Scale 20:1

L3 Fairing/LES Gallery

Upper Fairing under construction.
Note the intricate LES tower mating flange.

Good underneath view of the LES with all three nozzle sizes visible.

Complete L3 Fairing LES assembly. Keep reminding yourself that this is the smallest part of the vehicle (note the people at the bottom).

LES with the forward part of the Upper Fairing. Tower gripper plates and fairing separation motors plainly visible.

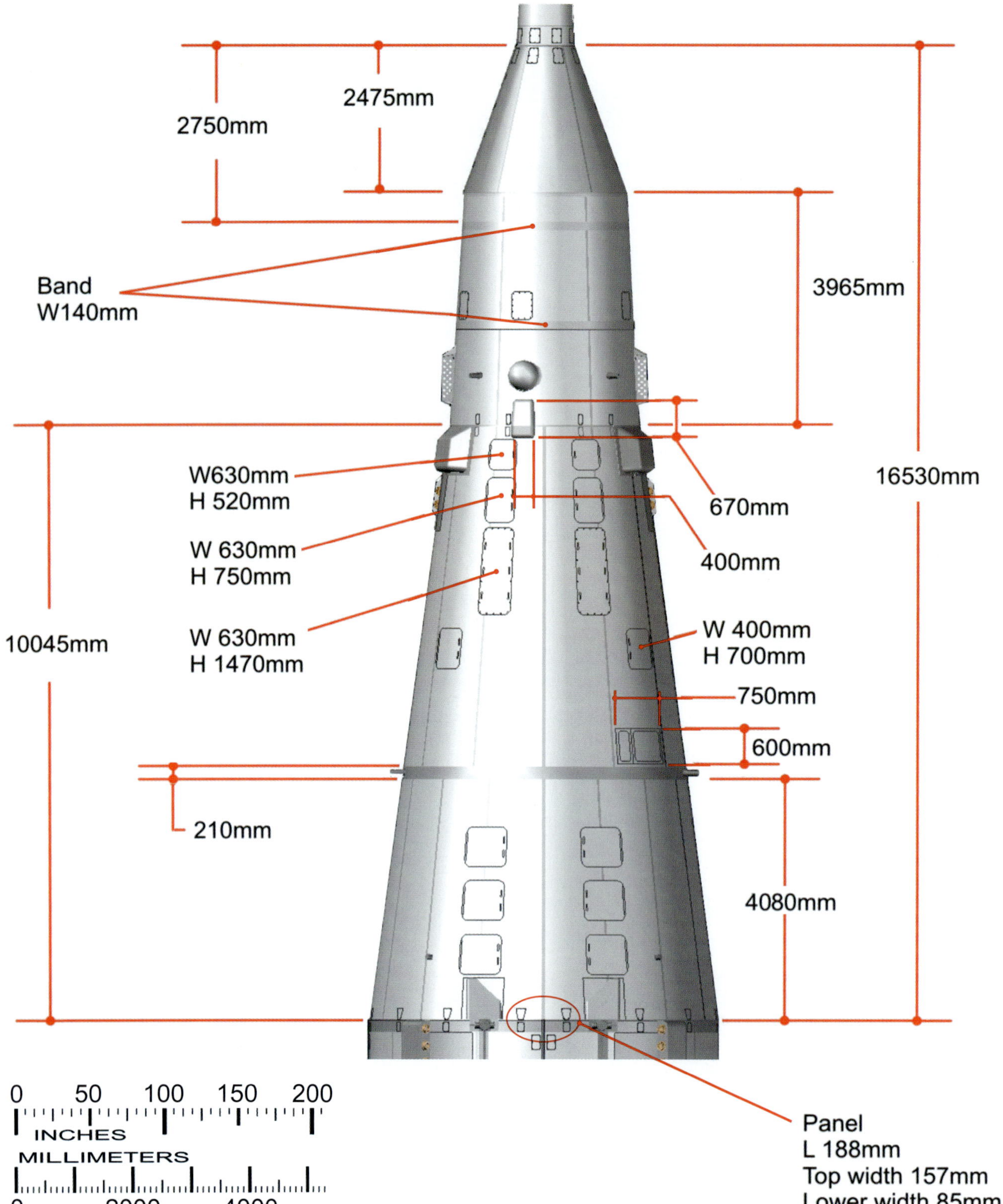

L3 Upper Fairing
0° Rotation, Scale 100:1

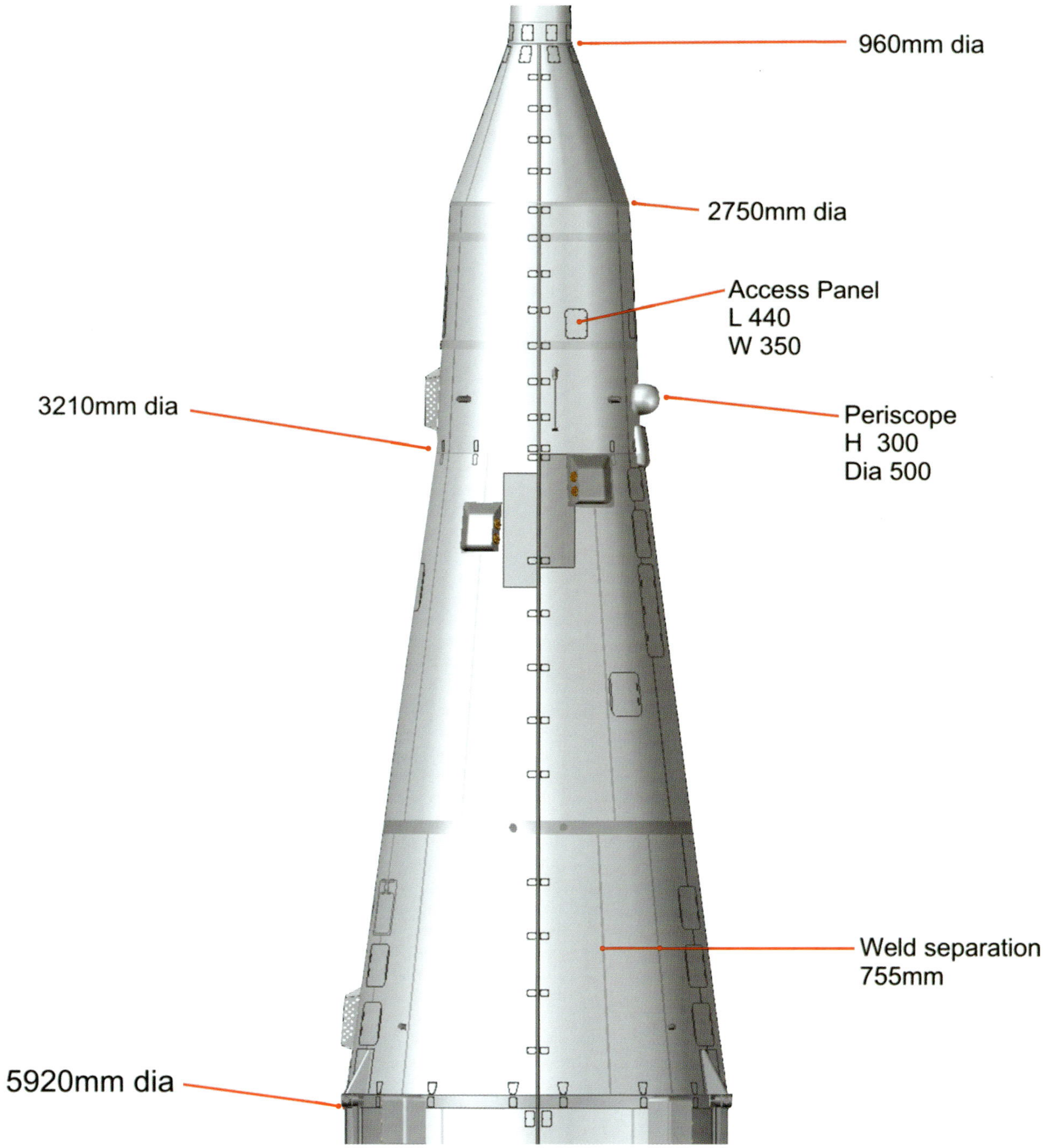

L3 Upper Fairing
90° Rotation, Scale 100:1

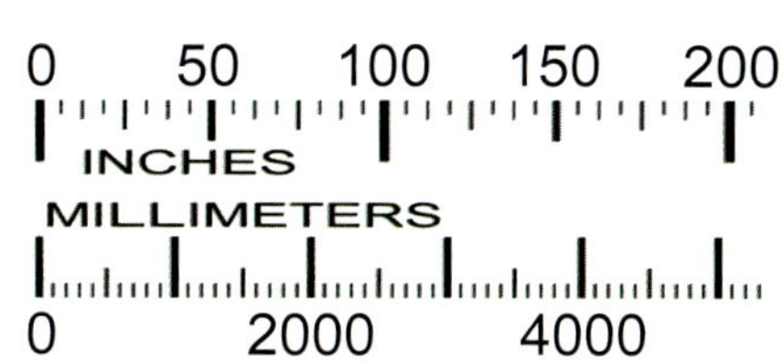

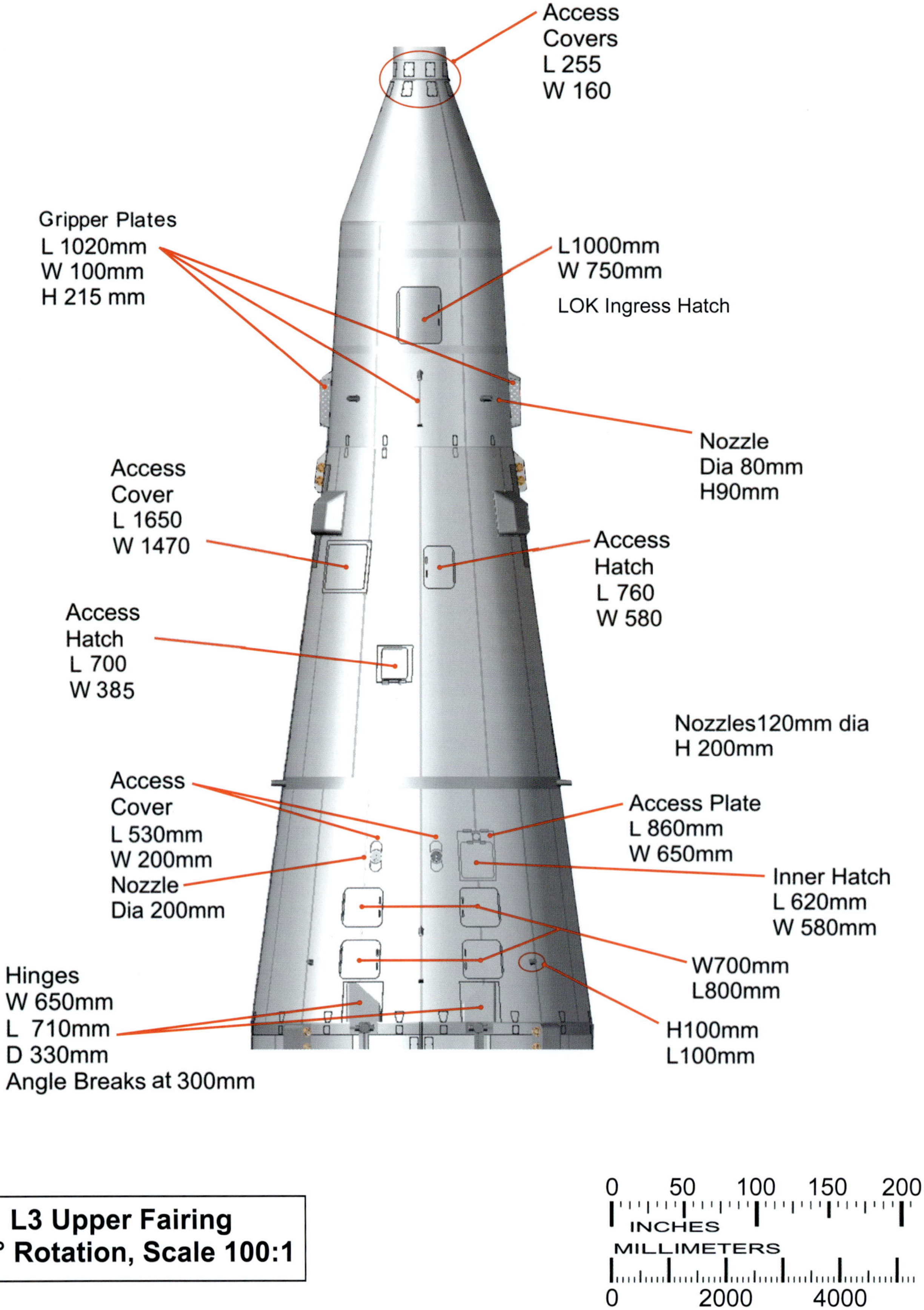

L3 Upper Fairing
180° Rotation, Scale 100:1

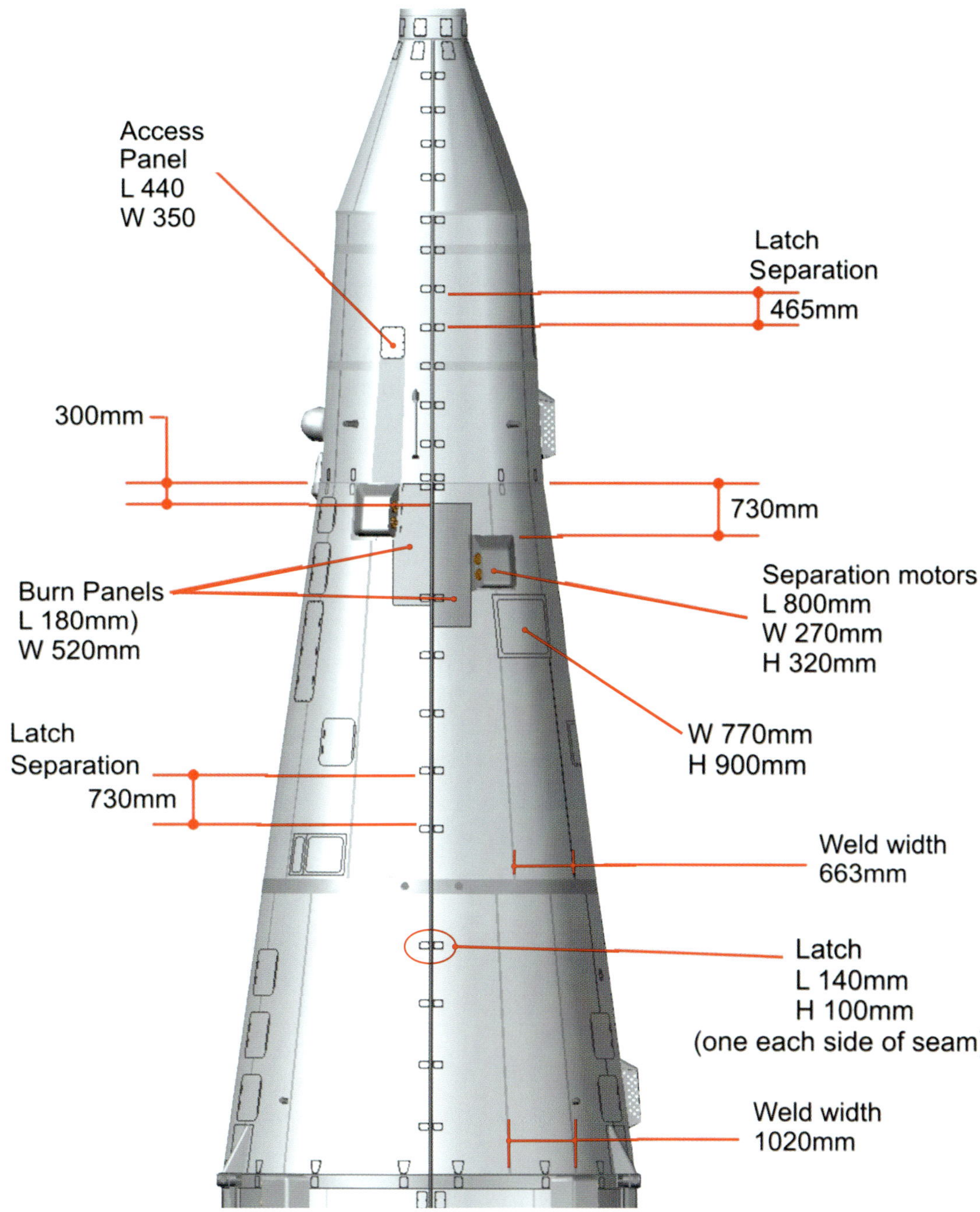

L3 Upper Fairing
270° Rotation, Scale 100:1

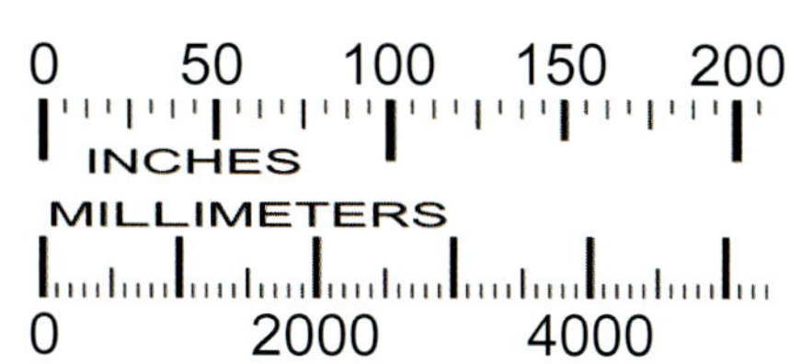

Launch Escape System

It took until the very top of the respective vehicles, but we finally have an example of where American designers used an open truss and Soviet designers used a closed monocoque, i.e. the Launch Escape Tower that connects the escape rocket assembly to the top of the spacecraft.

The main purpose of the LES was, of course, to pull the crewed section of the spacecraft away from the lower sections in the case of an emergency. Given this primary task, both the Soviet and American designers came up with remarkably similar but hardly identical designs as can be seen on the facing page.

To pull the spaceship modules away from an exploding booster at any point in the flight from sitting on the launch pad up to many kilometers of altitude and several times the speed of sound requires a lot of acceleration. This was provided by the large solid rocket motor attached to the top of the Launch Escape Tower. To keep the exhaust plume of this motor from impinging on the spacecraft it was saving, both designs used multiple nozzles angled outwards and arranged in a ring at the base of the motor. The Americans had four nozzles poking out between the tower struts generating a total of 670 kN of thrust, while the Soviets used eight to provide the greater total thrust (760 kN) needed to pull the heavier L3 Complex away.

One feature needed by an LES design is a way to get rid of the tower. This is necessary either after pulling the spacecraft away in an emergency or after it is no longer needed during a normal launch. To do this requires a second, smaller rocket motor mounted above the first. In the Apollo design, the Tower Sep Motor was literally right on top of the Launch Escape Motor and fired out of two large nozzles placed on either side of the motor housing. On the L3 the Sep motor was placed a bit further up on top of a second, smaller tower. This motor fired through a busy ring of 16 small nozzles with an interesting feature that we'll describe next.

A final ability needed by the system is to be able to pitch sideways. Whether pulling a spacecraft away from an expanding fireball created from thousands of tons of exploding propellants, or simply leaving on its own, it's best to get out of the way by pitching to the side. The Apollo LES did this by means of yet a third rocket motor small enough to fit sideways in the upper housing with its nozzle poking out the side. The L3 tower performed this pitch function in a somewhat more interesting manner. Of the 16 nozzles used in the Sep Motor, eight were larger and eight were smaller. This is clearly shown in Nick's rendering to the left. The unbalanced thrust caused the entire assembly to pitch over in the direction it was pushed by the larger nozzles.

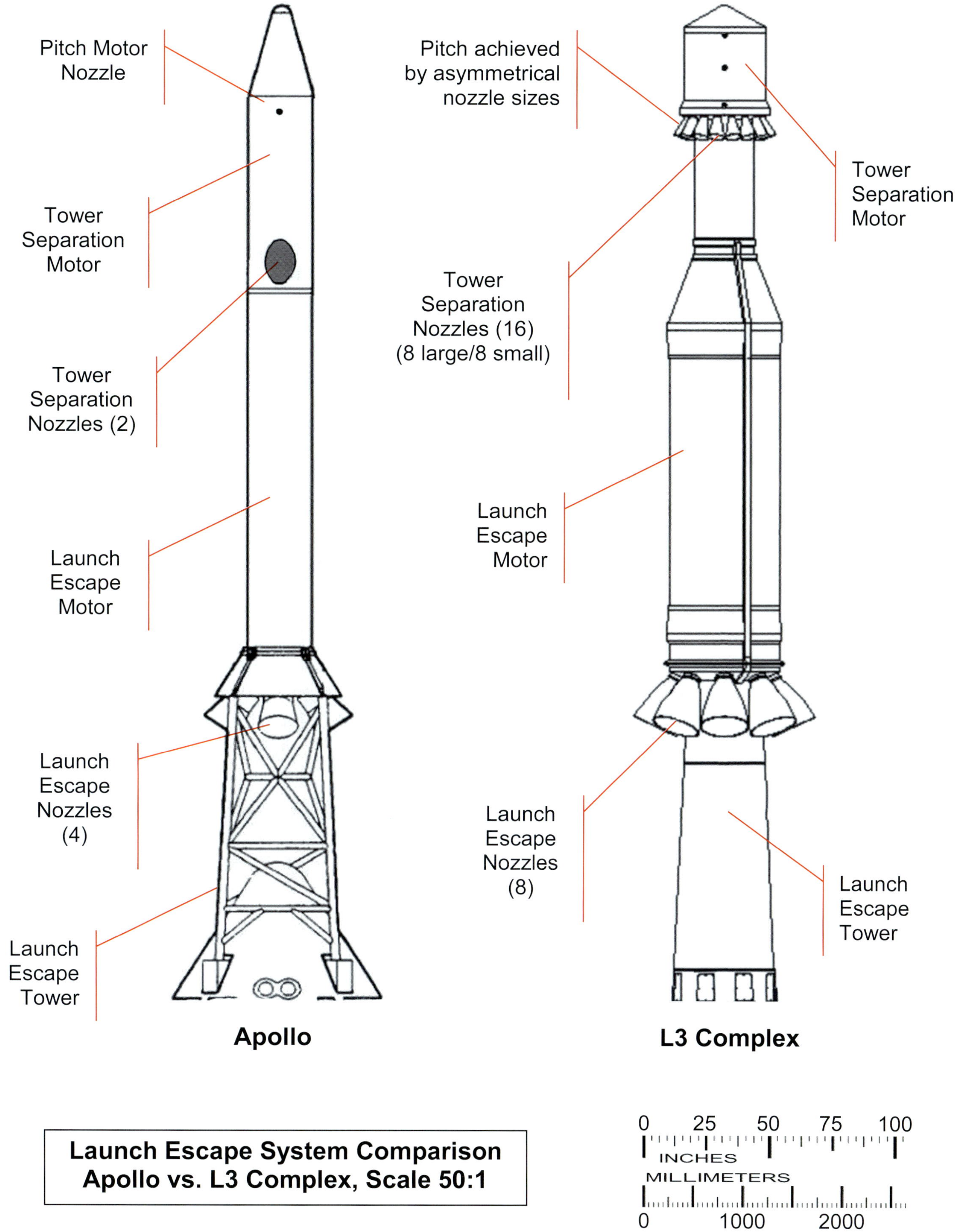

Launch Escape System Comparison
Apollo vs. L3 Complex, Scale 50:1

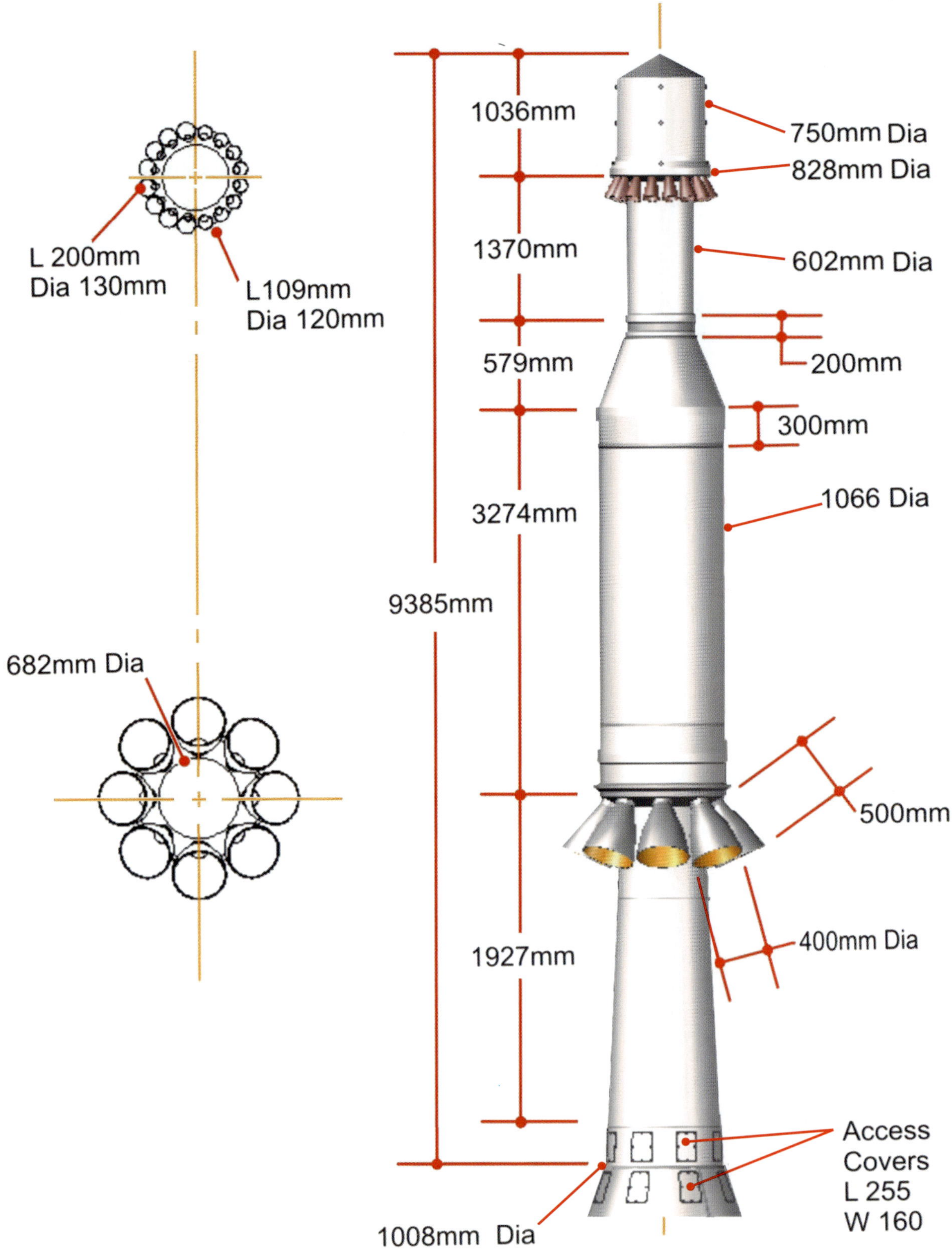

Launch Escape System
Scale 50:1

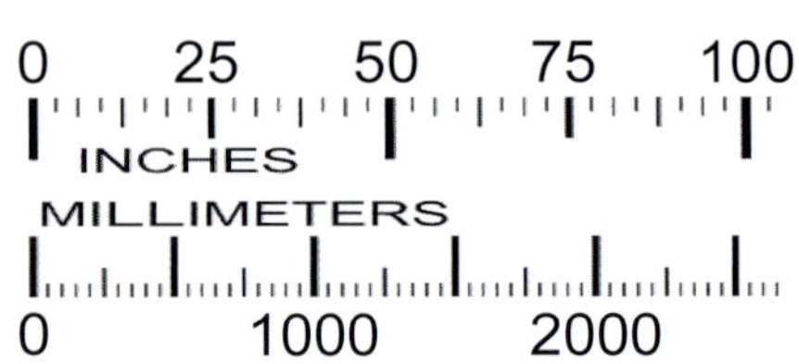

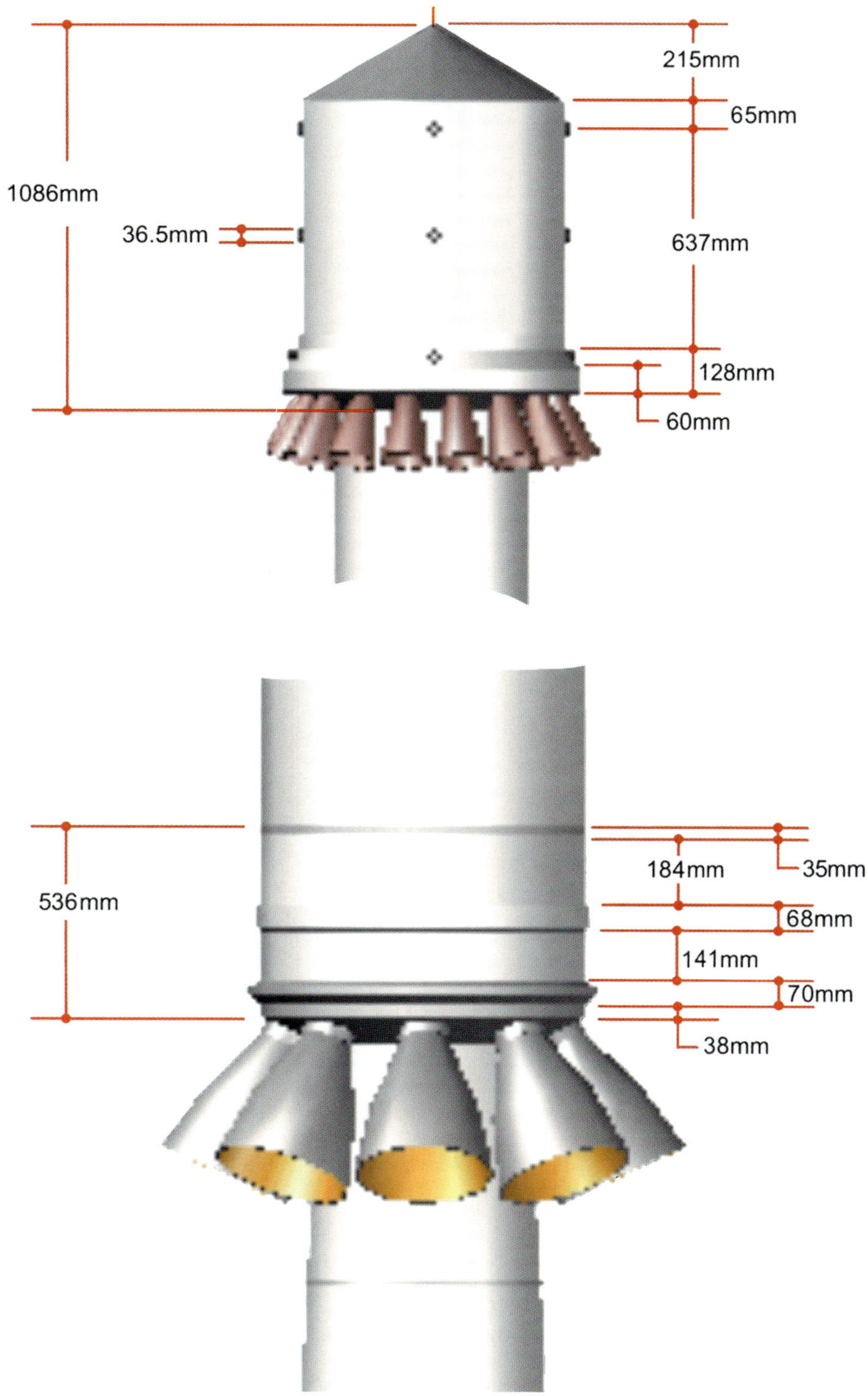

LES Nozzle Sections
Detail, Scale 20:1

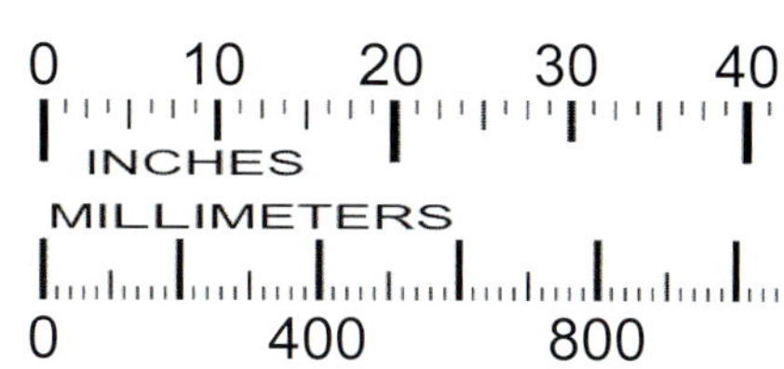

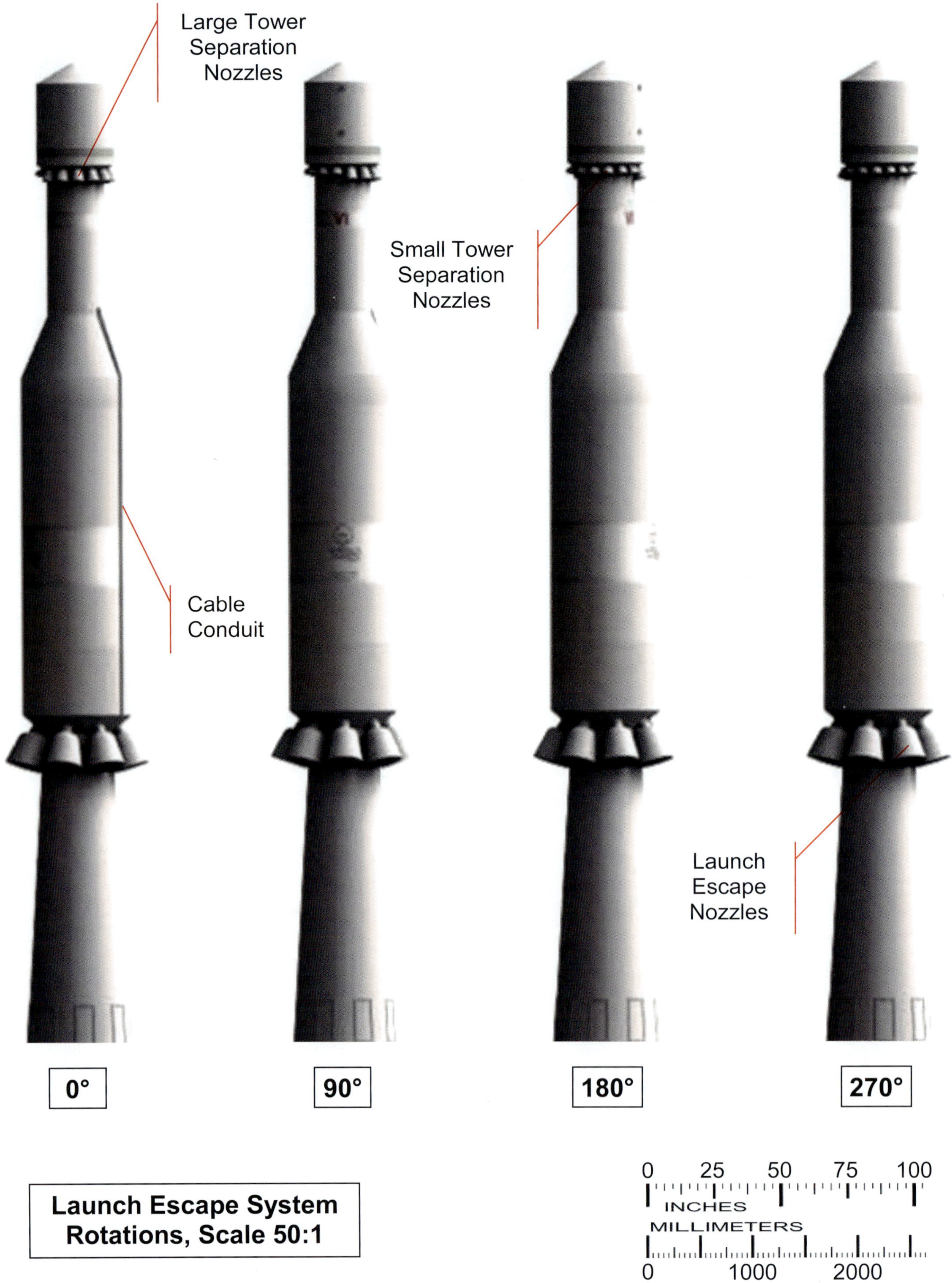

Launch Escape System Rotations, Scale 50:1

13: N-1 Launch Complex

13: Стартовый комплекс Н-1

CGI Renderings by NICK STEVENS
Dimensions Applied by MATT JOHNSON
Text by the Editor

As complicated as any space launch vehicle is, the launch complex that sends it into space is many times more so. With a vehicle as unprecedented as the N-1, the launch complex was unlike anything built before or since, with the exception of the equally groundbreaking Launch Complex 39 at Kennedy Spaceflight Center.

The N-1's launching facility had several different names depending on how it was being referred to. Most generically it was called "Site 110" at the Baikonur Cosmodrome, "site" being the designation for any group of structures dedicated to a specific task. Only a few of the sites actually do the launching, so sometimes you see it referenced as "Launch Complex 110" or simply "LC110."

Launch Complex 110 had two launch pads: The one to the west of the central transport line was known as 110 West. The other pad was, of course, 110 East. You also find them referred to as 110L and 110R (for left and right) respectively, but this is an English-language convention that was not used by the Soviets.

In this chapter we will be covering several aspects of LC110: the overall site plan, the launch pad design with its rotating tower and the "Grasshopper" transport that brought the N-1 vehicles from the Site 112 Assembly Hall to the pad.

Baikonur

Baikonur Cosmodrome is a vast, sprawling facility some 2,500 km (1,500 miles) southeast of Moscow in Kazakhstan. When the Cosmodrome was built, Kazakhstan was part of the USSR, but after the breakup of the Soviet Union in the early '90s, it reverted to being an independent country. Now Russia (which runs the space program) has to lease the facilities from Kazakhstan.

When built in the mid-1950's the facility was called the Tyuratam Test Range (after the local village and railroad station). Ironically, "Tyuratam" means "arrow burial ground" in the local language, very appropriate for a facility created specifically to test the R-7 ICBM. When that booster started being used to launch satellites, its role was expanded to include all aspects of spaceflight.

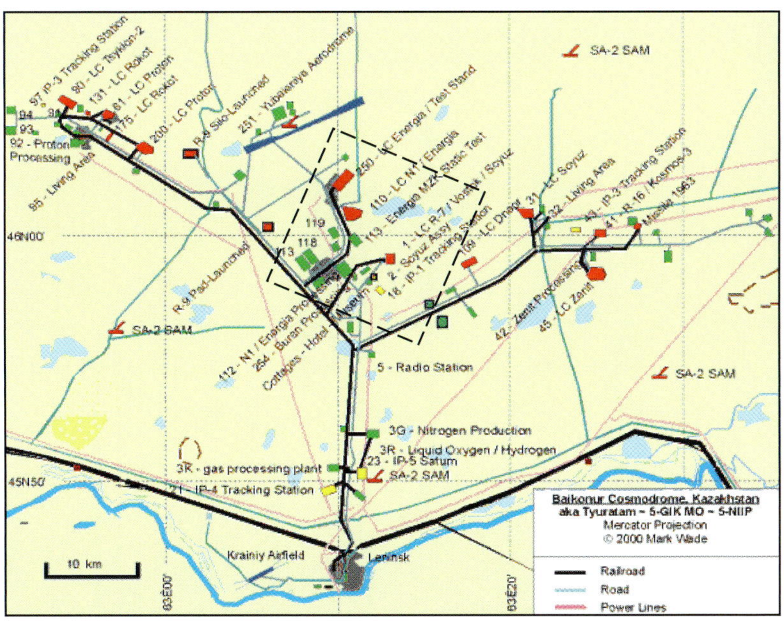

© Mark Wade

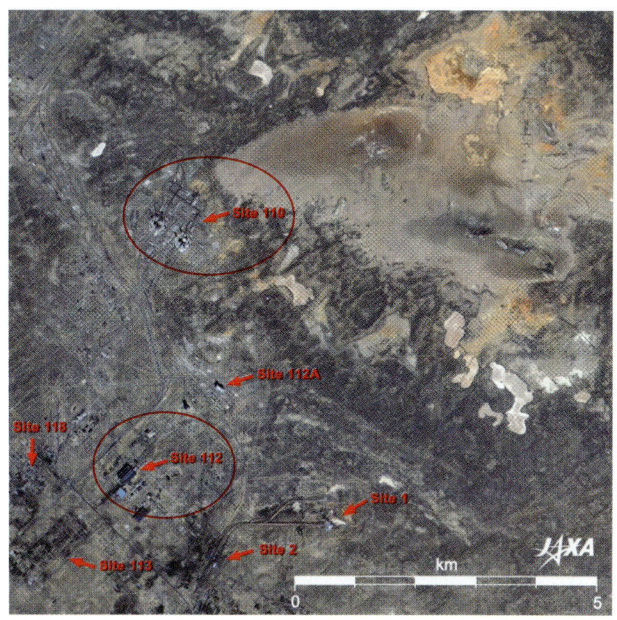

It is difficult to convey the vast size of the Baikonur Cosmodrome. The map on the left looks like any large industrial facility until you look at the scale and realize that it covers an area some 80 km (50 miles) wide by 60 km (37 miles) high! The dashed rectangle shows the area in the photo on the right taken by a Japanese Earth resources satellite. If, however, you include the downrange tracking and safety zones, the Cosmodrome extends 85 km from north to south, and 125 km from east to west, an area as great as the former Soviet Republic of Moldova. The two Sites to be discussed in this chapter are 110 and 112 (circled)

The name was changed to Baikonur after the *Sputnik 1* launch in 1957. Some claim that this name was chosen to deliberately misdirect the West to the small mining town of Baikonur about 320 km (200 mi) northeast of the launch center (in reconnaissance photos the large flame trenches under the launch pads do look a lot like open pit mines). Others claim that Baikonur was the name of the Tyuratam region long before the Cosmodrome ever existed.

A brand new town was built southward from the existing village of Tyuratam. At first it housed the workers that built the Cosmodrome, and now it's home to the people who run it and their families. Initially called "Leninsk," it was renamed "Baikonur" on 20 December 1995 by Nursultan Nazarbaev, the president of Kazakhstan. Building the modern highways and railroads from the manufacturing areas around Moscow out to this remote and desolate region was the largest public works project ever undertaken by the Soviet Union.

Site Plan

As we saw in Chapter 5, the west's first awareness of the massive N-1 project was through reconnaissance satellites, such as the CORONA image on page 56. To the CIA it was known as "Launch Complex J." If you want to act like a modern-day cold-war spy, you can find it yourself using any mapping software at 63° 18' 18" E longitude by 45° 57' 53" N latitude.

Launch Complex 110 consists of two identical launch pads (note, they are the same, not mirror image although some of the structures, like the lightning towers, are symmetric to the centerline of the Site). Between the pads is a large and very complex infrastructure (most of it underground) consisting of the giant pumps needed to fuel the vehicles, preparation rooms for the cosmonaut crews, etc.

The vehicles are assembled at Site 112 some 5 km away and transported to the pad horizontally on a dual-rail transport system called "The Grasshopper," as will be covered later.

Launch Pad

At the top of the next page is another of Charles Vick's analysis drawings, but this one was done in 1991 after information on the project began to be officially released. Ironically, it comes to us by way of being reproduced in a Russian article on the program, so the circle is complete. We'll start our analysis of the launch pad with the one thing you can't see in the illustration because the rocket is sitting on it, the Support Ring.

The middle figure opposite shows a portion of the Suppport Ring, the place where the Adapter Ring at the bottom of Block-A meets the pad. The Adapter Ring is another place where the Soviets and Americans diverged significantly in their design approach.

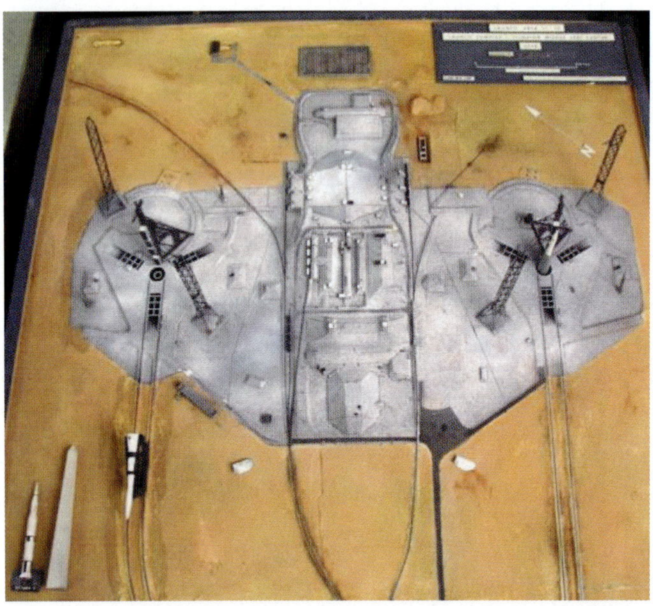

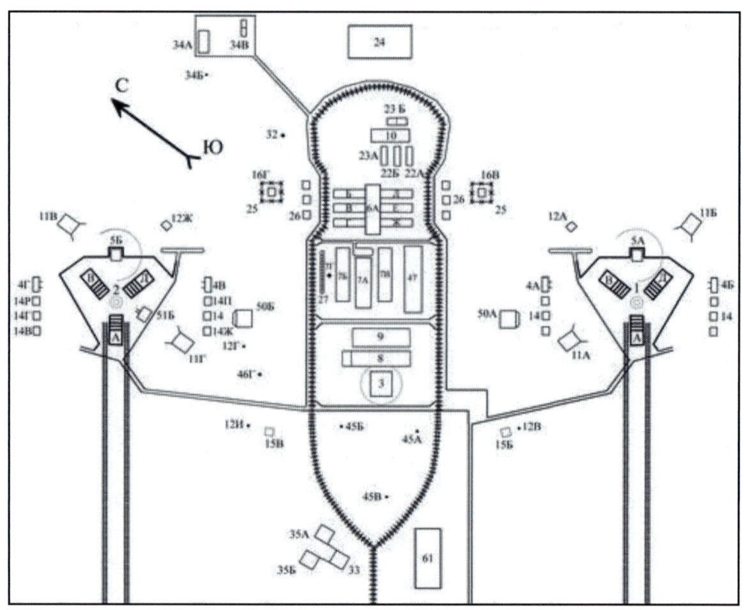

Based on interpretations of the CORONA images (such as those by Charles Vick's, see page 56) the CIA built a model (above left) of LC110. Compared to an actual Soviet site plan (right) it's pretty darned close. Note the Saturn V and Washington Monument in the lower left corner added for scale. This model is currently in the CIA museum in Virginia.

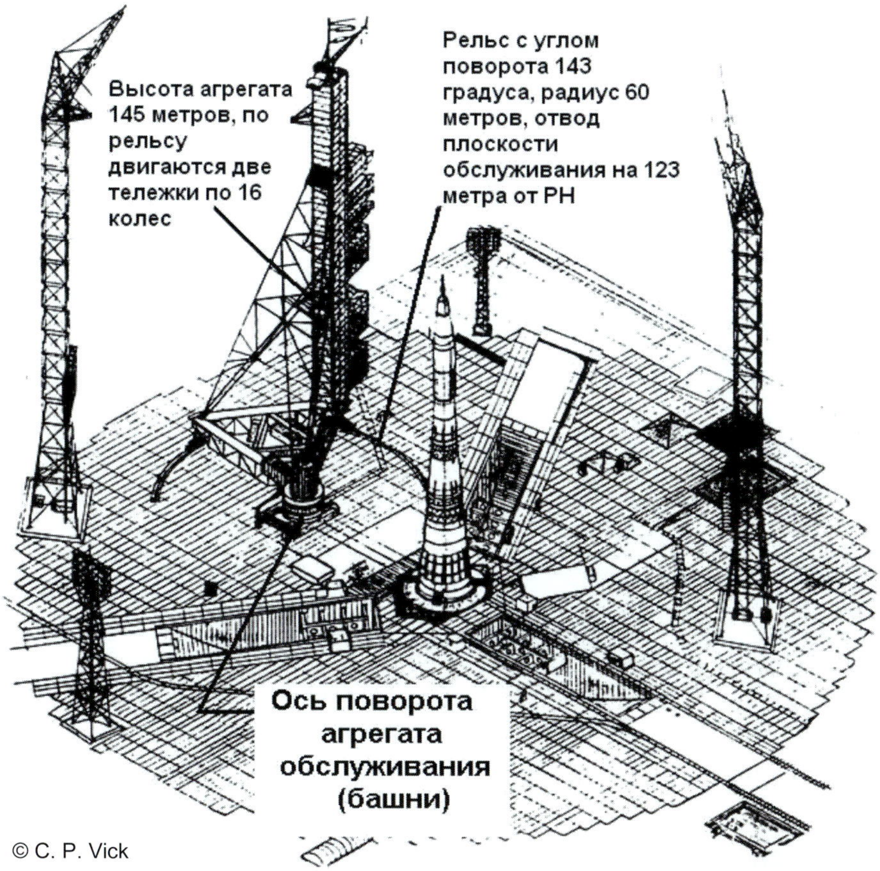

Left - Charles Vick's drawing of the N-1 Launch Pad.

Upper left caption: "Gantry Tower on rails, 145 meters (tall), moves on two trucks and 16 wheels"

Top center caption: "Circular rail with 143 degrees of rotation, a 60 meter radius, extending 123 meters from the LV (Launch Vehicle)

Bottom caption: "Gantry (Tower) rotational hub"

Below - A portion of the Support Ring and exhaust conduit. This trench is for the six central ring engines. The exhaust for the 24 outer ring engines blew by outside of the Support Ring.

Bottom - Detail of one segment (out of 24) showing the support pads and clamps rotated into the "engaged" position.

It's important that a rocket be attached securely to the launch pad. This is obviously needed during the days and weeks of preparation so that the rocket is correctly located relative to the Gantry Tower (see pp. 132/133 "Gripper Plate"). But less obviously, it is needed during engine startup to hold the rocket down until all engines are up to thrust and operating correctly.

On the Saturn V, this function was performed by four massive arms that clamped down onto "hard points" at the bottom of the first stage tail skirt and swung up and away at launch. The Soviets went with a much more belt-and-suspenders approach.

The Adapter Ring (see illustration pg. 104 and photo pg. 81) is a circular boxed structure that is attached to the bottom of Block-A. It is physically bolted to the Support Pads on the bottom of the stage (see pg. 88) using explosive nuts. Once the rocket is tilted up onto the Suppport Ring, two steel pads[1] in each of the 24 segments (48 pads total) support the weight of the rocket. Movable clamps (bottom image) rotate into position to engage the Adapter Ring and hold the rocket in place.

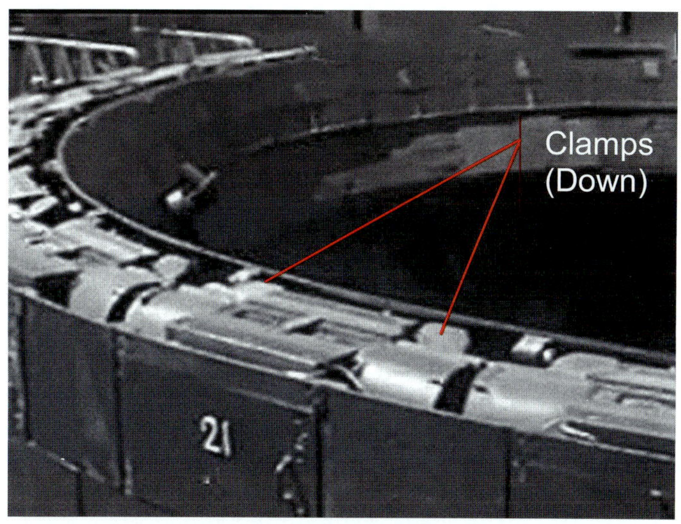

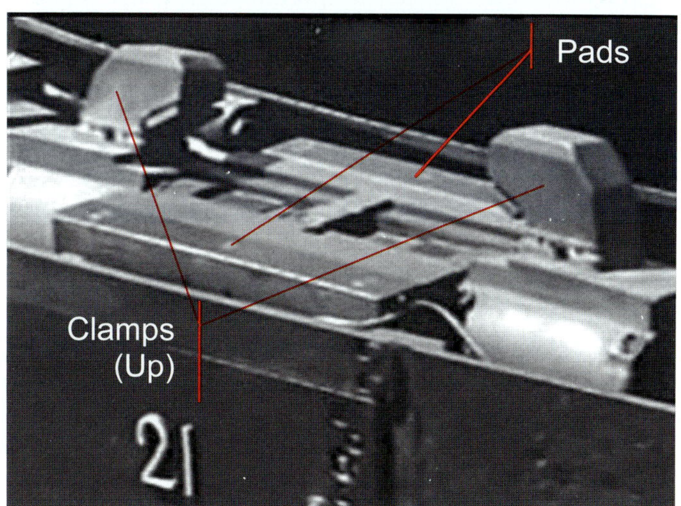

[1] These, unfortunately, also translate as "support pads." It's important to realize that there are support pads on both sides of the Adapter Ring; one set on the bottom of the booster and the others on the launch pad Support Ring.

Support and Adapter Rings after launch showing positioning rods

At launch, once the engines are started and the flight engineers are happy with their performance, the launch signal is given that fires the 48 explosive nuts and the rocket is away![2]

Inside the Support Ring is the exhaust conduit for the six engines of the Central Ring Group. The sides are beveled to form a venturi to draw in surrounding air to help keep the exhaust channeled correctly downwards as the rocket ascends. This bevel is continued upward by the Adapter Ring. The exhaust from the 24 External Ring Group engines travels into the conduit outside of the Support Ring but still within the much larger circle of the launch pad proper (with its own beveled edge). The exhaust travels 15 meters (50 ft.) down to a tetrahedron-shaped exhaust diverter that redirects it into the three large rectangular exit ports. Each port has to channel the exhaust from 10 engines and they create the distinctive look of the launch pad complex.

You would think that such a gargantuan exhaust conduit would be solid concrete, but amazingly, the area directly under the pad contains a five-story underground complex of support equipment as seen in this architects' model. The object in the center is the diverter and the rectangular opening is a section through one of the exit channels. One can scarcely imagine what it would be like to be in this area during a launch!

© Anatoly Zak

Above is an image of the Support Ring after a launch. The Adapter Ring is still attached and the magnified inset shows four of the positioning rods that aligned the Adapter Ring to the Block-A Support Pads. Some of the other rods are also shown.

Gantry Tower

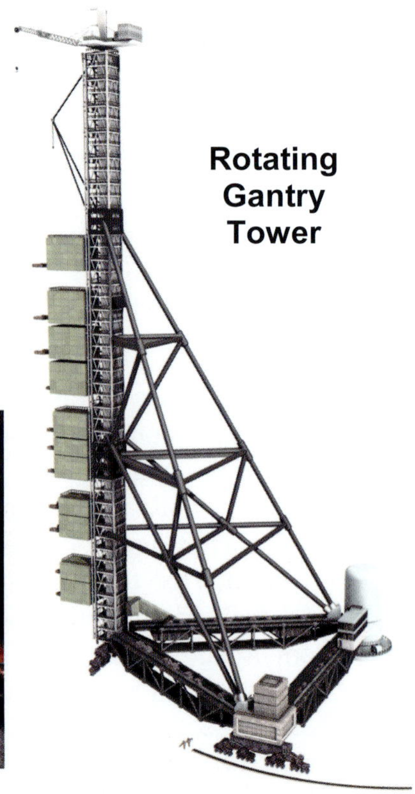

When preparing the N-1 for launch, engineers and technicians need access to all levels of the rocket. This is provided by the Gantry Tower shown to the right. Also known as the "Service Tower" or

[2] This is the hold-down scheme chosen for the Space Shuttle where four bolts with explosive nuts at the base of each SRB hold the vehicle to the pad.

"Service Structure" it is the largest movable piece of the N-1 system. As with most of the project, the tower was very complex and it helps to break it into its three functional subassemblies: the actual tower itself with its work platforms, the tower support structure and the machinery that lets such a gigantic assembly move.

The tower itself is, of course, the entire raison d'être for the system. The core tower consists of 26 individual levels, but even though the elevator may have 26 stops on it, they weren't all used in servicing the N-1. For that, there are 13 retractable service platforms along the rocket starting with Platform № 1 about mid-level of the rail carriage's main rocker (that will be explained in a bit). This "ground" level platform is actually some 3.5 m (nearly 12 ft.) off the ground.

The first three platforms are for servicing Block-A, the next three for Block-B, then three more for Block-V. Platform № 10 was for Block-G, Platform № 11 serviced Block-D, Platform № 12 for the LK and finally Platform № 13 for the LOK, which included cosmonaut ingress.

The location of the platforms, measured in height above the ground, is detailed in the photo, below, along with the top of the tower and the height of the lightning rod at the top, all in millimeters. For a nice detailed look at the sides and

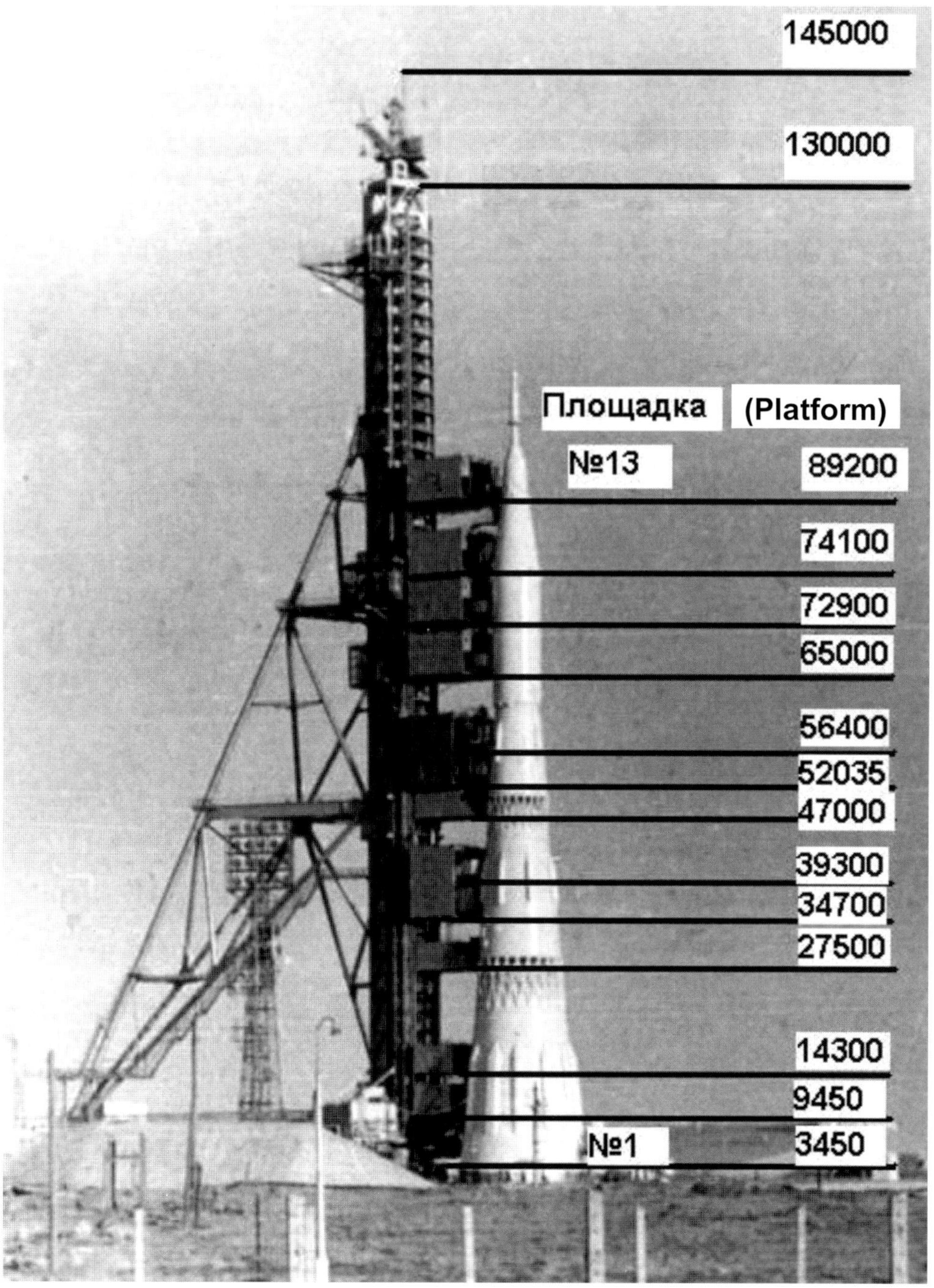

The height of the bottom edge of each platform, the tower and the lightning rod at the top (in millimeters).

open ends of the work platforms, see the image to the left. Some of the rounded guides on the platforms that match the rocket's diameter can be seen. These are all obviously retracted when the tower moves into place and then again when it moves away before launch. There are grapples to hold on to the Gripper Plates (see pp. 132/133) on the platforms that use them.

Before going on to the next section, it would be helpful once again to compare the N-1's hardware with the equivalent system on the Saturn V.

Left: The N-1 Gantry Tower with exposed Work Platforms.

Right: The Apollo 11 stack completely surrounded by the MSS (left) and LUT (right)

The part of the Saturn system that first comes to mind when looking at the N-1 Gantry Tower is the familiar red Launch Umbilical Tower (LUT) that's attached to the mobile launch platform. On the fold-out opposite is a comparison of the two to the same scale. The LUT accompanies the Saturn V all the way from the VAB where the rocket is assembled up to launch where it provides the umbilical connections until after the engines are started. Once the launch-commit is issued, the umbilicals disconnect and the arms swing to the side to clear the rocket as it ascends. Note that having the arms swing to the side is no substitute for the entire tower swinging away when launching rockets of this size. The Saturn LUT was a partially consumable item and extensive rework had to be done to the tower after every launch to repair the damage done by being that close to five F-1 engines as they passed by.

The LUT, however, was not the only service tower used by the Saturn. In the photo above you can see another large structure grasping the other side of the Saturn V stack. This is the Mobile

(continued page 185)

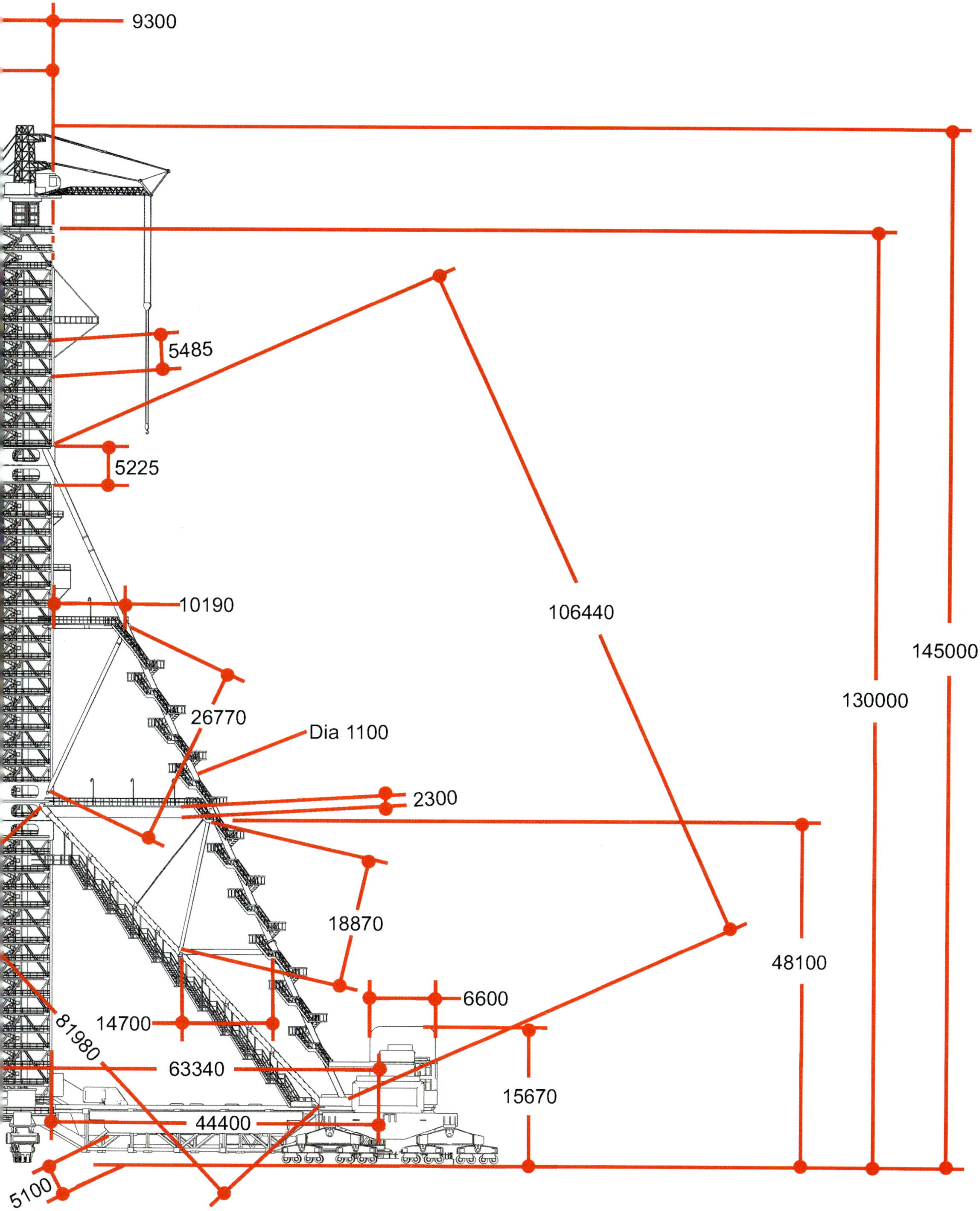
9300
5485
5225
10190
106440
145000
130000
26770
Dia 1100
2300
18870
48100
6600
14700
81980
63340
15670
44400
5100

Comparison: N-1 Gantry Scale

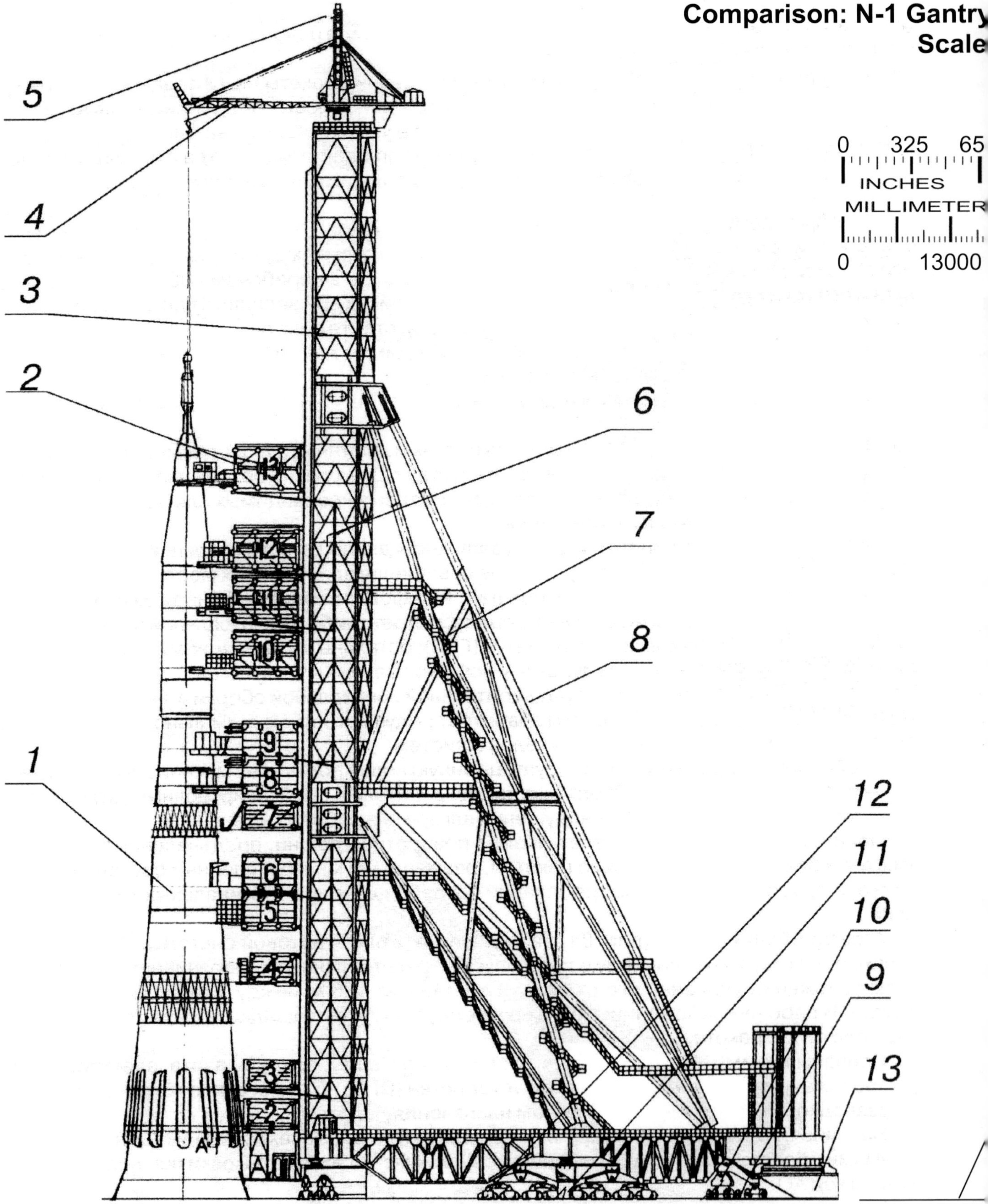

1 - ракета-носитель 11А52; 2 - площадка; 3 - ствол; 4 - монтажный кран; 5 - молниезащита; 6 - система сбора промстоков; 7 - аварийный трап; 8 - подкос; 9 - рама; 10 - холостой ход; 11 - противоугонное устройство; 12 - рабочий ход; 13 - опорно-поворотное устройство

1 - Rocket-Booster 11A52 (the N-1's official project designation); 2 - Service Platform; 3 - Tower; 4 - "Hammerhead" Crane; 5 - Lightning Rod; 6 - Effluent Collection System; 7 - Emergency Stairway; 8 - Strut; 9 - Base Frame; 10 - Idler Truck; 11 - Wheel Locking Units; 12 - Powered Truck; 13 - Rotation Hub

Tower vs. Saturn V LUT
650:1

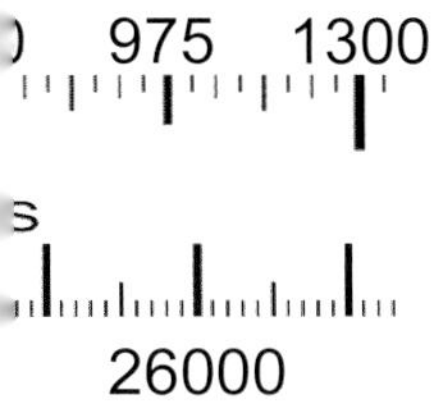

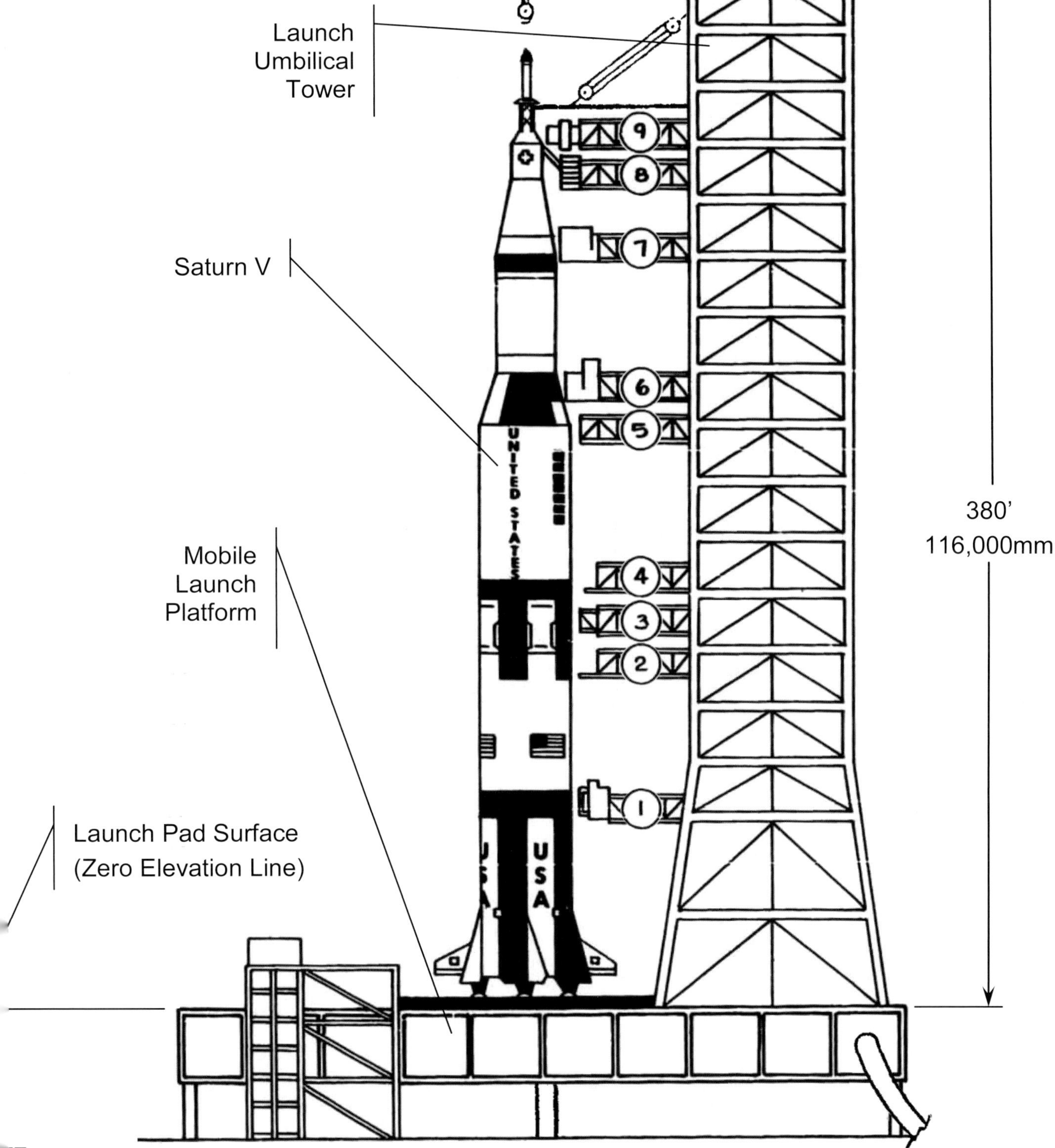

Launch Complex 110
Rotating Gantry Tower
Scale 700:1

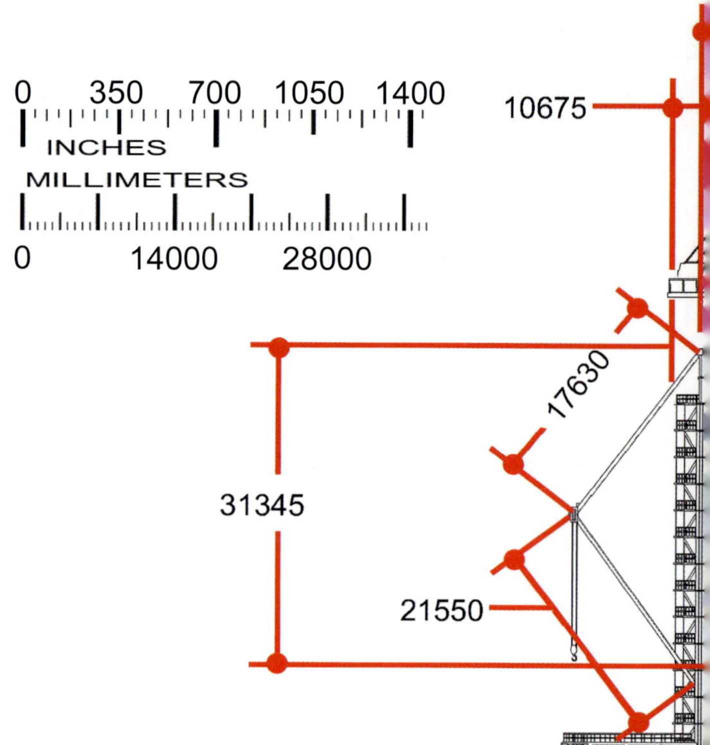

Modeling the Tower

I make no claims to have actual measurements; aside from the height of the Work Platforms and the tower itself, which are supplied from Soviet documents. The majority of the measurements shown are the result of extrapolating comparative drawings, photographs and CGI work from Nick Stevens measured with a micrometer. The end result was satisfactory for my 1/81.6 scale model (see Appendix D). I offer those measurements here to aid in any construction projects—Matt Johnson

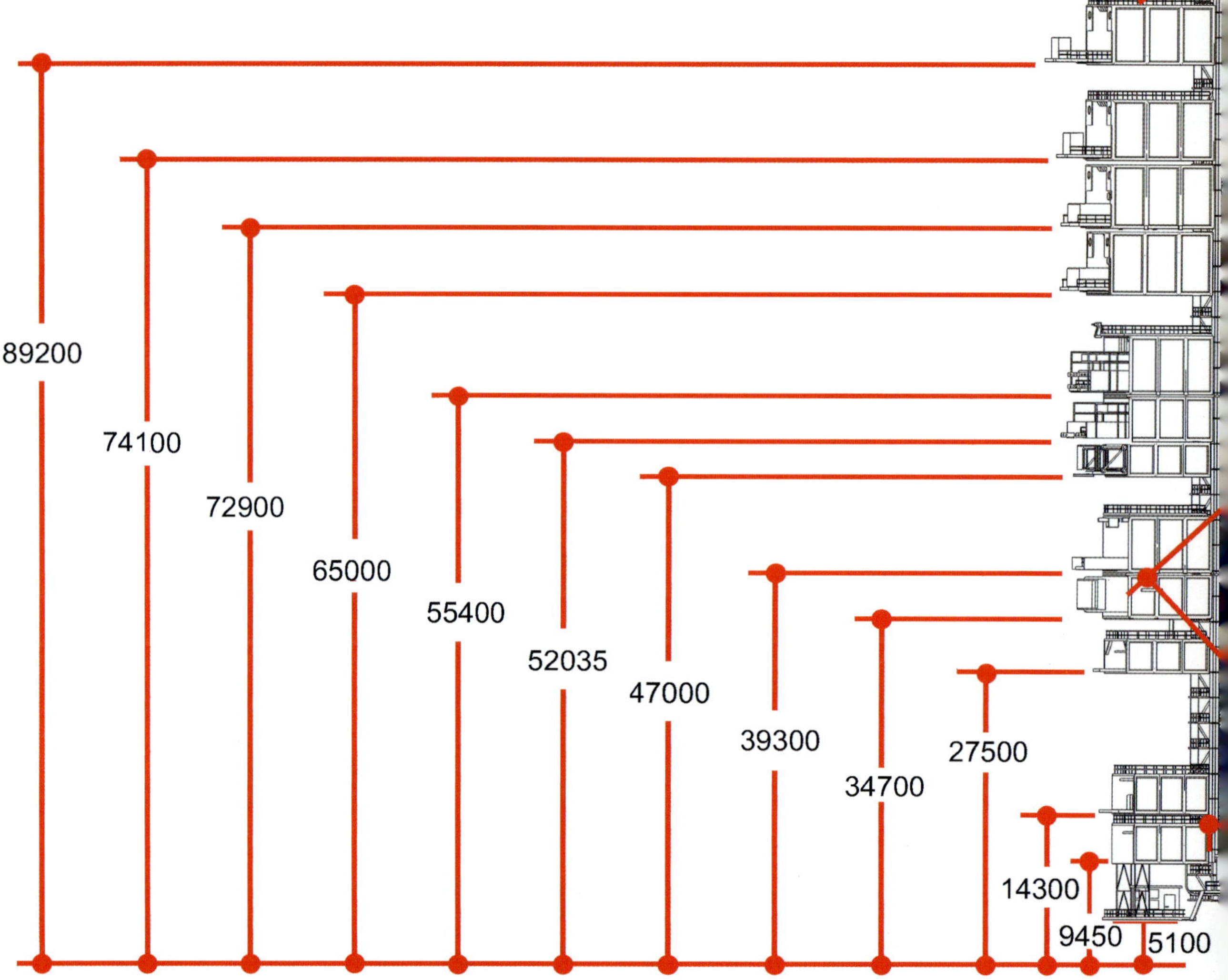

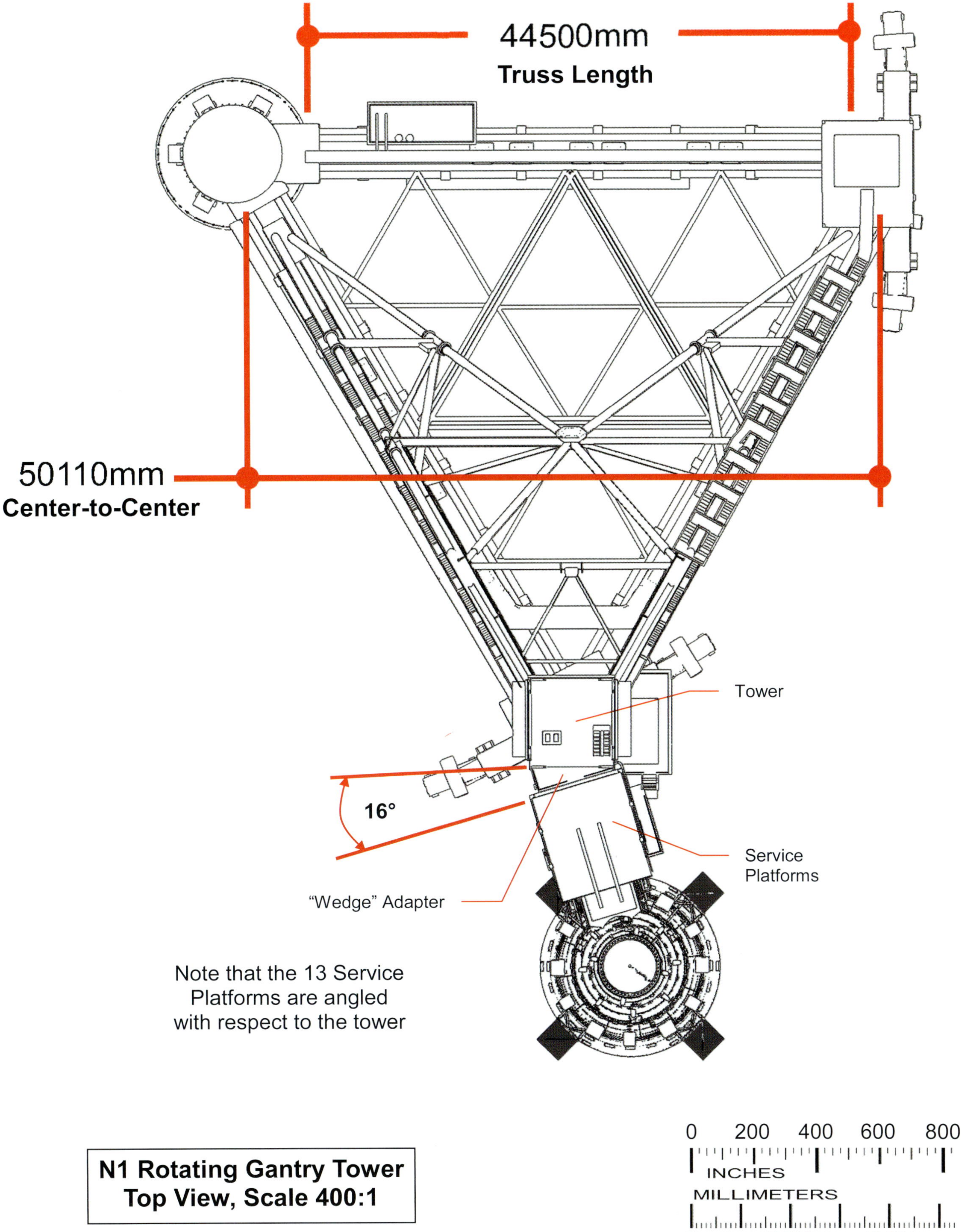

N1 Rotating Gantry Tower
Top View, Scale 400:1

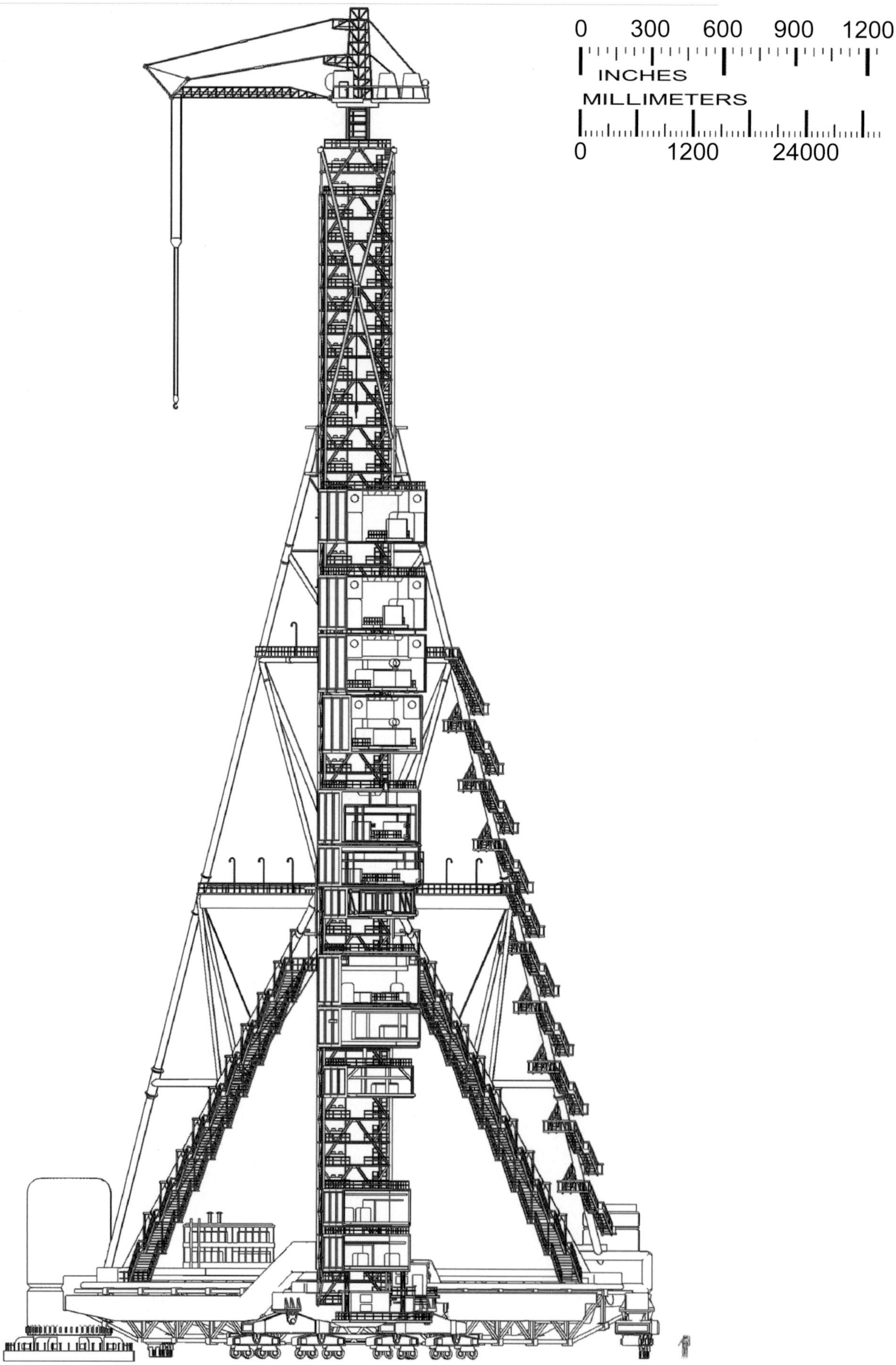

N1 Rotating Gantry Tower, 0° Rotation, Scale 600:1

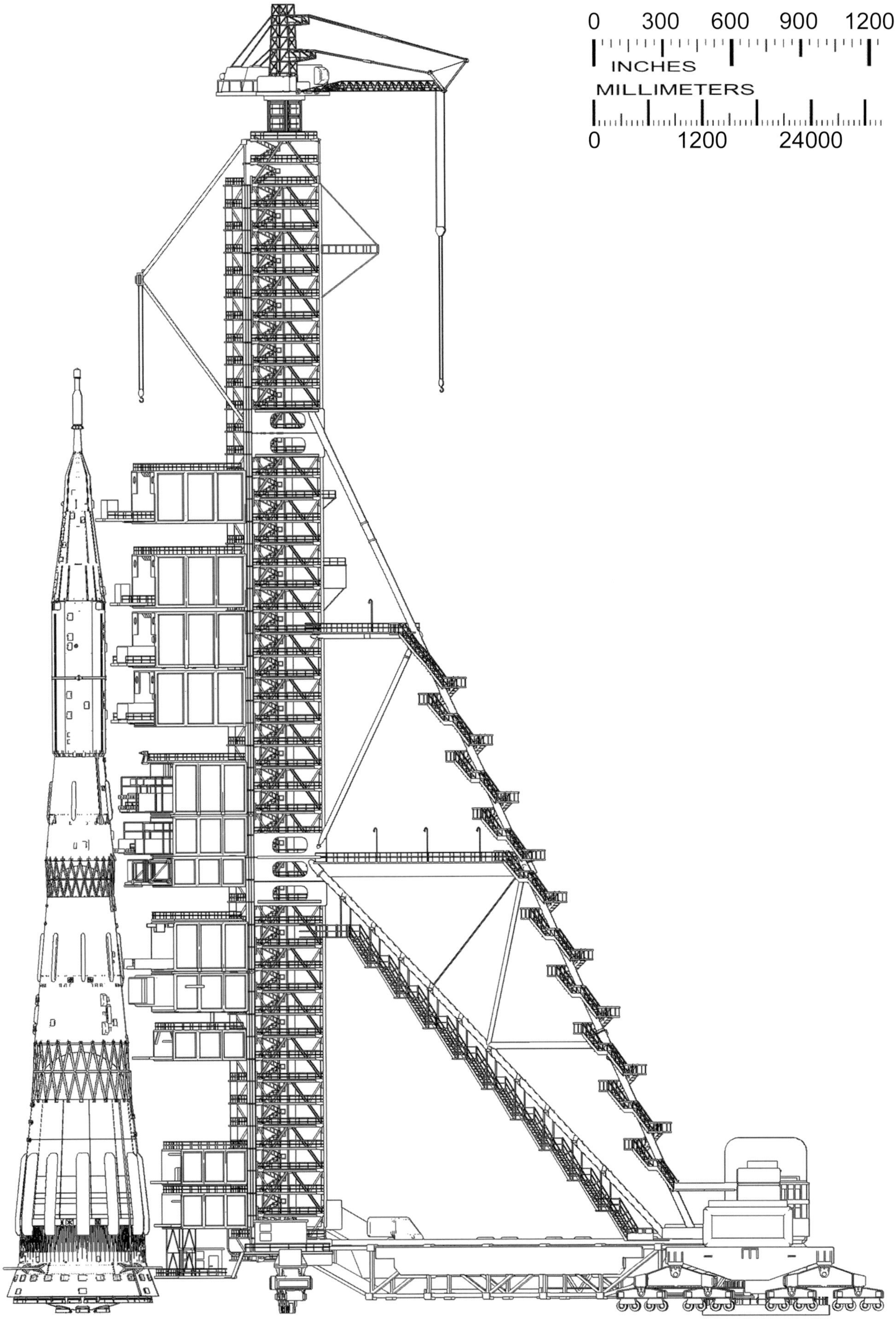

N1 Rotating Gantry Tower, 90° Rotation, Scale 600:1

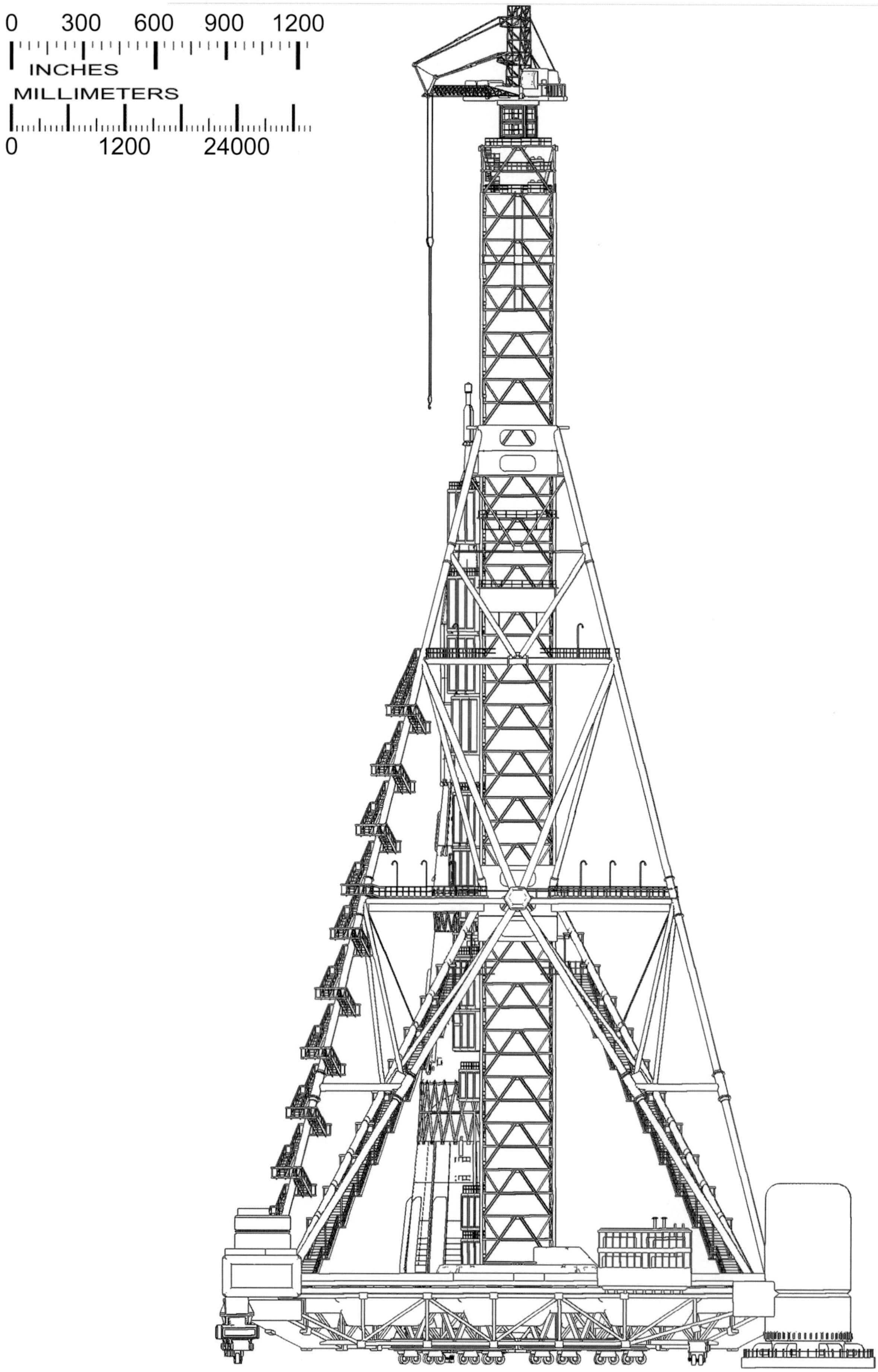

N1 Rotating Gantry Tower, 180° Rotation, Scale 600:1

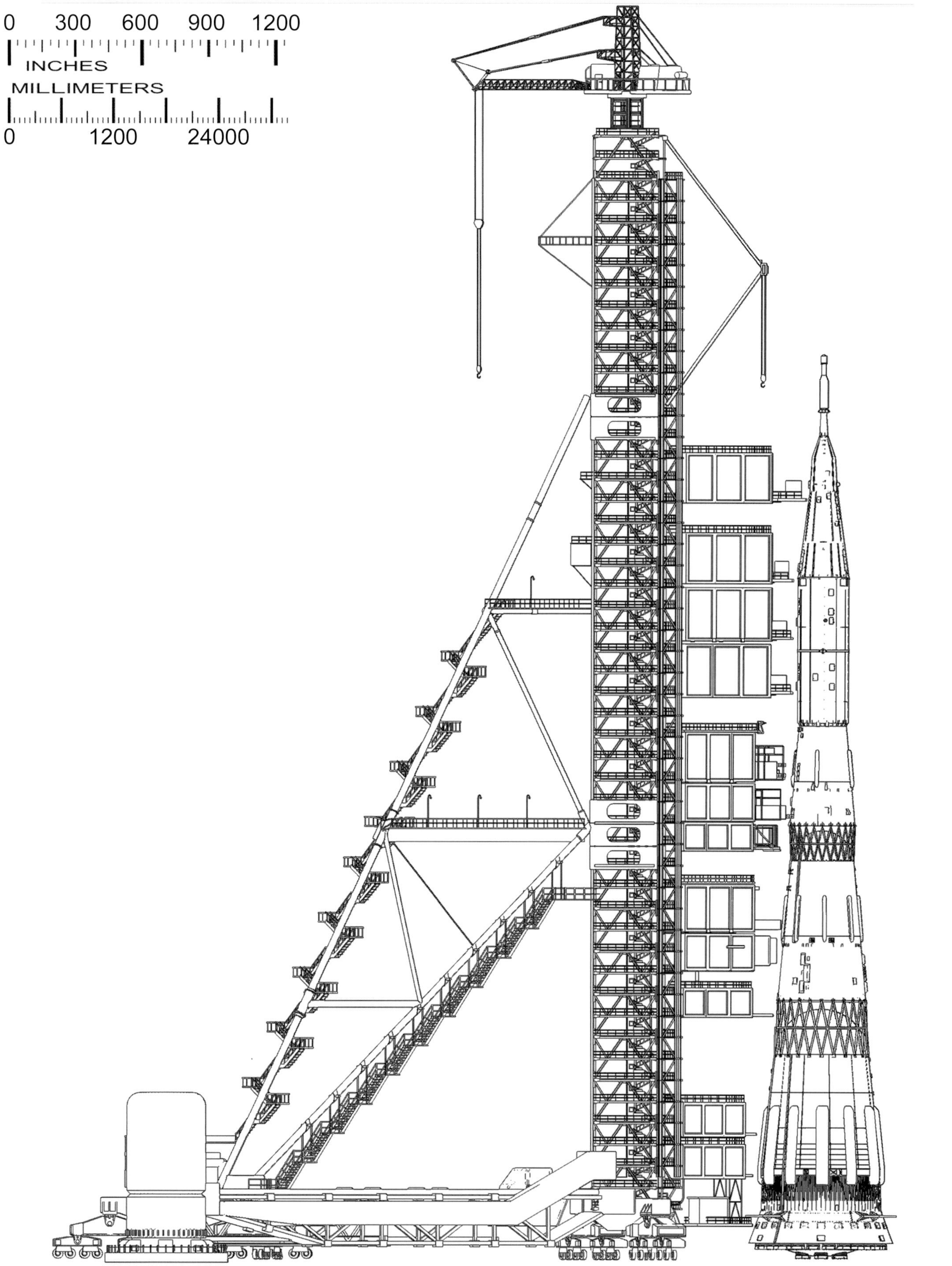

N1 Rotating Gantry Tower, 270° Rotation, Scale 600:1

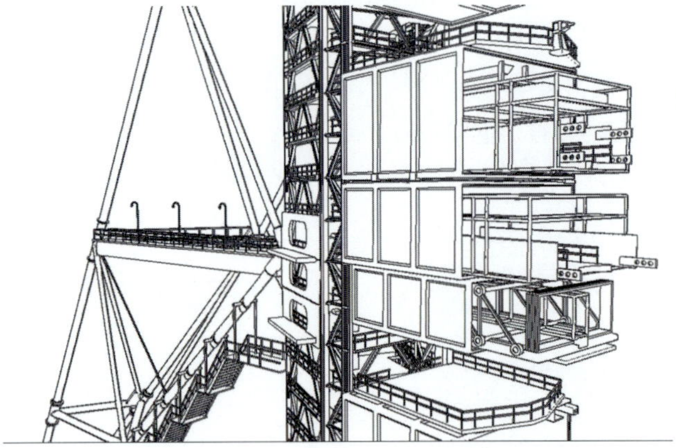

Typical Service Platform View

Crane Detail

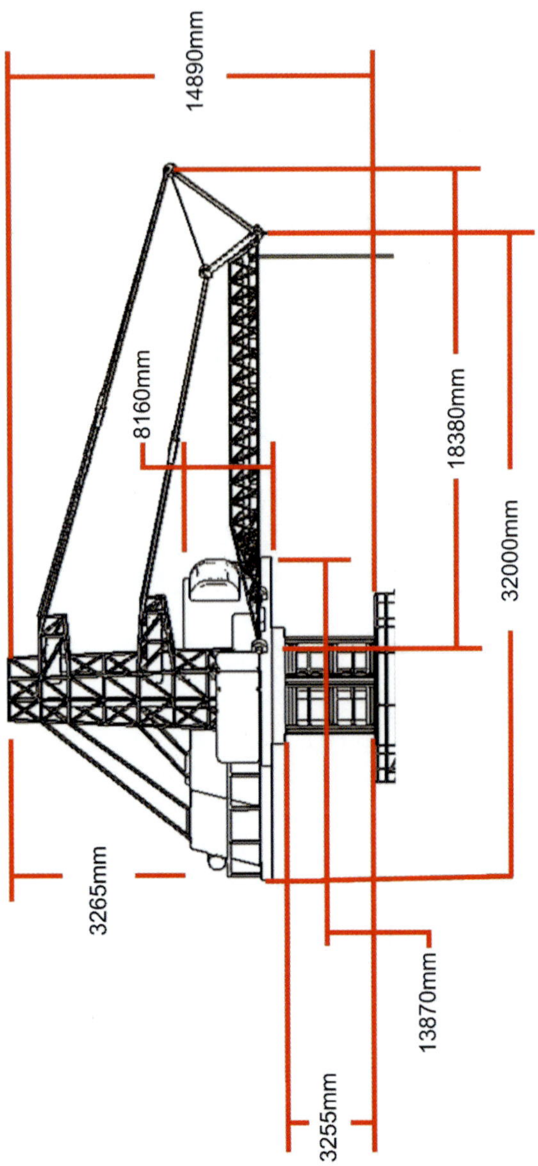

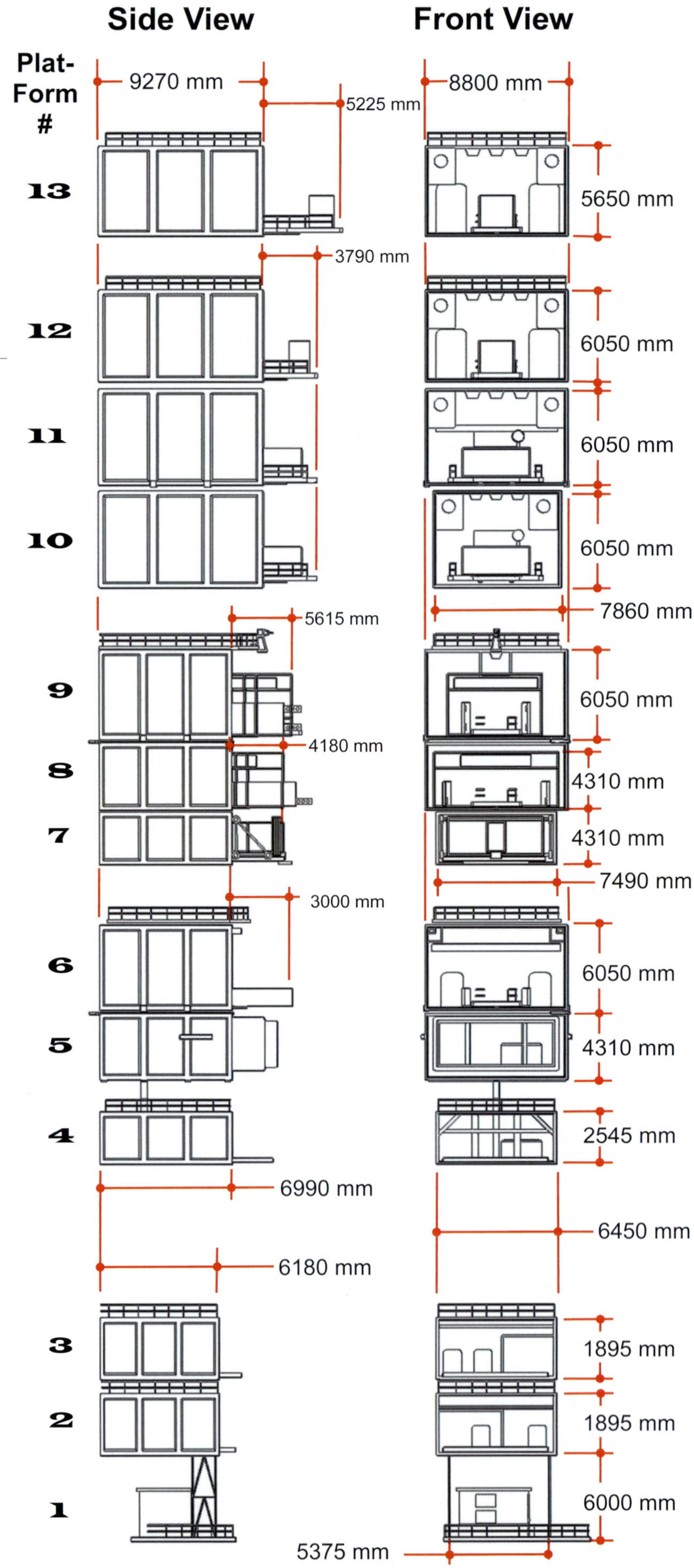

N1 Rotating Gantry Tower Details, Various Scales

Service Structure (MSS) whose use has an odd history. Originally called the "Arming Tower," its job was to provide a means to install all of the hundreds of pyrotechnic charges into the Saturn before launch. NASA safety rules consider it less of a risk to do this sensitive job in the open at the pad rather than in the controlled environment of the VAB and transporting a potentially armed rocket to the pad. To do this, though, required access to the entire circumference of the rocket at all levels, not just one side. With the Arming Tower providing that much access, more and more of the prep work was transferred there (hence the name change), reducing the LUT's role to largely umbilical connections and getting the crew into the spacecraft.

The MSS was delivered by the same crawler transporter that brought the mobile launch platform (with the rocket and LUT on top) to the pad. After dropping the platform onto its receivers, it went back along the causeway about halfway to where the MSS was stored and brought it to the pad. It returned the day before launch to remove the MSS and bring it back to the storage area.

The reason for going into so much detail on this other program is to show the amount of preparation required at the pad when launching rockets of this size. In this case the Soviets seem the model of efficiency, being able to do it all with just one tower that is able to get itself out of harm's way well before launch.

Tower Base Frame

If the tower were a permanent structure embedded in concrete, there would probably be no external framing required. But because it has to move and be located with a considerable degree of precision, a stiff supporting frame is needed.

The tower support consists of an equilateral triangular base frame (as seen in the top-view figure on page 179) topped by a support truss. Together they form an irregular tetrahedron (triangular pyramid) that's been pushed over so that one edge is completely vertical. This is the edge to which the tower is attached.

Each side of the base frame is a little over 50 meters (165 ft.) long, as measured between where the centerlines intersect:

The base sides are fabricated as open trussworks just under 45 meters (150 ft.) long joining the equipment at the three corners (tower, Central Hub and the "open" corner). Each frame truss would make a fairly high capacity bridge were they spanning a river rather than supporting a moon launch.

Tower Support Truss

The main support truss is unique among the ground support equipment in that it is made from round members rather than the more usual box section or I-beam elements. Each of the two main legs are themselves long, thin tetrahedrons that merge about ⅔ of the way up in a manner that allows the main members of both to continue to the top. Following classical design principles, the entire structure is heavily triangulated, probably best seen in Nick Stevens' CGI rendering on the next page.

Also very obvious in Nick's rendering are the ladders and stairways used to access the tower levels in the event of a tower elevator failure (or maybe if the crew just wanted a good cardio workout!).

View from the top of the winding staircase that wraps around the main tower leg

Tower Support Truss

Tower Rail Carriages

Each of the three frame corners is supported by a uniquely designed rail carriage in two different styles. The corner directly under the tower and the "open" corner both rotate on the 120 meter (400 ft.) diameter monorail outer track. These two corners are supported by a 16-wheel carriage arranged as shown in the following figure:[3]

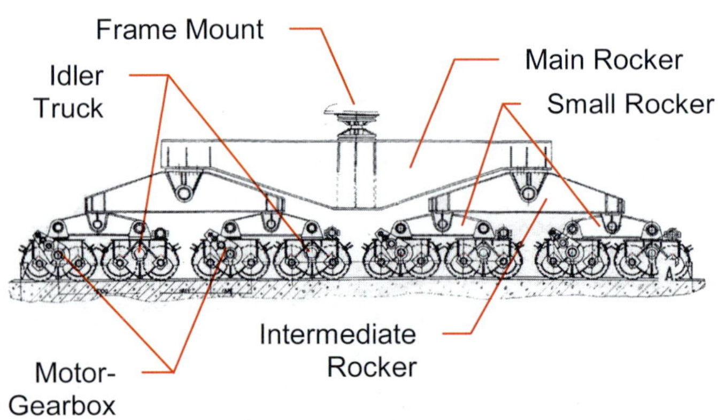

Monorail Carriage

As you can see, it is a well divided system consisting of a very large main beam (called a "Rocker") that supports the corner of the frame. The Main Rocker is itself supported by two Intermediate Rockers, which are in turn each supported by two Small Rockers. Each Small Rockers is supported by two trucks with two wheels on each truck. If you multiply that all out, you get 16 wheels per carriage. The two trucks on each Small Rocker are functionally different. The wheels in each of the left trucks in the figure are powered (you can see the motor and gearbox covering the space between the wheels) and the other truck is an idler.

Even the rail that these carriages run on had to be specially designed. With a loading of more than 160 tons per wheel, it far exceeded the load limit of any conventional railroad track. The rails were made from a high-strength alloy steel the Soviets called "G13p" and were built by the Novosibirsk Railroad Switches Plant. The Rockers, likewise, were built of AK25 steel, normally used in shipbuilding, despite the difficulties encountered in welding it. The rail was mounted to a solid concrete, load-spreading footing.

The corner of the frame next to the rotating Central Hub was completely different. First off, it wasn't a monorail, but a very narrow gauge double rail. This was needed since the radius of curvature of this track was only 10 meters (33 ft.) from the center of the Hub to the outer rail and even cutting the number of wheels to 12, they needed the carriage to be even shorter to make the geometry work.

The wheels were carried on three trucks with two axles (four wheels) each. Since 12 is not a power of 2, they couldn't use the elegant bifurcated suspension like on the other two carriages, so it was modified as shown in the figure on the facing page.

The Main Rocker still supports the corner of the frame, but only the right side, as seen from the Hub, is supported by a Sub-Rocker. The left side has an extended arm that goes down directly to one of the trucks. The other two trucks support the Sub-Rocker. The axle carriers at the end of each truck are angled so that all of the axles point to the center of rotation. Incidentally, none of the trucks on this carriage are powered. They are all idlers.

[3] Several pages of the original design report on the Rail Carriages and Hub are included (with translations) as Appendix C

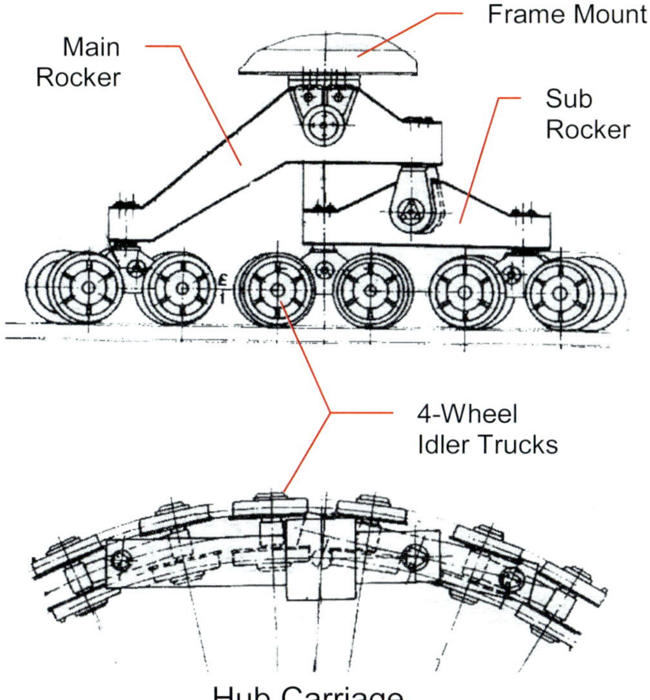

Hub Carriage

Central Hub

If the 44 wheels on the three carriages were not enough for you, there were 8 more in the system, although they ran sideways.

The Central Hub is more than just the axis of rotation for the tower assembly. It was also the central conduit of everything physical and informational that flowed between the rocket and the launch control center. Given the nature of the weather in central Kazakhstan (devastatingly cold in the winter, unbearably hot in summer) none of the utilities were constructed above ground. Power, communication, propellants, etc. were all sourced from the cluster of support buildings and underground tank farm between the two pad complexes. From there, an underground piping and wiring conduit network conducted everything to the Central Hub of each launch complex. Inside each Hub (see figure below) there was a solid half-core that provided the critical isolation between the fuel, oxidizer and power/communication cables.

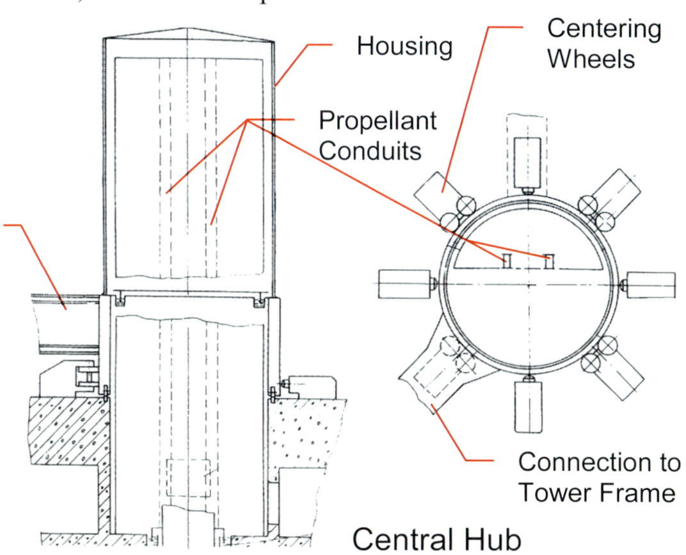

Central Hub

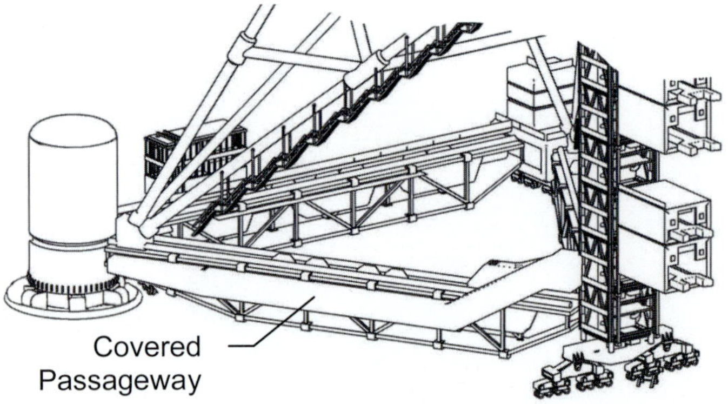

Surrounding the core was a protective steel shell that rotated with the tower. While the Hub was functionally connected to the tower frame, it did not support any of the weight, except for its own. The four sets of double wheels kept the Hub structure centered against any back-and-forth tugging by the tower frame as it rotated, pulling the Hub with it. To the right is the architects model of the Hub and the complex of support equipment beneath it. As with the complex under the launch pad (page 174) there is much more here than meets the eye (or spy satellite).

© Anatoly Zak

The huge propellant lines used to fill the rocket's tanks exited the Hub about midway up and were conducted to the Gantry Tower on a piping shelf, whence they travelled up one of the tower legs, splitting off at the different stages. Power and communication lines from the Hub travelled through a covered passageway attached to the frame side that went directly from the Hub to the Tower (figure left). The passage was possibly also used as an emergency escape route.[4]

Below are two frames lifted from a Soviet documentary on the program showing the Central Hub as seen from partway up the tower.

The left image is the dizzying view from the main truss leg beam (note the ladder rungs mounted directly to it). To the right is the same view after the camera had zoomed in on the hub, nicely resolving the Centering Wheel assemblies and other details.

"Grasshopper" Transporter

From the beginning of modern rocketry (~1920's) rockets have been built in assembly buildings and transported to the pad for launching. The assembling and transporting have almost always been with the rocket horizontal. This is for the obvious reason of gravity: Humans don't do "up and down" very well but we are pretty good at "sideways." This means that if you are building something long and thin, it's much easier if the whole thing is at one level where you can reach it. When transporting, it's a lot safer if the center-of-gravity of the vehicle is right down on the transporter rather than hundreds of feet in the air, and that it's securely attached along its length rather than balancing precariously on its tail. In other words, it can't fall over because it's already over.

However, after WWII when rockets started getting very large, unanticipated problems arose. For all their awesome power and impressive capa-

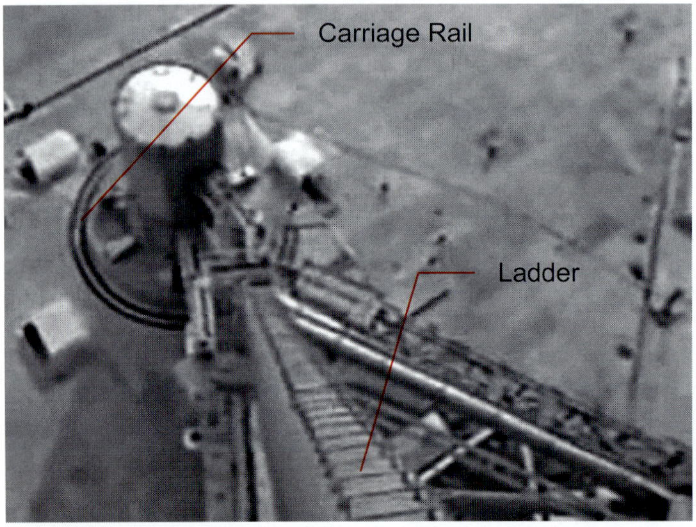

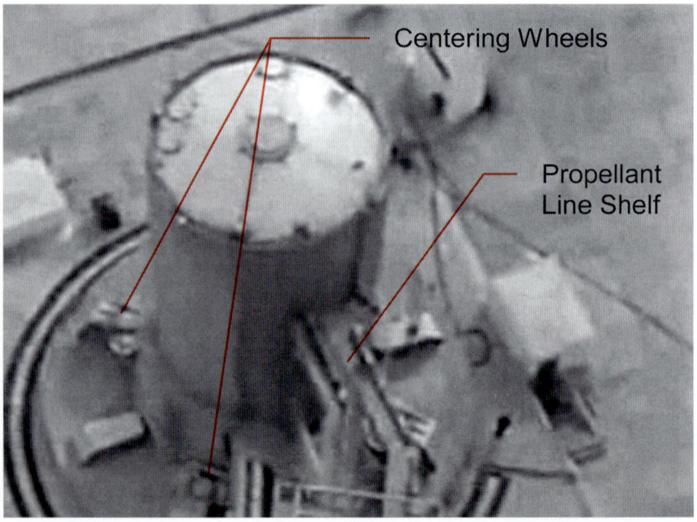

[4] This use of the covered passageway as part of the emergency escape system was found in documentation on the Buran program. No definitive document has been found establishing this use during N-1.

Two views of N1-5L leaving the Site 112 Assembly Hall. Left is a frame from a Soviet documentary, hence the lower quality. Note that the Locomotives are pulling the "Grasshopper" from the N-1's base end.

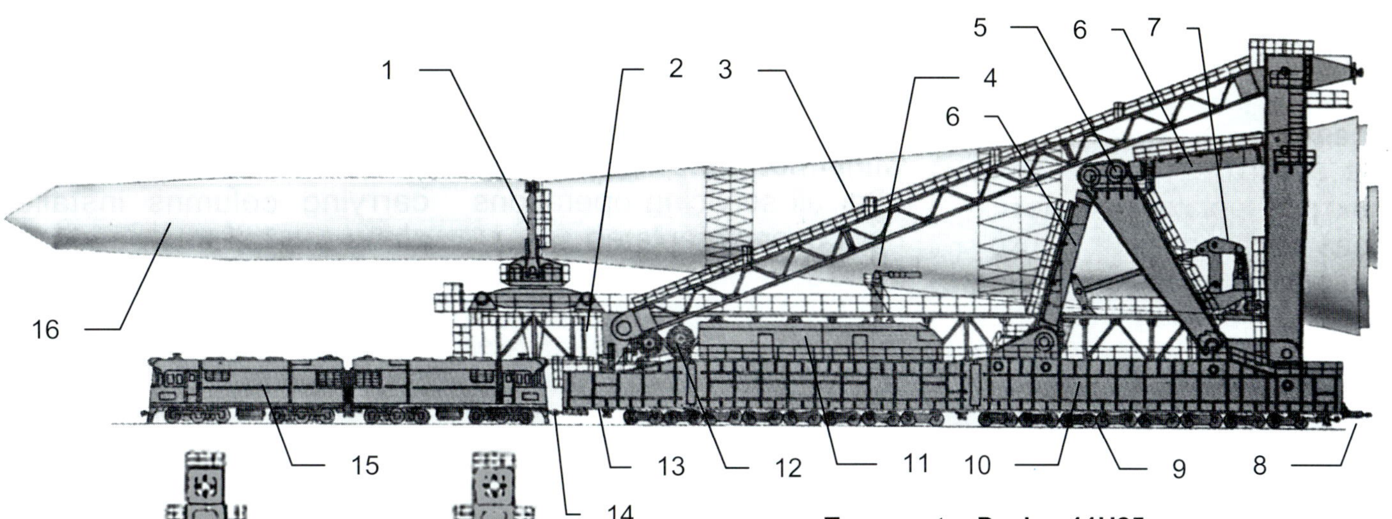

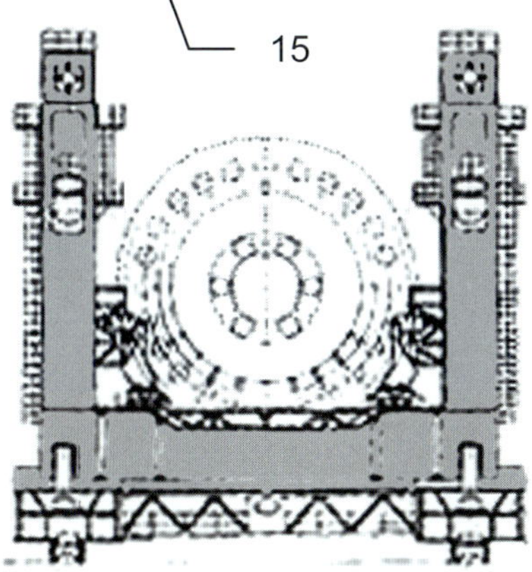

Transporter Design 11U25
(With Boom and Horizontal Positioner)

1 - Upper Support; 2 - Boom; 3 - Stairs and Service Platforms; 4 - Middle Support; 5 - Hoist Assembly; 6 - Hydraulic Cylinders; 7 - Lower Support; 8 - Alignment Mechanism; 9 - Chassis; 10 - Frame; 11 - Electro-hydraulic Pumps and Control Unit; 12 - Cable Reels; 13 - Hanging Mechanism; 14 - Coupler; 15 - Locomotives; 16 - Rocket-Booster 11A52 with Lunar Complex L3

Конструкция установщика 11У25 (со стрелой в горизонтальном положении):
1 - верхняя опора; 2 - стрела; 3 - комплект лестниц и площадок обслуживания; 4 - технологическая опора; 5 - механизм подъема; 6 - гидроцилиндр; 7 - нижняя опора; 8 - механизм доводки; 9 - ходовая часть; 10 - рама; 11 - контейнеры с электро- и гидрооборудованием и пультом управления; 12 - кабельные катушки; 13 - механизм вывешивания; 14 - автосцепки; 15 - тепловозы; 16 - ракета-носитель 11А52 с лунным комплексом Л3

The Grasshopper, about halfway between the Site 112 Assembly Hall (background) and Launch Complex 110 West (the twin rails for 110 East are in the foreground). Note that the locomotives have already been switched so that they're pushing from the N-1's L3 end.

bilities, rockets are essentially thin metal balloons. On average, the structure of the rocket represents about 10% of the liftoff weight and the payload about 1% (the other 89% is propellants). The designers want to use the absolute minimum amount of structure in the rocket because every 1% of structure saved from the rocket represents 10% more payload. Since the engines at the back push the rocket vertically, that's the direction that they are strong, but they can be pretty weak side-to-side. The problems arise when you consider that the transporter isn't holding the rocket along its whole length but just at a few points. This means that, like a rope being held only at its ends, its going to sag in-between the supports in its weak direction. It gets worse when the rocket is erected for preparation and launch since the cradle has to push more on the top end of the rocket, putting what engineers call a "bending load" on the whole structure (also in its weak direction).

The obvious solution to this problem is to stack and transport the rocket vertically so that it is never subjected to bending loads. This encounters the assembly and balance problems mentioned earlier which can only be solved by building an enormously tall assembly structure, the doors of which have to be even taller than the rocket, and transporting it very, *very* carefully. This is the solution that NASA pursued, lead by Saturn V designer Wernher von Braun (see sidebar, right).

A Brief History of Vertical Transport

The first person to recognize the problem of transporting very large rockets was, not surprisingly, the first person to work out the requirements of what it actually takes to get to the moon: Hermann Oberth, the original German Rocket Scientist.

In his 1923 book *Die Rakete zu den Planetenräumen* ("The Rocket into Planetary Space") Oberth proposed the *Modell E*, a rocket large enough to take a crew on a single loop around the moon (essentially the *Zond* mission profile, see page 48). He noted that the rocket would be too weak to transport sideways, but had no real solution until his follow-up book in 1927. That became the design used by Fritz Lang for the *Friede* in his 1929 movie *Die Frau im Mond* ("The Woman in the Moon") on which Oberth was technical advisor.

Oberth was von Braun's mentor and this concept "from the master" stayed with him. It was presented to the public on the first episode of Disney's *Man in Space* series, plus, of course, being the technique chosen to transport the *Saturn V*.

N1-5L approaches the launch pad. The 24 segments' support pads and latching clamps are visible on the Support Ring in the center.

When designing the N-1 system, Korolev's OKB-1 bureau did evaluate vertical transport because of its advantages in making the vehicle lighter, but were thwarted by the same thing that confounded Napoleon and Hitler: the Russian weather. While the central Florida coast does get several hurricanes every year, these come with a week or more of warning; plenty of time to move the vehicle back into the VAB if needed. The rest of the time is relatively calm and it very rarely dips below freezing.

Central Kazakhstan has, by comparison, very high and almost continuous winds which would make stabilizing a 100 meter tall rocket problematic at best. Combined with the difficulty (read "lack of funding") to build an unprecedentedly tall Assembly Hall, they bit the bullet and opted for horizontal transport. While the assembly building didn't have to be quite so tall, that was offset by the requirement to be extra long...and wide enough to accommodate two N-1 assemblies at a time. When finished, the assembly building was 240 meters (780 ft.) long by 190 meters (620 ft.) wide and 47 meters (155 ft., more than 15 stories) high making it the largest enclosed single volume in the Soviet Union.

Even if horizontal transport did away with the balance problem of the rocket standing upright, the rocket was still every bit as massive and needed an equally massive transporter. Nicknamed "The Grasshopper"[5] (for reasons that hopefully are obvious), it had to deal with the gigantic size of the first stage, which presented a balance problem of a different sort. The standard Soviet technique of transporting rockets on a modified railcar just wouldn't do. A rocket nearly 17 meters (55 ft.) in diameter required a transporter some 27 meters (90 ft.) wide. A vehicle that wide could not be safely balanced on a pair of rails less than two meters apart. A good stiff breeze (of which there were many) could easily blow it off the rails.

The twin-rail solution, with some 20 meters (65 ft.) total span between the outer tracks, seems obvious. It can even be seen in the still from *Frau* shown on the previous page. However, this overlooks the possibility of curves. Oberth's concept had a straight shot from the assembly building to the launch pad, as did the von Braun version used in the Disney film. The reality of the Baikonur site is that it's anything but straight, as can be seen in the photo at the top left of the previous page.

This made for significant difficulties for the drivers of the two locomotives that moved the whole massive assembly. When going around curves they had to vary the speed of each independently according to a geometric formula to keep from torquing the transporter left or right, which could lead to a derailment. Additionally, they had to uncouple the locomotives to switch ends partway along the trip. To pull the assembly from the Assembly Hall they were attached to the end of the Grasshopper below the N-1's base, but when pushing it onto the pad they had to be out of the way behind it, under the L3. Note that they are in front of the transporter in the photos at the top of page 189, but by the time the assembly had left Site 112, the locos had been moved to the back (photos on opposite page and above).

[5] Interestingly, the Space Exploration Corp ("Space-X") picked the same name for their vertical take-off and landing test vehicle.

All of this says nothing about the modifications to the N-1 itself to allow it to be transported sideways. In addition to having to strengthen the basic structure against bending, every single component, especially the giant propellant tanks and the extremely heavy engines, had to be mounted in a way that let them hang sideways. This made the mounts more complicated and that increased structured added weight.

Further, you couldn't simply lie the rocket down on its thin fragile skin; it would be crushed like an eggshell. The N-1 is supported horizontally at three very specific "hard points" on the three stages where there are special load-spreading structures installed. The cradle mounts for Block-A (called out on pages 83 and 97) and Block-B (pages 109 and 115) are attached through the skin to the inner frame that also supports the liquid oxygen tanks. The hard point for Block-V is not a dedicated frame-mount, but rather a special cradle at the forward end of the Grasshopper's boom which grabs the stage around the Payload Adapter Ring, already a reinforced structure.

Finally, you'll note that the Grasshopper's support ends with this forward cradle. The entire 30-tons (unfueled mass) of the L3 Complex with

shrouds and escape tower is hanging unsupported off of the front of the booster. Structural engineers call this a "cantilever" and they are difficult enough to do when it's a building made of steel and concrete embedded in a solid foundation. On the moonship this means that Block-G, Block-D, the LK lander and LOK all have to be able to hang sideways on their mounts and each module has to absorb the bending loads of all of the modules above it; then the entire 30 tons is cantilevered off of the Payload Adapter Ring at the front of Block-V.

The N1-M1 pad check vehicle. No reinforcing stringers at the bottom of Block-A

The final consequence of horizontal transport is the ability to erect the vehicle from horizontal to vertical at the pad. This is impressive when done with an R-7 or a *Proton*, so imagine the job it is to do it with something over 100 meters long and weighing several hundreds of thousands of kilograms, even when empty.

The hydraulic rams needed by the Grasshopper were far larger than anything that had been built previously in the Soviet Union. The biggest rams available at the time were, in fact, the ones developed for erecting the *Proton* booster. They had a hydraulic cylinder one full meter (40") in diameter and nine meters (30 ft.) long. The telescoping double-elements could extend up to 16 meters (52 ft.), but even that was too short. Engineers brainstormed different combinations of such devices and in the final design two pairs of such rams were used. They were mounted two on each side working against a central brace as seen in the photo opposite of N1-3L being raised. They were powered by on-board hydraulic pumps and controls.

As they moved the N1-M1 test vehicle onto the launch complex, they came up against one of the problems they were hoping to avoid. After tilting it up onto the pad, cracks were discovered encircling the base of Block-A right below the fuel line covers. This area is behind the aft-most cradle mounts as can be seen in the photo below. The 30 very heavy engines hang from their mounts cantilevered out unsupported and the bending moment

Reinforcing Stringers

The bottom of N1-5L Block-A on the pad showing the reinforcing stringers. The caption, pointing out the cradle mounts, reads: "Three sockets for attaching to the transporter supports, MIC and MIC cranes" (MIC is the Russian acronym for the Site 112 assembly building).

imposed on the structure cracked the skin and the inner frame. The solution was to add dozens of small reinforcing strips, called "stringers," around the base which, of course, increased the weight and reduced the payload. These were used on the first three N-1's launched, but by N1-7L they had improved the internal structure of the frame and lower skirt so the stringers were no longer needed.

Site 110 Legacy

With the many comparisons to the Apollo program, both in this chapter and earlier, the reader should have a pretty good understanding that Apollo and N1-L3 followed a parallel evolution throughout the 1960's and early 70's. Perhaps not surprisingly, this paralleling continued into the 1980's with both nation's reusable spacecraft programs capitalizing on the infrastructure built for their respective moon programs.

Buran lifts off on its one and only flight. The name seems quite literal when looking at the "snowstorm" of ice cascading off of the *Energia* booster.

For the Space Transportation System (the Shuttle's official project name), NASA reused the VAB, which could stack two shuttle assemblies at a time, and the Mobile Launch Platform/crawler-transporter to get it to Launch Complex 39, which could prep two shuttles at a time on pads A and B.

For *Buran* ("Snowstorm")[6] the Soviets reused the Site 112 assembly building, which could stack two Energia-Buran assemblies at a time, and the Grasshopper transporter to get it to Launch Complex 110, which could prep two shuttles at a time on pads East and West.

Of course, the use of these facilities by the new programs was hardly automatic. Extensive rework had to be done to the launch pad to accommodate Glushko's *Energia* booster (the rocket he developed after taking over OKB-1 and cancelling the N-1[7]). The rotating Gantry Tower was cut almost in half and a second, sliding tower was built on the other side.

The Grasshopper transport, too, was heavily modified. A combination of *Buran* being a shorter package and having a center of gravity closer to the end meant that the large diagonal braces on either side that gave the transporter its distinctive look (and its nickname) were not needed.

The *Energia* booster itself had to be modified for use with the shuttle. Originally, the four strap-

[6] In an interesting reversal, the Soviets went from a technical descriptive name for their moon program, to a lyrical-poetic name for their shuttle program while the Americans did just the opposite.

[7] Ironically, *Energia* does not use the hypergolic propellants that Glushko was pushing Korolev to use in the N-1, but rather LOX/kerosene (the N-1 propellants) in its four strap-on boosters and the much more efficient hydrogen/oxygen in its core vehicle. With the ability to put 105 tons into low Earth orbit, it actually has a lift capacity 10% greater than the N-1.

on boosters are spaced equally around the core vehicle, but the *Buran* orbiter takes up one of the positions, so they had to move the strap-ons around to make room. There was no room directly opposite the orbiter since that's where the cradle points on the core vehicle have to go. However, there was lots of width available thanks to the transporter being designed for the 16+ meter wide N-1 first stage. They decided to place the side boosters in groups of two on either side to take advantage of the generous space there. This had the coincidental effect of making the pair of strap-ons on either side look like a single unit and visually very similar to the US shuttle with the two Solid Rocket Boosters on either side of the central tank.

With the orbiter's shape admittedly copied from the US design (to save them the time and research on the aerodynamics), this made it appear that they had copied the entire STS system when, outside of the orbiter's basic configuration, they are really very different systems. All of the boosting effort comes from the *Energia* with the *Buran* basically unpowered except for the small orbital maneuvering engines. This actually gave the Soviet version several tons more payload capacity than the US shuttle since it didn't have to carry its heavy main engines into orbit.

On 15 November 1988 The *Energia* booster carrying the unmanned *Buran* lifted off from LC110 West for the shuttle's one and only flight. Three and a half hours and two orbits later, it landed by autopilot at the Yubileiniy airfield (Site 251) in Baikonur after an uneventful and completely successful flight (the only successful flight launched from LC110).

Buran and *Energia* being transported to LC110 (background) atop the emasculated Grasshopper

Epitaph

The collapse of the Soviet Union in 1989 meant the end of funding for both *Energia* and *Buran*. Before the shuttle flight, *Energia* had flown only once, in May 1987, carrying a military payload called *Polyus*. That flight was from the other *Energia* pad (Site 251) which was not man-rated. Both *Energia* and *Buran* were transferred to the new civilian Russian Space Agency, but it had no funds to fly such expensive programs, so all of the remaining *Energia* hardware, and the *Buran* shuttle that had made the flight[8] were mothballed in one of the N-1 assembly bays of Site 112. Finally, in the summer of 1993, at a very western-style press conference, the RSA announced that both programs were officially canceled. However, the stacked Energia-Buran in storage became a very popular "tourist stop" with journalists who would visit Baikonur for international launches. That is until it was destroyed when the roof collapsed in 2002 during a repair.

For the hundreds of millions (possibly over a billion) rubles spent on building, repairing and renovating/upgrading LC110, there were a total of only five launches flown from it: N1-3L and 5L (from LC110 East) and N1-6L, 7L plus the Energia-Buran flight (all from LC110 West). It is quite sad to see the state of decay into which the complex has fallen since being abandoned in the early 1990's. On the following page are a few images of the complex from around 2005.

---

[8] Depending on how you count them, there were a dozen *Buran* shuttles built, or at least contracted: five flight versions (two first-generation and three second-generation) and six or seven assembly test, drop test and pad-fit versions. The flight versions were:

- 1.01 *Buran* (Буран, "Snowstorm") - Originally called *Buria* (Буря, "Storm") since "Buran" was the program name, it was changed about one month before its flight.
- 1.02 *Ptichka* (Птичка, "Little Bird") - This was the nickname given to all the shuttles by the engineers during the program, but it stuck specifically to this orbiter.
- 2.01 *Baikal* (Байкал, "Nature") - First of the 2nd-generation orbiters. The design upgrades were based on lessons learned from the Orbiter 1.01 flight (mostly cockpit avionics and control systems).
- 2.02 Existed only as a partially completed airframe. Did not get far enough for a nickname. May have been destroyed when the program was cancelled.
- 2.03 Most parts were fabricated and delivered, but construction was not started.

Photo Gallery: The remains of Launch Complex 110

A - Approaching on the abandoned rail line

B - The Complex as seen from the Rotating Gantry Tower

C - The shortened Rotating Gantry Tower with the giant hinged service platform that surrounded the Shuttle;

D - The Sliding Service Tower built for *Buran*,

E - Reclaiming a bit of land under one of the Rotating Tower's edge trusses.

Appendix A

The L3 Moon Mission Renderings
by Nick Stevens

It all really happened by accident. I had just got the hang of some key techniques needed for digital rocket modeling so I started looking for a new project. I felt that US rockets had been done many times so I looked at Russian/Soviet ones. I found a couple of photos of the N-1, which I thought was distinctive and visually interesting.

I still remember the thought that went through my mind at this time: "There are only a few photos, so it will be easy to include every detail possible in the model, and make the best digital one ever built." So I kind of accidentally committed to a complete model!

And finding the information a bit at a time to start with meant it seemed like it was not going to be much more work for a long time. The scope grew slowly. The best bit was getting the information from Alex, and realizing just how good it could be. And the worst bit was when I realized I had messed up to the point where I needed to discard it and start from scratch.

Its sobering to think it took me longer to make this CGI model than it took the engineers to build the real thing!

– Nick Stevens

Nick standing in front of the Royal Observatory at Greenwich, the point defining the Prime Meridian of the globe

Clockwise from above:

Liftoff with all 30 NK-15 engines in Block-A running!

Staging to Block-B

Continuing to orbit using Block-V after the L-3 payload fairing jettisoned

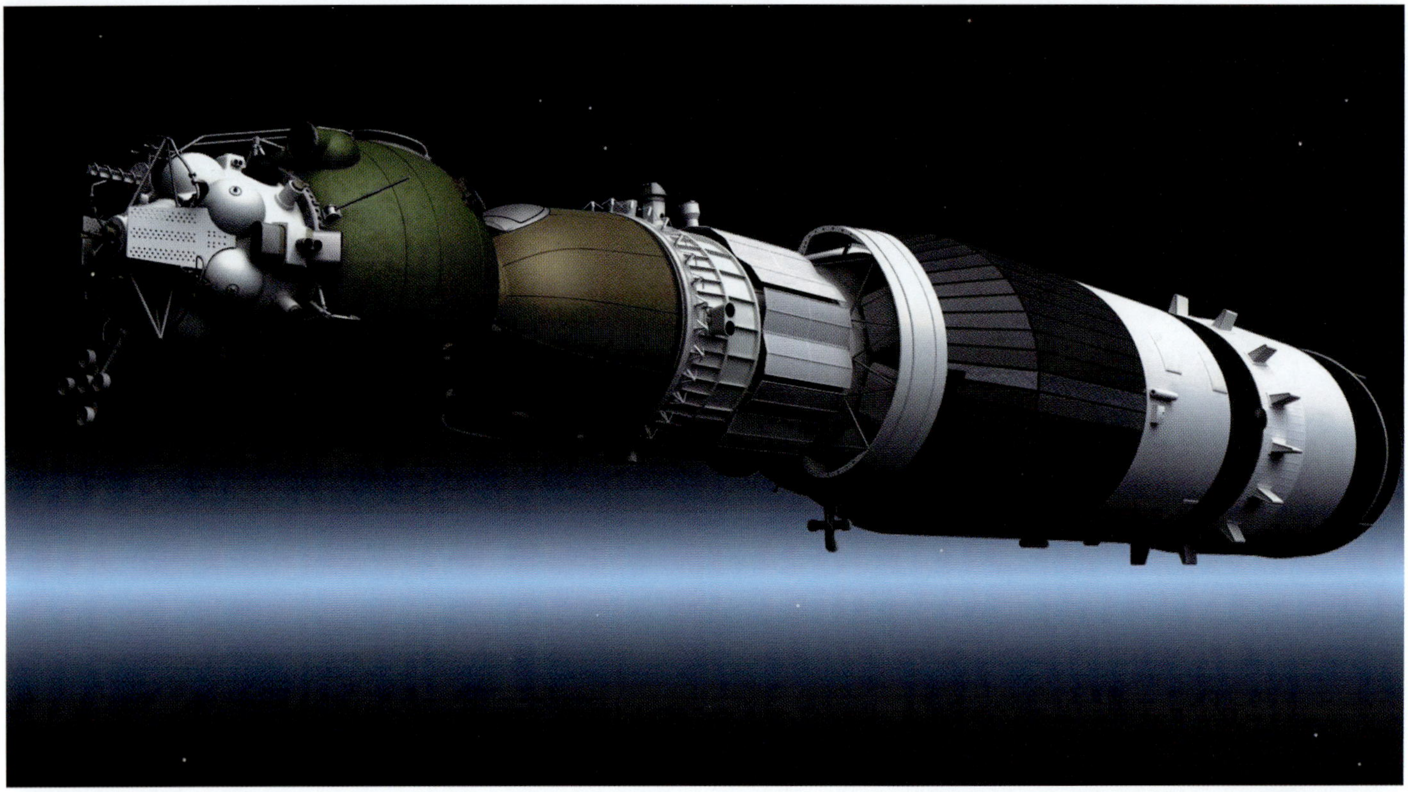

Above: the L3 Complex in Earth orbit getting ready to leave for the moon using Block-G

Below: Braking into lunar orbit using Block-D

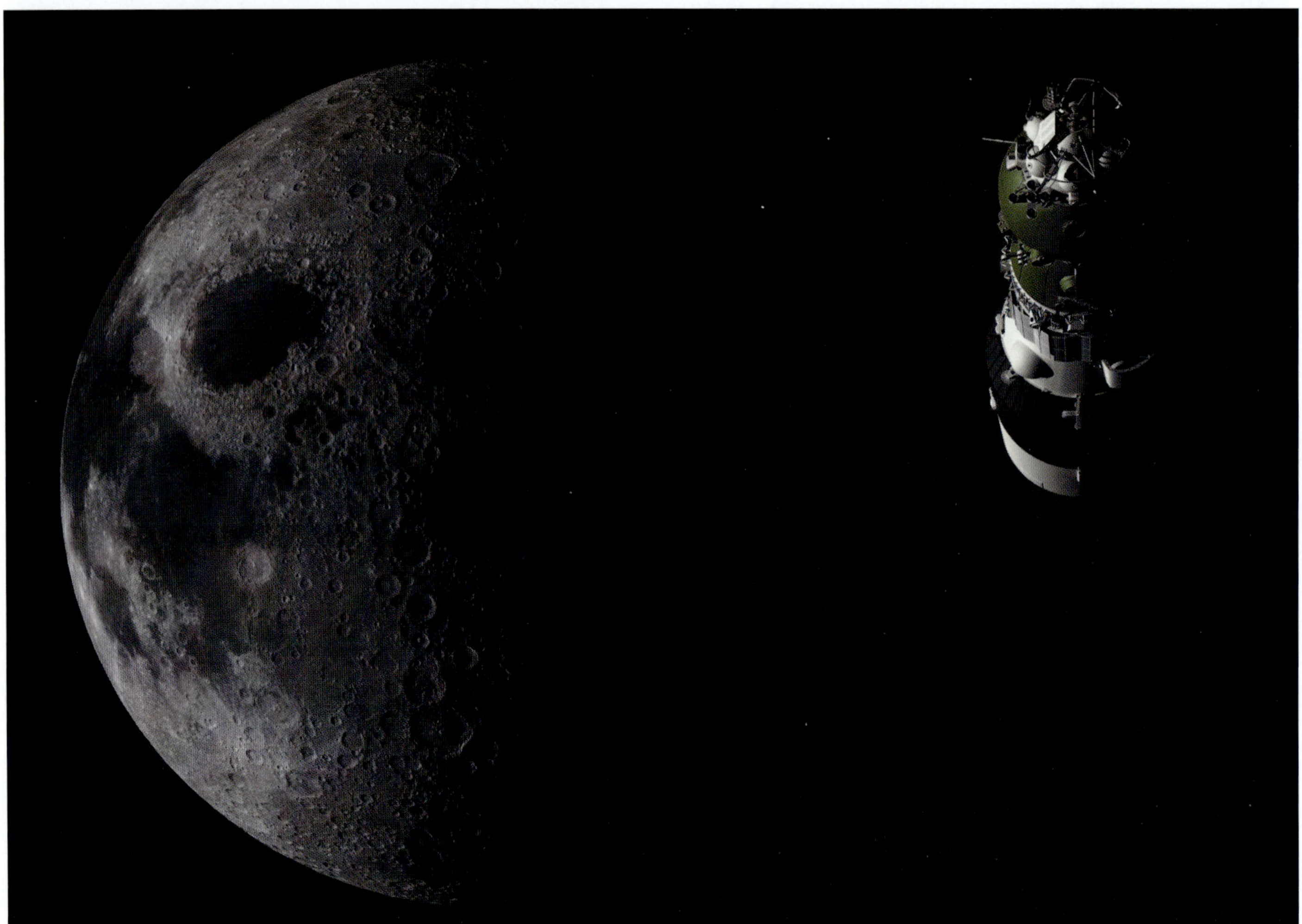

Left: Backing the LK out of it's shroud

Below: The connected Block-D and LK (Block-E) clear the shroud. The shroud itself will soon be jettisoned by the LOK

Left: Block-D fires for the last time, braking the LK towards landing.

Above: The LK finally gets to stretch its legs after jettisoning the spent Block-D. This occurs only 4,000 meters above the lunar surface.

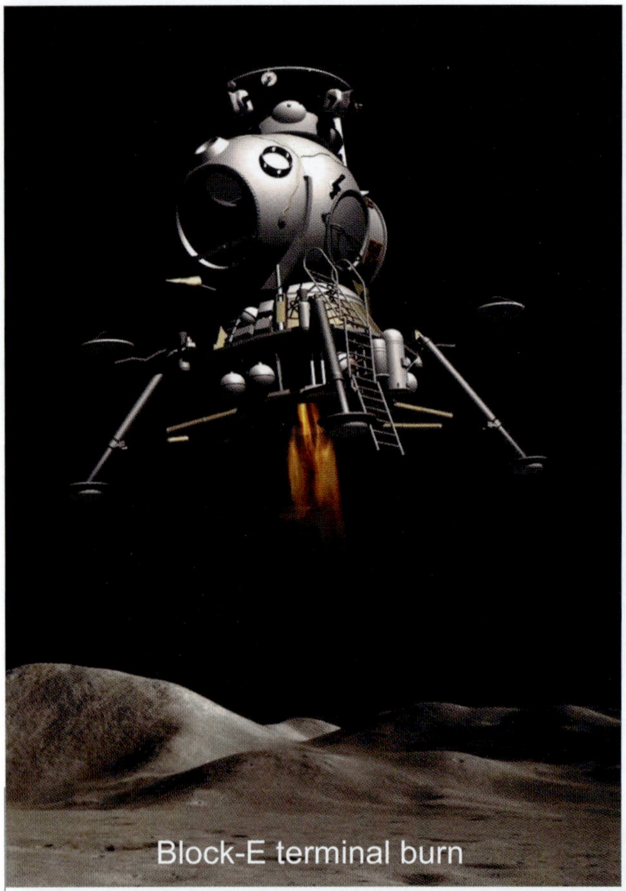
Block-E terminal burn

Touchdown! The Anchor Rockets fire to hold the ship in place

Triumph!

Above: The Lunokhod was delivered to the site a month earlier and the LK was guided in by the rover's homing beacon (striped conical antenna). If the LK is damaged in landing (note collapsed landing gear), the cosmonaut can use the Lunokhod to transport to the nearby RLK. It can also be used for exploration, but he won't get too far at 1.5 km/hr (1 MPH) with his limited surface time.

Below: The other crew member circles the moon in the LOK doing photographic mapping surveys and other scientific experiments

Clockwise from above left:

Liftoff using the same Block-E as for landing. Unlike Apollo, only the landing gear and some storage tanks are left behind

Back in orbit waiting for the LOK to find it

After docking the cosmonaut must do an EVA to transfer him-self and the sur-face samples back to the LOK

Clockwise from top: After jettisoning the joined LK and Utility Module, the Block-I engine at the back of the LOK is used to start the Reentry Module back to Earth

Less than an hour before reentry the Equipment Module is jettisoned and the cosmonauts prepare for a crushing "skip" trajectory down.

Home again!

Appendix B

The N-1 Table Medal

N1 TABLE MEDAL
1969

USSR

Photo of the N-1 table medal (obverse and reverse)

The 1969 N-1 table medal was most likely minted at the OKB-1/TsKBEM facility at the Baikonur Cosmodrome. The medal would have been designed by one or two selected military soldiers at the Cosmodrome who had shown artistic abilities.

N1 Table Medal Information:

- Commissioned for 1969's N-1/L3 launch.
- Issued to participants who constructed the N1 launch vehicle and launch facilities.
- Medal contains an alloy which has metal from an N-1 along with a copper interior.
- Between 50 and 100 were produced.
- Not an official medal. Never circulated due to secrecy of the N1/L3 project.
- Medals were ordered altered by the KGB.
- 72 mm in diameter and 5 mm in thickness.
- Believed to be the only item produced to commemorate the manned lunar program during the project's actual existence.

*Altered, in the case, means the reverse side (the text side) was completely stripped off of the medal. Some of these one-sided medals have survived as well as the complete two-sided medals.

HISTORY OF THE N1 TABLE MEDAL

The 1969 N1 table medal was commissioned for the contractors, engineers and participants involved in the building of the N1 lunar launch vehicle and its launch facilities.

The N1 was built to launch Soviet cosmonauts to the moon as part of the N1/L3 manned lunar program. The rocket was nearly 345 feet tall and capable of 9.6 million pounds of thrust at takeoff generated by 30 engines. Twelve launch attempts of the N1 were scheduled. All four of the development test flights launched experienced engine failure before reaching the second stage.

According to Alex Panchenko and Michael Rose, experts in Soviet/Russian space collectibles, the N1 table medal was not recognized as an official medal and none of the samples created were ever circulated.

Mr. Panchenko was informed by Vladimir Katayev, a former Soviet rocketry general, that only 50 pieces of the medal were custom ordered and presented to the recipients.

According to Victor Nikolaev, an employee of the Russian space agency and also a space collectibles expert, the medal was commissioned by the military and the Ministry of General Machine Building, approved by the KGB and presented to the State Commission. He believes up to 100 medals were made.

General Katayev also revealed to Mr. Panchenko that the N1 table medal consisted of an alloy containing metal from an N1 launch vehicle. Based on cut versions of the medal, Mr. Nikolaev concluded the interior was made of copper.

Mr. Nikolaev also discovered that the KGB later ordered the medals altered* in an effort to remove evidence of the program's secret existence. How many survived intact is unknown.

The front of the medal shows the profiles of three men looking at a N1 rocket on its launch pad. Mr. Nikolaev confirmed the symbolism of the men represents the Soviet workers, soldiers who worked on the N-1 project and the cosmonauts who would venture to the moon.

The Soviet Union's hammer and sickle symbol and a stock of wheat (to symbolize economic wealth) are also on the medal's front artwork.

The back of the N1 medal has text in a circular pattern which translates as "In Honor of Completion of the Construction of the Object N1 1969". A small star (symbolizing communism) is located above the letters N1.

This collectors page is reproduced through the courtesy of Doug Smith, the owner of the medal. Doug is only the second owner in the US after buying it from Michael Rose (mentioned in the write-up) who had bought it originally from a Russian dealer many years ago.

Appendix C

Tower Design Details

Notes by the Editor:

While working on Chapter 13, part of the research involved reviewing some pages of a design report on the tower rail carriages, Central Hub and some of the service platforms, forwarded to me by Nick Stevens. He said he had received them from *Energia* expert Vadim Lukashevich as background material.

I originally intended to just review the illustrations since they were, I thought, detailed enough to explain how the various components worked. The more I studied them, though, the more questions were raised on just what the different parts did. I decided to translate the drawing callouts using my usual tedious method of building up a Russian text file one letter at a time using the Cyrillic alphabet characters in the "Symbol" menu of Microsoft Word and then translate it using Google Translate. This usually gets me close enough that I can figure out what each piece does and what I can call it to be understandable to an English-speaking audience.

Not this time. For example, item #11 called out on the drawing opposite translated as "Anti-Theft Device." Anti-Theft? Really? They were afraid of someone coming in and stealing a 7,000+ tonne structure? Even if they managed to disconnect it at the Central Hub, how could they sneak it out the front gate? Obviously something more was needed to understand this.

After a bit of searching, I found an on-line site that could run OCR (Optical Character Recognition) on Russian text. So I created some clean text-only images of the book pages, ran it through the OCR, corrected the mistranslated characters and then ran it through Google Translate. Getting closer! I called what came out of the translator "Russlish" since the syntax was distinctly Russian and some parts of speech (like adjectives, adverbs and articles) lacked the context of the original or were missing all together. Still, it was close enough that after some study, and having some idea going in of what they were trying to describe, I was able to come up with the descriptions you see in the rest of this Appendix.

This was a heck of a lot of work just to get some callouts on drawings in the main text, so I decided to display the fruits of my labors here for all of you who enjoy the finer details.

By the way, the "Anti-Theft Device" is actually the locking mechanism that prevented the tower from moving in the wind as is explained at the bottom of page C-5.

- Jack Hagerty

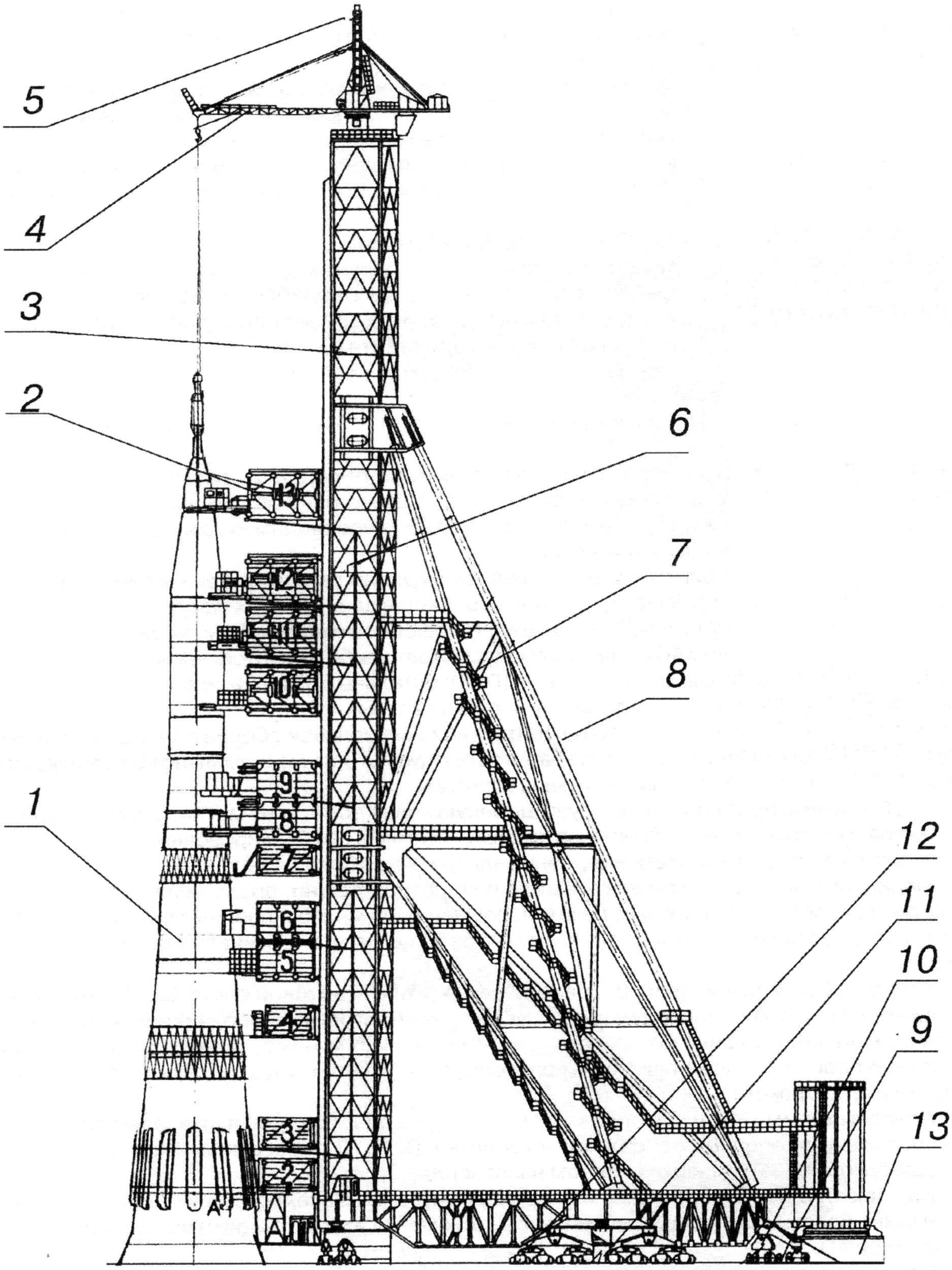

Рис. 84. Агрегат обслуживания 11У212:
1 - ракета-носитель 11А52; 2 - площадка; 3 - ствол; 4 - монтажный кран; 5 - молниезащита; 6 - система сбора промстоков; 7 - аварийный трап; 8 - подкос; 9 - рама; 10 - холостой ход; 11 - противоугонное устройство; 12 - рабочий ход; 13 - опорно-поворотное устройство

Fig. 84. Service Tower 11U212
1 - Rocket Booster 11A52, 2 - Work Platform, 3 - Tower, 4 - Construction Crane, 5 - Lightning Rod, 6 - Effluent Collection System, 7 - Emergency Stairway, 8 - Brace, 9 - Frame, 10 - Idler Truck, 11 - Locking Device, 12 - Powered Truck, 13 - Rotation Hub

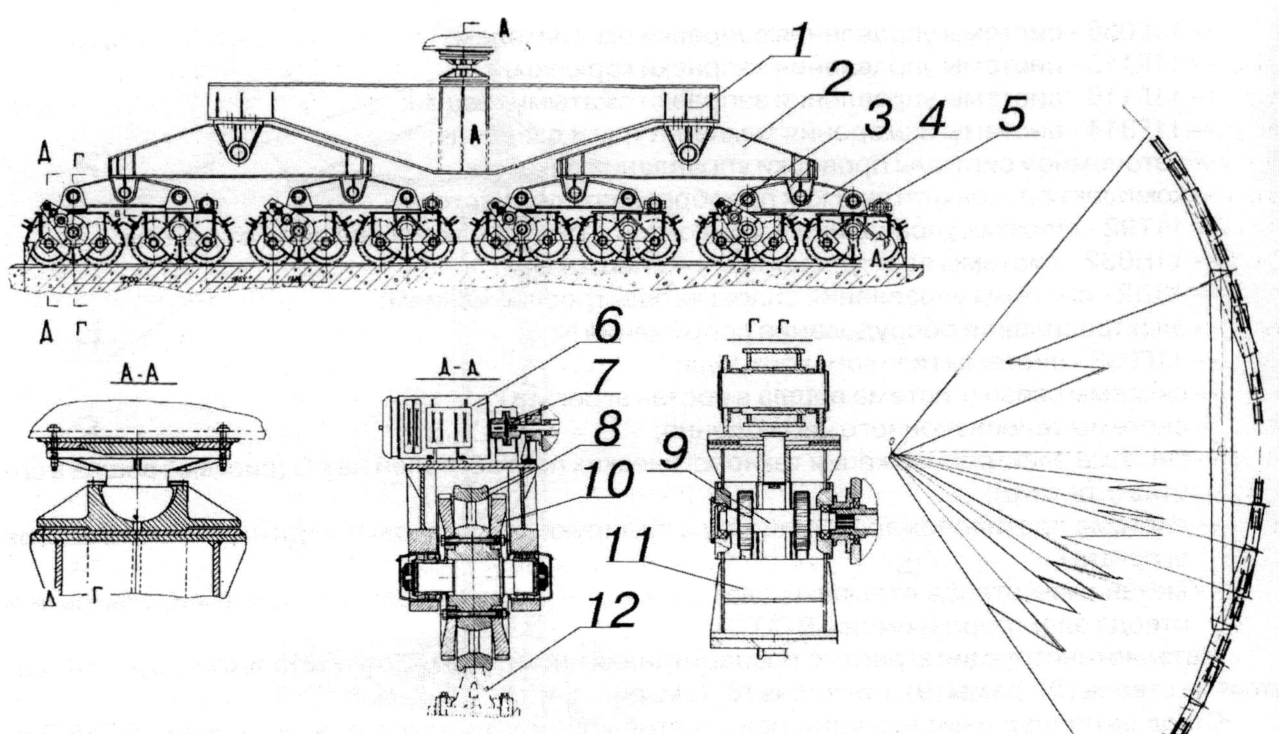

а
(a)

в
(c)

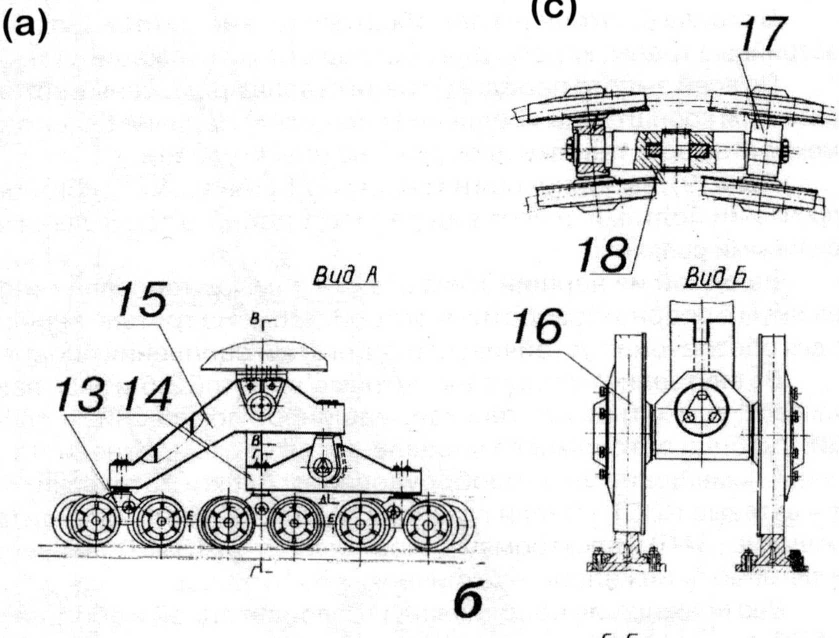

б
(b)

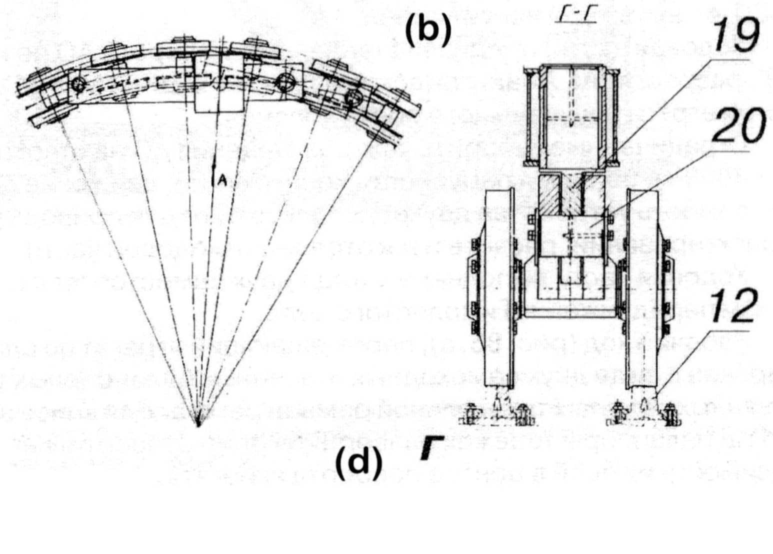

(d)
г

Fig. 85. Chassis Assembly, Service Tower 11U212:

a - Drivetrain, b - Idler Trucks, c - Movement of Powered Trucks on the outer circular rail, d - Movement of Idler Trucks on the inner circular rail

1 - The Main Rocker, 2 - Intermediate Rocker, 3 - Gearbox, 4 - Small Rocker, 5 - Driven Monorail Trolley, 6 - Motor; 7 - Clutch, 8. Drive Wheel, 9 - Double Spur Gear Shaft, 10 - Ring Gear, 11 - Bogie Frame, 12 - Rail, 13 - Idler Truck, 14 - Lower Rocker, 15 - Upper Rocker, 16 - Wheels, 17 - Transverse Cross Member; 18 - Longitudinal Traverse, 19 - Ear, 20 - Horizontal Axis

Рис. 85. Ходовая часть агрегата обслуживания 11У212:
а - механизм передвижения (рабочий ход); б -холостой ход; в - схема вписывания механизма передвижения в наружный круговой рельс; г - схема вписывания холостого хода во внутренний круговой рельс;
1 - главный балансир; 2 - средний балансир; 3 - редуктор; 4 - малый балансир; 5 - приводная однорельсовая тележка;
6 - электродвигатель; 7 - муфта; 8. ходовое колесо; 9 - сдвоенная вал-шестерня; 10 - зубчатый венец; 11 - рама тележки; 12 - специальный рельс; 13 - неприводная тележка; 14 - нижний балансир; 15 - верхний балансир; 16 - колеса; 17 - поперечная траверса; 18 - продольная траверса; 19 - ухо; 20 - горизонтальная ось

Только оси колес (8) двухколесных приводных тележек (5), имеющих каждая привод кранового типа, общий на оба колеса, направить в центр поворота не удалось, так как не нашлось предприятия, на котором можно было нарезать приводные зубчатые венцы и шестерки с углом конусности ~ 58'. Вследствие этого оси колес остались параллельными, но столь незначительный угол отклонения колес от направления на центр поворота особых осложнений в движении агрегата не вызывал.

Для создания рельсового пути под рабочий ход в ЛИИЖТе был разработан, а на Новосибирском заводе стрелочных переводов изготовлен 3-метровыми секциями специальный рельс. Рельс имеет коробчатое сечение из стали $Г13_п$ и допускает нагрузки на колесо 166 т, что превышает нагрузку на самые тяжелые типовые железнодорожные рельсы. Рельс установлен на специальном сплошном бетонном фундаменте балочной конструкции.

Балки балансиров рабочего хода из-за высоких нагрузок на ходовую часть выполнены из стали АК25, применяющейся в судостроении. Большие трудности возникли при сварке деталей балансиров из этой стали, а также при испытании самих балок балансиров.

Холостой ход (рис. 85, б) с радиусом поворота 10 м выполнен в виде трех неприводных 4-колесных тележек, установленных на несимметричной трехзвенной балансирной системе.

Геометрические оси вращения колес сходятся в центре агрегата.

Противоугонное устройство предназначается для предотвращения перемещения заторможенного АО под действием ветра.

Устройство состоит из двух заделанных в бетонный фундамент упоров и с пазами, в которые входит выступ рычага, установленного на кронштейне, закрепленном на раме агрегата.

Кронштейн опускается и поднимается электромеханическим приводом. Электрическая блокировка привода исключает возможность передвижения агрегата при опущенном рычаге.

On the drive wheels (8) and two-wheeled idler trucks (5) (Fig. 85, a), each having a drive-line, it is not possible to align each wheel to the center of rotation (Fig. 85, c) as there was no process to cut the drive sprockets and ring gears with a cone angle ~ 58°. As a result, the rotational axis of the wheels remain parallel, but so slight is the deflection angle from the wheel axels to the center of rotation that this did not cause any complications in the movement of the unit.

A special track was developed for the LIIZHT. It was manufactured by the Novosibirsk Railroad Switches Plant in 3-meter special rail sections. The rail has a box-section of G13p steel and allows a wheel load 166 tons, which exceeds the load on the heaviest standard railroad tracks. The rail is mounted on a special solid concrete foundation construction.

Due to high loads on the chassis, the beams of the Rockers are made of AK25 steel, as is used in shipbuilding. Great difficulties have arisen in the welding of parts made of this steel, as well as testing the beams themselves.

The Idler Trolley (Fig. 85, b) with a turning radius of 10 m is in the form of three non-drive four -wheeled carts that are installed on an asymmetrical three-tier system.

Geometric axis of rotation of the wheels meet in the center of the unit (Fig. 85, d).

The locking device is designed to prevent movement of the Service Tower in the wind. The apparatus consists of two levers embedded in a concrete foundation with matching grooves, which are a projection of the lever, mounted on a bracket secured to the frame. The bracket is lowered and raised an electromechanical drive. The Electric Lock actuator eliminates the ability of the unit to move when the lever is down.

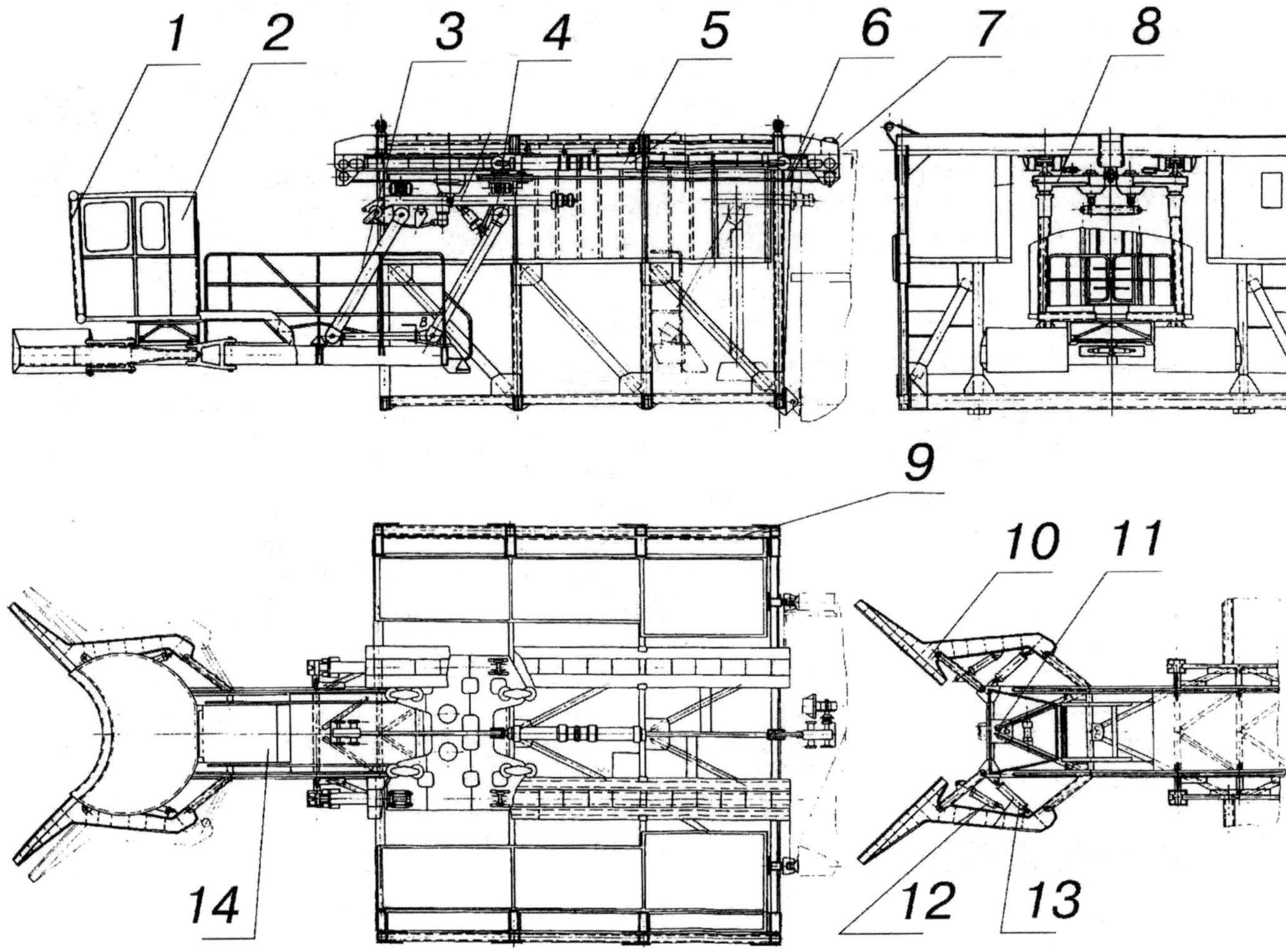

Рис. 86. Следящая площадка обслуживания N 13:
1 - уплотнение; 2 - кабина; 3 - тяга пантографа; 4 - комплект пружин возврата пола; 5 - пневмоцилиндр каретки; 6 - цепь; 7 - редуктор с маховиком; 8 - каретка; 9 - металлоконструкции площадки (стационар); 10 - ловитель; 11 - захват; 12 - пружина возврата ловителя; 13 - гидроамортизатор; 14 - подвижный пол

Fig. 86. Retractable Service Platform M13:
1 - Seal, 2 - Cabin, 3 - Pantograph, 4 - Retracting Spring, 5 - Air cylinder carriage, 6 - Chain, 7 - Gearbox with flywheel, 8 - Carriage, 9 - Metal pad, 10 - Catcher, 11 - Grip, 12 - Spring return safety device, 13 - Hydraulic shock absorber, 14 - Walkway

К стационарным элементам относятся:

— укрепленная на фундаменте сегментообразная труба (см. рис. 88) с двумя силовыми колоннами (3) и вертикальной стрелой. Эти элементы разделяет внутреннее пространство ОПУ на два отсека - кислородный и керосиновый - для раздельной прокладки коммуникаций этих компонентов и предотвращения смешения паров горючего и окислителя;

— воспринимающие горизонтальные нагрузки опорные тележки (4) с каретками и на случай выхода их из строя страховочные упоры;

— корыто жидкостного разделителя для изолирования кислородного отсека от паттерны с электрокоммуникациями.

Подвижными элементами являются: опорное кольцо с круговым рельсом под опорные тележки и с элементами болтового соединения ОПУ с рамой агрегата; цилиндрический ствол (2) (купол) с крышкой (1) и площадками для обслуживания выходных элементов коммуникаций.

Fixed elements are:

- A foundation of embedded reinforced pipe (see Fig. 88) with two power conduits (3), and the vertical arrow. These elements divide the inner space of the GTC into two sections - oxygen and kerosene - to separate the components and prevent mixing of fuel and oxidizer
- Connector from the horizontal load bearing carriage (4) to the rail carriages and to the case rotation stops;
- Separator to isolate the oxygen compartment from the electro-communications.

Rotating elements are:

- The support ring with a circular rail trucks connected to the frame assembly, Body Cylinder (2) with Cover Dome (1) and the utility shelf for connecting pipes and communications.

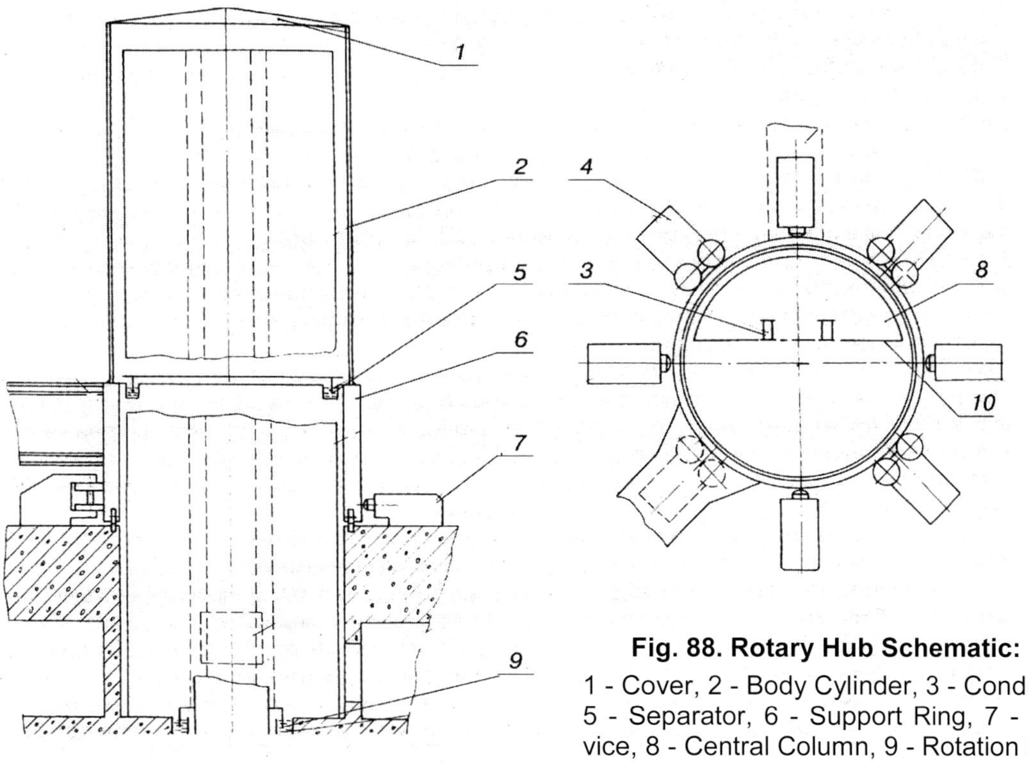

Fig. 88. Rotary Hub Schematic:

1 - Cover, 2 - Body Cylinder, 3 - Conduits, 4 - Truck, 5 - Separator, 6 - Support Ring, 7 - Centering Force Device, 8 - Central Column, 9 - Rotation Limit Stop, 10 - Wall

Рис. 88. Схема опорно-поворотного устройства:

1 - крышка; 2 - ствол; 3 - колонна; 4 - тележка; 5 - разделитель; 6 - кольцо; 7 - упор; 8 - труба; 9 - разделитель; 10 - стенка

Сравнительно большое внутреннее помещение ОПУ (диаметр по куполу 8 м и по опорному кольцу 9,5 м) насыщено расходящимися от него коммуникациями, их компенсационными устройствами и элементами обслуживания.

Управление механизмами АО предусмотрено с трех мест:

— автономного пульта управления на агрегате;

— дистанционного пульта управления № 1, установленного в сооружении № 3, где находится ЦППП (центральный пульт подготовки и пуска ракеты);

— дистанционного пульта управления № 2, установленного в сооружении № 103 - главном командном пункте, откуда поступает команда на пуск ракеты.

The relatively large interior space of the GTC (diameter of the dome of 8 m and support ring 9.5 m) is filled with a wide range of communications equipment, actuators and control elements.

The mechanisms can be operated from three places:

- Stand-alone control unit to unit;

- Remote control № 1, set in Building № 3, which is the TSPPP (Central Control for Preparation and Launching of Missiles);

- Remote control № 2, at Site № 103 - the main Mission Control, where he launch command is issued.

Appendix D

Author Bios

Matthew Johnson

Grand Rapids, Michigan, USA

Matt Johnson, a long time science fiction and space program enthusiast and model builder, became a Born Again Rocketeer* in 2006, after the death of his father spurred him to recover his childhood. He began researching the N-1 in 2008 in order to create a thirty engine flying scale model for entry the National Association of Rocketry's Scale Competition. This competition requires the assembly of a "data pack" documenting the real rocket being modeled and requires that the model weigh less than 1,500 grams (3 lbs. 5 oz.). He hoped to complete it in time for the 50th version of the NAR's Annual Meet (NARAM) in 2009.

That research put him in contact with Nick Stevens, CGI designer in London and the late Mike Dorffler of Estes. Mike was the American contact for N-1 historian Alex Shliadinsky in St. Petersburg, Russia; internationally known as the first and best historical expert on the N-1. Matt's research provided the necessary details of the N-1 to create a 1/81.6 scale model with thirty flight engines and the creation of a data pack to support his findings.

His 2009 version failed to meet the weight restrictions of the NARAM contest, but Matt continued on, building three more N-1 models of increasingly lower weight, and a scale launch tower to match. Finally having a contest-qualified N-1 for NARAM 52, Matt did enter and fly in that contest.

The untimely death of Mike Dorffler in 2010 left Alex Shliadinsky without a contact in the US, and Matt offered to continue working with him. With the enthusiastic support of Jack Hagerty at ARA Press, that offer lead directly to the creation of *N-1: A Reference Guide to the Soviet Superbooster*.

Ready to roll the tower ← back for launch!

To the Moon! ↑

↑ Matt and his creation

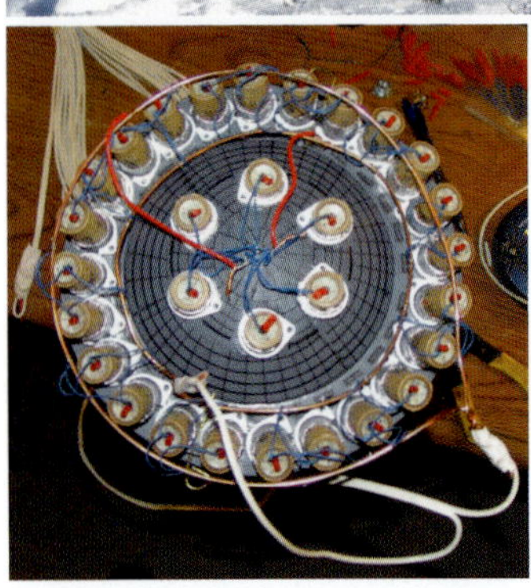

← The modeling equivalent of the Adapter Ring to wire up 30 model rocket motors

Ignition! →

*Born Again Rocketeer is a phrase Rocketeers use to denote a middle-aged person who gains the urge to re-enter model rocketry after a long absence, usually with a vengeance. The cycle of life often finds such people in a psychological and financial position which both affords and justifies the move to high powered rocketry.

Nick Stevens

London, UK

Nick Stevens was born just in time to catch the start of the space age, and studied Astronomy at University College, London. He founded and ran single-handed the UK's first online astronomy service "Starbase 1" back in the days of dial-up (before the world wide web). Starbase 1 also functioned as an ad hoc computer section for the Society for Popular Astronomy.

He has long been fascinated by the possibilities of using computer graphics to accurately recreate unflown space exploration projects, starting with his first ZX Spectrum computer.

Currently Nick is a Computer Capacity & Performance specialist in the day job, and is an Artist member of the International Association of Astronomical Artists, and an administrator of the International Space Art Network. You can find more of his work here:

http://spaceart1.ning.com/profile/NickStevens

Vladimir Antipov

Moscow Region, Russia

Vladimir Antipov worked at the Baikonur Cosmodrome for 27 years. As an engineer, he was engaged in the testing of ground equipment and participated in spacecraft launches.

He arrived at the spaceport in 1980 and immediately noticed a crater and a lot of debris from the N1-6L accident, a rocket that was unknown to him at that point, as well as the abandoned cyclopean structures for its operations.

He eventually learned about the N-1 and watched with sadness and regret the irrevocable decay of its fragments and the dismantling of the technical equipment in order to launch the *Energia* rocket.

Around 1990, in the period of Glasnost and Perestroika, when the first data on the N-1 appeared in the press, he began to collect information and reconstruct the history of the lunar rocket. To accomplish this task, and promote the history of the space program, he arranged a children's "study group" for the N-1 rocket. With the approval of the authorities, they visited the Baikonur facilities associated with the lunar program of the Soviet Union, photographing and describing the fragments of the N-1 and L3 ships, and talking with the veterans of Baikonur.

The collected research is stored at the N-1 L3 museum. Vladimir has written articles for newspapers and magazines, began to teach children in the International Space School of Baikonur and wrote a textbook for the school. Alexander Shliadinsky helped Vladimir in the preparation of drawings for the modeling of the N-1.

Following the deterioration of his health in 2008, he moved his family from the Baikonur Cosmodrome in Kazakhstan back to Russia where he currently lives outside of Moscow.

Alexander Shliadinsky

St. Petersburg, Russia

N-1 historian Alex Shliadinsky is internationally known as the man who first researched N-1 after records started being unsealed in the late 1980's. His meticulous engineering drawings of the rocket, first by hand and then in AutoCAD, have been reproduced in professional aerospace histories of the program.

His work has appeared in the magazines *Astronautics News* and *Russian Space and Spaceflight*. His early drawings of the N-1 were published in *Wings of the Motherland,* № 8 and № 9 in 1993. Most notably, he provided the N-1 drawings for Boris Chertok's history of the Soviet space program, *Rockets and People*, Volume IV, *The Moon Race*.

Working with Mike Dorffler in the US, Shliadinsky's AutoCAD drawings of the N-1 were published on CD between 2005 and 2007 and in an enhanced New Edition shortly before Mike's untimely death in 2010.

Alexander Shliadinsky (L) and Peter Druzinin with models of the US and Soviet Space Shuttles, and the V-2 based V-1Ye research rocket

Peter Druzinin

Tallinn, Estonia

While not a direct contributor to this book, Peter was none the less instrumental in its coming to be.

It was Peter who introduced Nick Stevens to Alex Shliadinsky, which led to Nick's CGI project on the N-1. That, in turn, led to Matt Johnson contacting Nick for dimensions and details when building his flying model.

Peter was born and raised in Keila, Estonia. His formal education was at TTY (Tallinn Technical University) in Tallinn, Estonia where he eventually moved in 2004 and currently resides.

He has been fascinated with astronautics and spaceflight since childhood and first heard of the N-1 from a newspaper article in *Izvestia* in 1988. Intrigued, his research led him to Vasily Mishin's book, *Why We Never Landed on the Moon* in 1990. His "breakthrough" (as he calls it) on the N-1 happened when he joined the online forum hosted by professional journal *Novosti Cosmonautica* (*Astronautics News*). There he met Alex Shliadinsky and Vladimir Antipov.

After 25 years as a radio ham, he switched hobbies to aerospace modeling in 2003 so that he could create miniatures of the vehicles that have fascinated him his whole life. He has made pilgrimages to aerospace facilities all around the world, including both the Baikonur Cosmodrome and Kennedy Spaceflight Center.

He met Nick Stevens in 2007 and started supplying him with detailed sketches (from his own model making projects) and photos to supplement the general information from Alex. Some of these photos are reproduced in this book. However, for you, the reader, Peter's greatest contribution was in reviewing the text. He cast an uninvolved, but expert eye on it, picking out things that the rest of us missed by being too close.

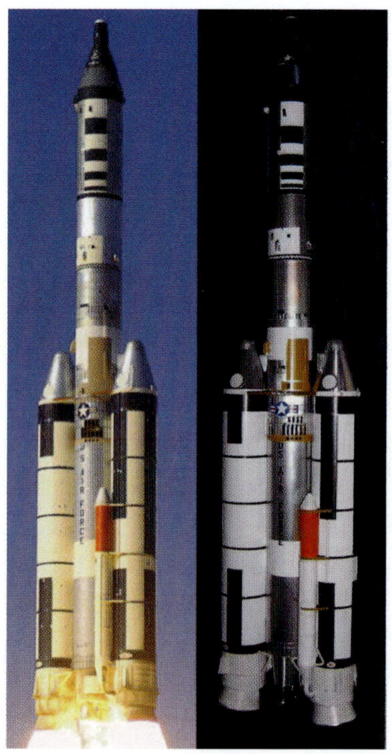

Titan III-C Round 6 carrying the mockup MOL (left) and Peter's stunning model of it

Igor Bezyaev

Obninsk, Russia

Igor was born in 1964 and as an eight year old schoolboy, he entered an art school. The teacher there was an artist specializing in fictional subjects and he used the fantastic paintings of Leonov and Sokolov in the lessons. He would get the class fired up about space travel and Igor was especially excited by the dream of interplanetary flight. All through school he was fanatically interested in space technology, astronomy and dreamed of becoming a cosmonaut. He loved to draw and designed the spaceships, which he then planned to build. Today this raises a nostalgic smile, but at the time he was serious about it!

For a number of reasons, it was not possible for him to work in the space industry, so he began his career in aviation equipment (at least a little closer to the cosmos). He worked as a technician in the industry for about 10 years while he earned the Russian equivalent of an MBA and now works as a production manager for specialized aviation equipment. But he's never forgotten his dreams of space.

He first learned of the existence of the USSR's moon and Mars programs in the early 2000s. As he was already well versed in 3-D graphics programs, he immediately had the idea to recreate all the famous Soviet interplanetary ships as three-dimensional models. His childhood dreams of building a spaceship came true! (Even if only virtually)

Since then, he has mastered working in three-dimensional graphics and has collected a large library of materials related to lunar and Martian ships. He continues to work on the simulation of spacecraft and the space environment to create a scene. He is thrilled to be able to participate in this book on the rocket masterwork of Sergey Korolev, the rocket that was intended to carry out the long-standing dream of mankind: travel to other planets!

Jack Hagerty

Livermore, California, USA

Photo by Tom Evans

Your humble editor and publisher started ARA Press nearly 20 years ago to provide modelers interested in science fiction and real aerospace subjects access to accurate, high-quality reference books at reasonable prices. After publishing his own, now classic titles *Spaceship Handbook* and *The Saucer Fleet*, he broadened ARA's scope by publishing other authors' work such as *2010: A Scrapbook Odyssey*, and two titles on *Lost in Space* history and hardware. ARA also provides continuing publication of books to the rocket hobby that had been abandoned by their publishers such as *Rockets of the World* (and all of its supplements), and the *Extreme Rocketry* technical series of booklets.

In 2010 he was a scale model judge at NARAM 52 (see page D-2) where he saw Matt Johnson's flying version of N1-L3. He was astounded at the level of information Matt had compiled on this project that had supposedly been completely destroyed. A deal was quickly struck, the result of which you're holding in your hands.

When not word-smithing in the world of rockets, Jack is a senior R&D test engineer for a company that builds artificial hearts.

Special Thanks

The top of the Special Thanks goes to all of the wonderful Kickstarter backers who made this book possible! We couldn't have done it without you!

On the production side, in addition to the specific people mentioned below, the authors would like to thank all those, too numerous to mention, who chipped in with comments on forums, observations, images they found, and above all, encouragement to keep us all going throughout this long, long, project.

From Matt:

To Vicky Dorffler for her kind permission to continue working with her husband's material after his untimely death.

From Nick:

To everyone who helped me to understand the Manned Soviet Lunar program, Particularly all those on the "Novosti Cosmonavtiki" forums, who put up with my computer translated Russian questions!

To Peter Druzinin for all his helpful sketches, and Sergey Andreev for his support.

To the crew at Foundation 3D for their encouragement and advice on modeling techniques.

From Jack:

To Ari Krupnik for reviewing my machine-translated chapter titles and hardware callouts thus saving me from some embarrassing mistakes.

Since Nick mentioned Peter Druzinin again above, let's share a couple more images of Peter's work. In the top image we see a collection of his scratch-built models, an eclectic mix of Soviet/Russian, American, western European and Chinese rockets and spacecraft with his "trophy wall" extending to the right.

Below we see a close-up of the wall. At the bottom are signed photos of Yuri Gagarin and Neil Armstrong. Above those, photos of his visit to Kennedy Spaceflight Center. To the lower right a rather sad photo; the shuttered doors of the Site 112 Assembly Hall guarded by two leviathans; the abandoned "Grasshopper" transporters—Ed

Index

Index Legend

Bold Entry—Text Reference
Italic Entry—Image Reference
Letter Prefix—Appendix Reference

— E —

— F —

— G —

— H —

— I —

— J —

— K —

— L —